Innovations in Measuring and Evaluating Scientific Information

J. John Jeyasekar
Tamil Nadu State Forensic Sciences Department, India

P. Saravanan
Lekshmipuram College of Arts and Science, India

A volume in the Advances in Standardization Research (ASR) Book Series

IGI Global
DISSEMINATOR OF KNOWLEDGE

Published in the United States of America by
IGI Global
Information Science Reference (an imprint of IGI Global)
701 E. Chocolate Avenue
Hershey PA, USA 17033
Tel: 717-533-8845
Fax: 717-533-8661
E-mail: cust@igi-global.com
Web site: http://www.igi-global.com

Library of Congress Cataloging-in-Publication Data

Names: Jeyasekar, J. John, 1965- editor. | Saravanan, P., 1971- editor.
Title: Innovations in measuring and evaluating scientific information / J. John Jeyasekar and P. Saravanan, editors.
Description: Hershey PA : Information Science Reference, [2018] | Includes bibliographical references.
Identifiers: LCCN 2017017433| ISBN 9781522534570 (hardcover) | ISBN 9781522534587 (ebook)
Subjects: LCSH: Scientific literature--Evaluation. | Scientific literature--Data processing.
Classification: LCC Q225.5 .H36 2018 | DDC 507.2/1--dc23 LC record available at https://lccn.loc.gov/2017017433

This book is published in the IGI Global book series Advances in Standardization Research (ASR) (ISSN: 2472-7296; eISSN: 2472-730X)

British Cataloguing in Publication Data
A Cataloguing in Publication record for this book is available from the British Library.

For electronic access to this publication, please contact: eresources@igi-global.com.

Advances in Standardization Research (ASR) Book Series

Kai Jakobs
RWTH Aachen University, Germany

ISSN:2472-7296
EISSN:2472-730X

Mission

Today, the technical, economic and regulatory importance of standards is no longer questioned. Standards allow complex technical systems to interoperate, they extend or even create markets and have been an important tool in the establishment of the European single market. The underlying standardization process provides a platform for pre-competitive co-operation and can support open innovation.

The **Advances in Standardization Research (ASR) Book Series** seeks to advance the available literature on the development, characteristics, deployment and value of standards. Highly multi-disciplinary standardization research provides deeper insight into the ways standards shape the environment within which they function and how, in turn, their development is shaped by the respective institutional and procedural environment. The series thus helps to better utilize standards for the improvement of organizational processes in both the private and the public sector. It also contributes to a better understanding of standardization processes and should thus enable a more effective and efficient development of more useful standards.

Coverage

- Emerging Roles of Formal Standards Organizations and Consortia
- Technological Innovation and Standardization
- Standards in the Private Sector
- Standardization for Organizational Development
- Case Studies on Standardization
- Multinational and Transnational Perspectives and Impacts
- Conformity Assessment
- Intellectual Property Rights
- Open Source and Standardization
- National, Regional, International, and Corporate Standards Strategies

IGI Global is currently accepting manuscripts for publication within this series. To submit a proposal for a volume in this series, please contact our Acquisition Editors at Acquisitions@igi-global.com or visit: http://www.igi-global.com/publish/.

The Advances in Standardization Research (ASR) Book Series (ISSN 2472-7296) is published by IGI Global, 701 E. Chocolate Avenue, Hershey, PA 17033-1240, USA, www.igi-global.com. This series is composed of titles available for purchase individually; each title is edited to be contextually exclusive from any other title within the series. For pricing and ordering information please visit http://www.igi-global.com/book-series/advances-standardization-research/154747. Postmaster: Send all address changes to above address.

We dedicate this book to our parents.
The joys of parents are secret, and so are their griefs and fears.
- Francis Bacon

Table of Contents

Detailed Table of Contents

The present chapter analyses the information literacy of the faculty members and also the relationship between information literacy and growth of literature. The survey was conducted among the faculty members of both Engineering and Arts & Science Colleges in Tamil Nadu. The study revealed that the majority of the faculty members have information literacy at average level. Academic productivity is more among the members of the faculty of Arts and Science Colleges than Engineering Colleges. Academic productivity shows information literacy significantly contributes to the growth of literature in their respective field. The comparison of information literacy among the faculty members shows that they differ significantly on information literacy based on age, type of college and familiarity of computer; and do not differ significantly based on gender, locality, experience, discipline and familiarity of internet.

This chapter is intended to present briefly the open educational resources initiatives in India especially related to growth of scientific literature. Scholarly communication can be defined as the system through which research and other scholarly writings are created, evaluated for quality, disseminated to the scholarly community, and preserved for future use. An old tradition and a new technology have converged to make it possible as an unprecedented public good. The old tradition is the willingness of scientists and scholars to publish the fruits of their research in scholarly journals without payment, for the sake of inquiry and knowledge. The new technology is the internet and web 2.0 tools. Thus, the scholarly communications are available in the public domain in the open access platform. The system includes both formal means of communication, such as publication in peer-reviewed journals, and informal channels in electronic media of open access. Also presented is an overview of all available open source resources in Indian context.

This chapter describes how health is the most precious component for the happiness and all-round development of human being in the society. Alongside this, information is an important resource for individual growth and survival. Therefore, a Health Information System (HIS) is a system for collecting/processing of data from various sources, and using the information for policy-making and management of health services. This chapter discusses health literacy and its association with health information needs and health information literacy, the status of women in India and their influence in the well-being of the family, and the impact of HIS. This chapter also proves that with enough data obtained from the women of Kanyakumari District in India, the government's policy needs to be directed towards women for the success of its health-care programmes.

The present chapter explores the essential stages in designing and developing the questionnaires and offers a basic introduction to beginners who are at the preliminary stages of their research. The study was approached purely from a theoretical perspective and various issues and challenges have been addressed including the length of questionnaires, how to distribute a questionnaire and how to analyze the data. The study concluded that researchers often rely on the previously developed questionnaire rather than developing the new one because it requires the knowledge and expertise. For novice researchers, developing entirely a new questionnaire requires a basic understanding of the format of the questionnaire, length of the questionnaire and how to distribute a questionnaire. Therefore, the chapter presents a systematic way to develop a questionnaire keeping in view all the issues and problems faced by researchers who are engaged in study for a thesis or in other research projects.

The quantum of information being produced has made the information scientists to devise various ways and means to effectively organize and disseminate them when needed. Bibliographical, citation and abstracting databases are the some of the tools. This chapter intends to describe the conceptual base of bibliographical databases, their genesis and impact in library science perspective and their importance in measuring information. The chapter also presents the science of measuring information using various quantitative and qualitative approaches like bibliometrics and scientometrics.

The chapter attempts to identify the information needs of educationists based on the citations in Master's degree dissertations in Education. The source for the study is the Master's degree dissertations submitted to the Manonmanium Sundaranar University during the period 1996 - 2006 from N.V.K.S.D. College of Education. The citation-based approach is followed in the study. Number of citations, cited articles, authors, frequency and percentage distributions were the tools used to analyze the data. Findings showed that journals were the most utilized reference materials in the dissertations. Indian periodicals are ranked first, followed by that from the USA. Mostly cited periodicals are published in the year 1993 with 181 citations (14.23%) and mostly-cited books were published in the year 1996 with 71 (13.95%) citations. Moreover, highest number of citations was in the year 1997 with 702(10.19%) citations and the lowest number of citations was during the year 2001 with 337(4.89%) citations.

This study outlines a significant relationship between a spurt of research activity in the field and the outbreak of the Zika disease, for the purpose of visualizing the phenomenon. By using bibliographic research, it is possible to correlate data with this disease. To identify the hotspots of research the data is evaluated using the following parameters: literature growth since 1975; most prolific authors and their contributions; top contributing institution, country, and journal; identify co-authorship clusters of the authors and institutions; and finally, to create and examine co-word maps of the keywords. It is also found that under-developed countries like Senegal and Zambia have contributed to Zika research along with the USA, France and Germany. Forty-five per cent of the contributions are concentrated in six journals.

This chapter outlines how scientometric tools and techniques are increasingly used to find the growth and direction of research. Forensic sciences are the application of science and technology in the administration of law and justice. Forensic Sciences encompass a wide range of subspecialties which are applied in adjudication of civil and criminal legal issues. Five subspecialties, viz., Forensic Anthropology (FA), Forensic Genetics (FG), Forensic Medicine (FM), Forensic Psychology (FP), and Forensic Toxicology (FT) were comparatively analyzed to find the growth of literature, source documents contributing to the literature growth and also the keywords used to find the research trends in these subspecialties. Data pertaining to these subspecialties documented from 2006 to 2015 were retrieved from the Science Direct platform and studied using MS-Excel and VOSviewer. The cumulative growth of literature showed a linear trend. The study found Forensic Science International as the top contributing journal. Significant association was observed between FA, FG, FM and FT literature while FP was not.

Science and technology are continuously changing as new discoveries and inventions are made. Research funding agencies, project directors, and individual researches need to keep a tab on these dynamics. This chapter tracks the research directions of forensic science for a period of thirty-eight years starting from 1975. Data for the analysis was obtained from SCOPUS bibliographic and citation database. Over the study period there was an exponential growth of forensic science literature and documentation. The United States of America contributed about one fourth of the research papers published while the most prolific author was Bruce Budowle (University of Texas). The majority of the contributors were from the non-governmental sector. The Journal of Forensic Sciences was the most productive journal during the study period in terms of number of published papers. It was also found that internationally collaborated papers attracted more citations.

Studying the communication pattern of scientific information produced as a result of academic and industrial research using mapping techniques is part of scientometric studies. In this chapter, the bibliographic data related to computer forensics retrieved from SCOPUS is subjected to mapping and visualizing. The results showed that the computer forensics literature grows exponentially. USA ranks first among the countries contributing to the literature growth. The University of South Australia is top ranking institution and K.K.R. Choo is the top-ranking author. Conference proceedings are the dominant form of communication flow. The scientograph of keywords reveals six clusters and the keywords with high frequency of occurrences are computer crime, security of data, network security, electronic crime countermeasures, digital forensic, internet, cryptography, cloud computing, cyber security, malware, and intrusion detection.

The present chapter evaluates the growth of literature of Toxicology in Homoeopathy. The Scopus database is used as source for the study. Data pertaining to toxicology in homoeopathy for the period 1963 to 2017 is retrieved and analyzed. The study reveals that toxicology literature grows rapidly in the later stages than in the earlier period. The United States of America is the leading contributors at the global level and India is in second position. Among the contributors, E.J. Calabrese, Professor of Toxicology, University of Massachusetts is the most prolific author for Toxicology in Homoeopathy literature. University of Kalyani and University of Massachusetts are the leading institutions, majority of the records are published in the journal 'Homoeopathy', which is published from the USA and major contributions are in the subject domain 'medicine' (65.4 per cent). Most frequently repeated keywords in toxicology of homoeopathy literature are: homoeopathy, humans, non-human, unclassified drug and animals.

This chapter creates the scientometric profile of Vikram Sarabhai Space Centre (VSSC) for the period from 1973 to 2016. The scientific productivity of the scientists is analyzed using scientometric indicators based on the data retrieved from the Scopus database. The analysis revealed that linear growth is prevailing. It also shows that the scientists are actively involved in disseminating their research findings in scholarly journals and G.V. Rao secured first position among the top-ranking authors. The Indian Institute of Science is the leading collaborating institution with VSSC and the United States is topmost collaborating country. Most of the scholarly communications of VSSC are published in the 'Journal of Sound & Vibration' and a majority of the records are of an article type. Most of the articles are published in the inter-disciplinary subject of Engineering. The study also found that the total number of citations received by the documents published from VSSC are 17395 and out of the 1783 documents published and 1385 documents received citations.

Nephrology is the adult and pediatric study of the kidneys and associated diseases. This chapter deals with the research productivity on Nephrology. Since the study is based on scientometric approach, the data has been retrieved from Web of Science database. The objectives of the study are: to identify the growth rate of research productivity, to trace the types of documents in Nephrology, to know the authorship pattern and language wise classification, and to find out the institution and geographical area wise distributions. Software packages such as Histcite and VOSviewer have effectively been used to analyze the data. In addition, various scientometric tools were also applied. As far as the contribution is concerned, the USA leads other parts of the world in Nephrology. English language dominates the globe in terms of publications.

The chapter brings out the results of a bibliometric analysis of the journal titled Asian Journal of Diary and Food Research for the period from 2011 to 2015. The data was downloaded from the journal's website. This study aims at analyzing the research output performance of co-authorship index and collaborative index. The analysis covers mainly the publication of articles, year-wise distribution of articles, degree of collaboration, institution-wise distribution, authorship pattern, etc. The results show the merits and weakness of the journal which will be helpful for its further development.

The concept of academic humility has often been ill-defined or ill-conceived by members of the Nigerian academia. It argues that this concept has objectionably been associated – nay confused – with "academic subservience", "academic hypocrisy", "false modesty" and "yesmanism". Such misconception could partially be attributed to the prevalence of the "seniority syndrome". In effect, the seniority syndrome has often wanted that pertinence be ascribed unto a research idea or thesis not necessarily on account of the robustness of the methodology that led to its enunciation, but principally on account of the "seniority" (credentials) of the researcher who authors the thesis or idea in question. This tendency has generally stemmed from the faulty assumption that a senior researcher is theoretically more knowledgeable than his or her junior counterparts; and that the "junior researcher" must manifest unconditional reverence for his or her senior colleagues. This scenario is most often observed during Ph.D. these defenses and similar forums aimed at evaluating research in Nigerian universities.

Foreword

Scientometrics, ever since it came into existence as a field of study, has given new dimensions to the two main scientific principles – Objectivity and Reproducibility. With modern techniques at hand, it has now become possible for us to measure the impact of scientific outputs and the outcomes. Perhaps, librarians and information scientists, have been the first ones to recognise the importance of big data when they started analysing the impact of science, understanding citations and production indicators, and the implications and the ultimate purpose of science. Thanks to Eugene Garfield, Derek J de Solla Price and the thousands of information scientists such as the ones that have authored papers in this publication, Scientometric results have had an influence on the way scientific literature and other outputs have been further harnessed by mankind in recent years, due to which materials such as science citation index, current contents and other citation databases became treasured possessions in libraries.

Today, we witness a range of big companies developing algorithms for analysing and influencing 'consumer behaviour' and 'knowledge seeking behaviour' in the modern age of Artificial Intelligence, Machine Learning and Augmented Reality research. The multi-armed bandit problem models tested to optimise the use of existing knowledge, called 'exploitation' and the attempt to acquire new knowledge, called 'exploration' are the new steps that extend the idea of measuring science and its outcomes. We are in the age of big data analytics. Recommender systems that are likely to replace the functions of a search engine have started to predict, manipulate and influence user preferences towards any kind of information, thus, going a step above the objectivity of measuring a scientific output to filtering and influencing users' information and knowledge seeking behaviour. Many a times, in this process, the objectivity is lost; it is the idea of 'influence' and market-driven competitiveness that dominates these new developments in the 'post-truth' world. This is the reason why the inventor of the World Wide Web, Sir Tim Berners-Lee is seeking transparency in algorithms so the ideas of 'objectivity' and 'openness' with which he gave the web freely to the world are maintained.

While the world of computing is advancing its field with the above techniques, it becomes imperative for the library and information science community to also defend science by advancing their own proven techniques such as scientometric, bibliometric and citation analyses, and the enterprise of measuring productivity in precision. The thrust librarians and information scientists place on principles such as normalisation, fairness and pro-quality metrics that are averse to manipulation and improper validation should combat the 'unfair' systems that have come to influence the way knowledge is sought, accessed and promoted. In this context, one must pay tribute to the man the world has lost this year, Sir Eugene Garfiled, by advancing the field of scientometics that he has left behind. I commend my friend and colleague, Dr J John Jeyasekar, Dr P Saravanan, and the scores of authors who have done their bit to advance the field that upholds the scientific principles of 'objectivity' and 'reproducibility' through their articles in this publication.

B. Shadrach
Alliance for Affordable Internet – London, UK

Preface

Since Second World War enormous quantity of scientific data are being produced. These data when collected, collated, and communicated in a meaningful way over a formal or informal medium becomes scientific information. The scholarly communication of scientific information changes the entire landscape of science and due to this, the knowledge domains keep on changing. Since information plays a vital role in the development of policies and programs of a country, the researchers as-well-as authorities and policy makers are challenged with the cognitive trajectory of the science dynamics.

The dynamics of science has opened a vista of issues to be solved by the information scientists and also by those engaged in research and publishing. The problem of '*Information overload*' started right from the days of Gutenberg. With the advent of modern computers, the doubling of scientific information happens at an alarmingly exponential rate. It was found from studies reported during 1970s that information in science doubles every five to ten years and in social sciences every eight to twelve years. Later, a study reported in 2001 shows the number of documents on the Internet doubled from 400 million to 800 million from 1998 to 2000. In the beginning of the twentieth century librarians began to organize the huge quantities of information to put them into use and the later developments in the Information and Communication Technology (*ICT*) made the organization of information as a distinct system.

While information is assimilated, organized, and transmitted, a need arises to increase the information literacy level of the end-users about the organized information so that it is put to optimum use. Chapter 1 deals with the association between information literacy levels of the faculty and the growth of scientific information. Library resources in the information era are so complex and information literacy has direct influence on academic productivity. Academic success and excellence are the objectives of the higher education system. Hence evaluation of the information literacy levels among the academic community is imperative for the institutions to design their methods of pedagogy.

The epistemology of scientific metadata has several facets including the communication of science. The invention of modern printing system had transformed the mode of scholarly communication. Further, with the availability of new electronic resources such as World Wide Web and web 2.0 tools the entire dimension of this system has been revolutionized. Information is freely available anywhere and anyone with access to internet and a computer system or hand-held electronic communication device can access it, albeit with the inherent dangers. One of the major dangers of free availability of information is its authenticity. Many semi-literates tend to interpret the incomplete data and become ill-informed ones. The second chapter defines what scholarly communication is in the context of modern *ICT* era, its transmission to the public domain and how to evaluate it. As the thrust towards open access to scholarly communications continues, libraries invest significant resources in setting up and maintaining institutional repositories. This chapter provides plenty of information on the open access educational resources available in India.

The governmental funding for various programs are on the decline as the society has started to move towards consumerism. The economics behind this is pushing for a need to make use of the available resources with great care. The information systems (*IS*) are focused on making difficult value judgments on what decisions are to be taken by government, agencies, research centers, and educational institutions. Businesses have to decide where to invest their limited funds, researchers have to decide which problem to study, and institutions have to decide on the direction of their educational programs. Similarly, health agencies have to make the right policy decisions in formulating and implementing their programs towards the betterment of the society with least cost and without compromising on the quality. Chapter 3 evaluates the health information systems and their impact on health-care. It is also focused on the health information needs of the women confined to a geographic area though with wider application.

Sherlock Holmes, the protagonist of Sir. Arthur Canon Doyle, states that it is a capital mistake to theorize before one has all the data. Primary data is the fundamental unit of any research project. Collection of primary data is an art, which requires sound techniques backed by strong methodology. The fourth chapter clearly brings out the characteristics of a questionnaire, a data collection tool. When the universe of the study is large and geographically wide collecting data through questionnaires is very cost-effective. When it is well structured and presented it becomes a very effective means of raw information eliciting tool.

Knowledge begets knowledge, as money begets money is a famous saying. Similarly, information also begets information. The cumulative nature of technical and scientific data and information gave birth to the need of some kind service which points to earlier documents. The scientific literature available when provided to the scientists at the right time leads to further inventions and knowledge. Sir Isaac Newton rightly said that he was able to see the farther things by standing on the shoulders of the giants. Bibliographic information is a pointer to previously published documents and is made available through bibliographic databases. This information are being increasingly used for measuring and evaluating scientific research though the original purpose was to aid the librarian in selecting the journals for subscription and to assist the reference librarian in locating and retrieving the needed information. The fifth chapter is an exhaustive study on the bibliographic databases and their role in the bibliometric as-well-as scientometric studies. The meso, macro, and the micro level indicators that evolved as a specific answer to a specific problem in the measurement and evaluation of information are also given due consideration.

The sixth chapter is a citation based study to assess the information requirement of the particular group of academicians, namely, educationists. Scientometrics can be simply stated as the quantitative study of scientific communication. Scientometrics uses a multitude of sophisticated techniques, some borrowed from bibliometrics including citation analysis. Citation analysis techniques provide a measurement of the relationships and also the flow of information between people, groups, or organizations. Citation analysis is based on two basic concepts. They are (i) an author's reference to a published work indicate the relationship between the present work and the earlier (cited) work; and (ii) works which cite the same work have some subject relationships with each other. Hence by studying the citations of the source data available in the Masters Degree dissertations, the author is able to trace the information flow and demonstrate to the educationists the highly useful reference materials.

The seventh chapter is visualization and scientometric mapping of the research literature in a micro level viral disease namely, *Zika*. Domain visualization is a relatively newer research front. The idea is to use information visualization to represent large amounts of data in research fronts (Chen, et al., 2001). This allows the viewer to look at a large corpus and develop deeper insights based on a high-level view of the map (Card, Mackinlay, & Shneiderman, 1999). The authors have retrieved data from *PubMed*, the largest database of medical literature and have used the visualization of similarity techniques to analyze the network between various players engaged in *zika* virus research. The findings of the study have far reaching consequences among the health policy makers and health workers.

The eighth chapter is a study on the research trends prevailing on the knowledge domain of forensic sciences. Forensic science is a multi-disciplinary subject which encompasses a variety of scientific knowledge in the adjudication of legal matters. The author has selected five subspecialties of forensic sciences and analyzed the literature growth, and source documents. The keywords used in the published literature have also been subjected to visualization of co-word similarities and a thesaurus of each sub-specialty has been created. The study is novel because it has used the bibliographic data extracted from Science Direct and also the comparative analysis of keywords shows the association and dissociation between the subspecialties of forensic sciences. This chapter is of immense use to the researchers and research funding agencies in forensic sciences.

Science is not static but is dynamic. With every new discovery, the domain grows and sometimes makes amends to the earlier theories and laws. The scientists working in a field have to keep them abreast of the latest developments due to the ephemeral nature of science. The ninth chapter systematically captures the dynamics of Indian forensic sciences research literature. This chapter tracks the research direction of Indian forensic sciences for a period of thirty-eight years beginning from the year 1975. This chapter is aimed at providing a base for the knowledge creators, knowledge workers and knowledge managers of Indian forensic sciences.

The concepts of text mining and visualization are being increasingly used to identify and analyze the topical coverage of large text corpus. This process, also known as scientometric mapping, is a nascent technique used predominantly by library and information science professionals. The practical applications include the means to understand the scholarship documented in a bibliographic database, how closely different terms are placed forming a cluster of co-occurring terms, and how these clusters represent a research area. Data visualization technique is a useful way to explore a set of documents, uncover their latent patterns, and pose new questions for further analysis using additional methods. Computer forensics is relatively new subspecialty of forensic sciences, which deals with legal application related to computer or digital crimes. The tenth chapter documents the growth of scientific literature, the countries contributing to the literature growth, and the prolific authors in computer forensics as seen through *SCOPUS* database. The study also creates cluster maps of computer forensics to visualize the subject.

Carizzo-Sainero (2000) defines bibliometrics as "the ensemble of methodological knowledge that will serve the application of quantitative techniques in or to evaluate the process of production, communication and use of scientific information". Homoeopathy as a complementary and alternative system of medicine plays a key role for effective treatment of diseases and it is widely used around the globe. Homeopathy became spectacularly popular in the United States and Europe in the 1800s and its strongest advocates included European royalty, American entrepreneurs, literary giants, and religious leaders. Toxicology is the branch of science that deals with poisons, and narcotic or toxic substances. Chapter 11 gives an account of publication count year-wise, country-wise, author-wise, institution-wise, etc. and also network maps of collaborating countries, and institutions arrived from *SCOPUS* database. At the time when alternative system of medicine is gaining credence the study would be a useful one for health information system managers apart from the policy makers and researchers.

Research institutions' heads, funding agencies of research projects and also governments, all face for one reason or other, the same evaluative question: how can one define, recognize and compare 'science excellence' as objectively as possible? The quality related parameters are used to assess the research productivity of research institutions. The resultant indicators benchmark cross-country and cross-discipline and help one to identify and classify institutions as Centers of Scientific Excellence. Vikram Sarabhai Space Centre (*VSSC*), Thiruvananthapuram, India is the major space research centre of Indian Space Research Organisation (*ISRO*), actively involved on rocket and space vehicle for India's satellite programs. Twelfth chapter gives an account of *VSSC*'s research productivity, author-wise contribution, institutional level research collaboration, and international collaboration, articles with high citations, and a keyword cluster map.

Chapter 13 is related to research contributions of *Nephrology*. *Nephrology* is the adult and pediatric study of the kidneys and associated diseases. The software packages *HistCite* and *VOSviewer* are used to analyze the data downloaded from *Web of Science* provided by Thomson Reuters Institute for Scientific Information (ISI). The *Web of Science* provides seamless access to current and retrospective multidisciplinary information from approximately 8,700 of the most prestigious, high impact research journals in the world. *HistCite* tool automatically generates chronological tables and historiographs of topical paper collections. It assists researchers and librarians in the following areas: identifying core papers on a topic in question; understanding the impact of specific authors, papers, and journals; and making sense of the history of old and new research topics.

Chapter 14 is a single journal bibliometric study in the domain of food sciences. Single journal bibliometric studies result in the creation of a portrait of the journal under study. It is an indicator of quality and productivity of the journal. In general, only very popular journals of the field are selected for such studies. *Asian Journal of Dairy and Food Research* is the official publication of the Agricultural Research Communication Centre. It provides a forum for the scientific community to publish their original research articles in the field of Dairy and Food Science. Hence this journal was subjected to a bibliometric study focusing on the growth of literature and collaboration.

Measurement and evaluation of research activities including the productivity of the author or institution is fundamental in assessing their prestige and eminence. Academic and research institutions, however, while striving to attain excellence are plagued with certain problems like 'academic humility or subservience', 'ghost authorship', and 'seniority syndrome'. These problems cripple the effective functioning of the institutions. Chapter 15 highlights these problems while discussing the remedies as experienced in Nigeria, while this experience can be seen everywhere throughout the developing countries.

Scientific research is a large enterprise and the substance of it is so complex and specialized that personal knowledge and experience are no longer sufficient to quantify and evaluate the scientific information produced the researchers and institutions. Peer review was the standard approach to evaluate a research project until partially replaced by scientometric tools. Scientometric tools became popular because of the flaws like bias and favoritism inherent in the peer review system. There have been arguments for and against publication counts being used for research evaluation. Prof. Subbiah Arunachalam, eminent professor of chemistry and information science, states that the total number of scientific research paper can be used as a science indicator. He further argues that rejecting publication counts is like throwing out the baby with the bathwater (Arunachalam, 2003). Though there are some disadvantages in using scientometric techniques for evaluating scientific literature / information, the advantages are far more.

The government agency which assesses the research need, or a funding agency which makes a funding choice, or an institution which has to assess a candidate for appointment or promotion, or a library facing collection related issues, or a researcher pondering on a topic of research and later for the publication of the results have to depend on the evaluation of scientific research performance based on scientific information and its surrogates. This book will be a valuable addition to the above categories as it attempts to make use of scientometric techniques for the betterment of the civil society.

J. John Jeyasekar
Tamil Nadu State Forensic Sciences Department, India

P. Saravanan
Lekshmipuram College of Arts and Science, India

REFERENCES

Arunachalam, S. (2003). Use of SCI-based publication counts. *Current Science*, *85*(10), 1391–1392.

Card, S., Mackinlay, J., & Shneiderman, B. (1999). *Readings in Information Visualization: Using Vision to Think*. Burlington, MA: Morgan Kaufmann.

Carizzo-Sainero, G. (2000). Towards a concept of bibliometrics. *Journal of Spanish Research on Information Science*, *1*(2), 5986.

Chen, , Paul, R. J., & O'Keefe, B. (2001). Fitting the jigsaw of citation: Information visualization in domain analysis. *Journal of the American Society for Information Science and Technology*, *52*(4), 315–330. doi:10.1002/1532-2890(2000)9999:9999<::AID-ASI1074>3.0.CO;2-2

Acknowledgment

We thank God Almighty because we believe that every idea and the strength to execute the idea come from Him.

We take this opportunity to thank all our well-wishers and those helped us complete this project, one way or other. We also thank all the contributors and those who doubled-up as reviewers as-well. We thank the editorial advisory board members who gave their valuable suggestions for the improvement of the quality of the book and also for their comments as reviewers of the chapter.

Special words of thanks are due for Dr. B. Shadrach, Asia Coordinator, Alliance for Affordable Internet (London), for devoting his time to go through the manuscript and writing the foreword. We also thank our respective institutional authorities for permitting us to take up this project and also for their encouragement.

The first editor thanks his wife, Mrs. Jacquline and daughter, Dr. Judith for sacrificing their time.

The second editor too, thanks his family members and colleagues for their encouragement and motivation to complete the task.

J. John Jeyasekar
Tamil Nadu State Forensic Sciences Department, India

P. Saravanan
Lekshmipuram College of Arts and Science, India

Chapter 1
Association Between Information Literacy and Growth of Scientific Literature

P. Saravanan
Lekshmipuram College of Arts and Science, India

ABSTRACT

The present chapter analyses the information literacy of the faculty members and also the relationship between information literacy and growth of literature. The survey was conducted among the faculty members of both Engineering and Arts & Science Colleges in Tamil Nadu. The study revealed that the majority of the faculty members have information literacy at average level. Academic productivity is more among the members of the faculty of Arts and Science Colleges than Engineering Colleges. Academic productivity shows information literacy significantly contributes to the growth of literature in their respective field. The comparison of information literacy among the faculty members shows that they differ significantly on information literacy based on age, type of college and familiarity of computer; and do not differ significantly based on gender, locality, experience, discipline and familiarity of internet.

INTRODUCTION

Generally, the library and information centre is an integral part of the higher educational institution, which satisfies the information requirements of the academic community consisting of students, teachers and scholars. It acts as a hub of the teaching and learning activities where students, teachers and researchers interact with print and non-print materials to fulfil their day-to-day information requirements. In the traditional library of print sources, the concept of information searching becomes more tedious and users have to spend more time, energy when compared to latest emerging technology of online sources. Advancements in Information Communication Technology (ICT) based technology and its implications to the library made the information search become more complex. The information search can be made easy through information literacy programmes which in turn help to enhance the academic productivity of end users.

DOI: 10.4018/978-1-5225-3457-0.ch001

The impact of ICT in the information age is rightly pointed out by Swain & Panda (2009) as

...the rapid advancement of information and communication technology has brought a revolutionary change in the information scenario giving rise to a number of options to handle varied information sources conveniently and effortlessly as a result of which e-resources have become the most sought after modern library's reserves in satisfying varied needs of students, teachers, and researchers with minimum risk and time.

The impact of technologies such as CD-ROMs, multimedia, computer networks, Internet etc., have lead library into a paperless library. Changes in technology have dramatically altered how information is accessed, stored, preserved and disseminated. Now-a-days, the word *e* is associated not only with library resources such as e-resources but also with every field. In this context, the effective retrieval of information by the academic community depends on their information literacy on e-resources.

INFORMATION LITERACY

Information literacy is the ability of the individual to recognize when information is needed and have the ability to locate, evaluate and use effectively the information needed. It is the ability to use information from various sources in an effective and efficient manner to promote learning and scholarly communication.

Being information literate requires knowing how to clearly define a subject or area of investigation, select appropriate terminology that expresses the concept, formulate search strategy that takes into consideration different sources of information and various ways that information is organised; analyse the collected data for its value, relevancy, authority, quality, reliability and subsequently turn information into knowledge. In the changing environment of the library, the information literacy skills or information competency becomes an essential one. It is noteworthy to mention the definition given by the American Library Association (ALA) Presidential committee on Information Literacy, "To be information literate, a person must be able to recognise when information is needed and have the ability to locate, evaluate, and use effectively the needed information".

Information literacy skills are closely related to information technology skills, especially when information sources and access tools are increasingly based on digital technologies. As more digital technologies spread, an appropriate level of education should be developed to ensure that people are able to use technologies.

REVIEW OF LITERATURE

Literature review is concerned with the survey of a particular work or work published in a particular field or topic. Some of the studies related to information literacy are briefly presented here.

Adeleke and Emeahara (2016) studied the relationship between information literacy and use of electronic information resources for academic purposes by postgraduate students of the University of Ibadan, Nigeria. Low level of usage of electronic resources, in particular, full texts data bases is linked to lack of search techniques skills by many postgraduate students of the university to access the myriad of e-resources. There was significant relationship between information literacy skills and use of electronic information resources at $\alpha = 0.05$ ($p = 0.34$). Lack of adequate skills of postgraduate students resulted into the non-encouragement of the expected research-led enquiry in using electronic information resources (EIRs) in this digital age. Information and Communication Technology (ICT) course be part of the curriculum of every postgraduate programme, with emphasis on e-information literacy.

Bibina and Kabir (2016) investigated the digital literacy among science-research scholars of University of Kerala. Their study revealed that a great majority of the respondents are familiar with web portal, online databases, digital library/archives, open access e-books/e-journals and search engines. Moreover, majority of them have awareness about Google Scholar and web 2.0 based services. A good percentage of the respondents used desktop, laptop and smart phone for internet access. The study also revealed that all respondents browse internet daily. They are in need of digital-literacy programmes. The constraints faced by the science-research scholars while using digital information resources. Some of them are: slow access speed, difficulty in finding relevant information and then frequent power failure.

Catalano and Philips (2016) investigated the impact of library instruction on information literacy (IL) skills. The study also intended to determine the relationship between information literacy scores and research experiences with student success outcomes such as retention. The respondents of the study are graduate and postgraduate students of mutli-discipline. The outcome of the study was there were no significant differences between those students who had instruction and those who did not on the IL test, a regression analysis revealed that experience writing research papers that required library resources and an individual's use of library books throughout their academic career demonstrated significant, positive relationships with whether a student passed the information literacy test.

Similarly, the studies of Khan (2015) revealed that 25.6 per cent of the PG scholars do not know the meaning of information literacy and information literacy is related to use of e-resources; and Slauddin (2015) discloses that digital literacy influences the usage of e-resources. Similarly, Prince and Saravanan (2015) conducted a study on use of open access electronic resources among the academic community in the higher education system in kanyakumari district, in which the role of information literacy is anlaysed on the usage of open access resource and in another study on teacher approach to internet information sources (Saravanan and Mary, 2007). The closer analysis of available literature shows that studies on information competency and its effects on growth of scientific literature is scanty, and hence the present study is pioneer one.

NEED AND SIGNIFICANCE OF THE STUDY

With the advancement of Information Communication Technology (ICT), most of the traditional sources are being changed into electronic resources such as e-journals, e-books, e-databases, e-newspapers, internet resources, online database and CD-ROM databases. Retrieval of information is also become more and more complex in e-environment when compared to traditional print media. Thus, information literacy skills of the faculty towards e-resources will differ when compared to print sources. Moreover, information literacy skills helps faculty to access latest information in their field. These skills in turn help them to contribute to scholarly communication in the form of journal article, publication of books, and undertaking minor and major projects. In other words, information literacy helps the faculty to enhance their academic productivity, which leads to growth of literature. In this context, an attempt is made to analyses the information literacy of the faculty members and the effect of information literacy on academic productivity. Several studies have been conducted on information literacy or its related factors but no systematic study has been conducted so far on the effect of information literacy on growth of scientific literature, and hence the present study.

STATEMENT OF THE PROBLEM

The present study is intended to evaluate the information literacy of the faculty based on selected personal variables and to determine the effect of information literacy and academic productivity. Hence the problem for the present study is entitled as "Effect of Information Literacy on Growth of Scientific Literature: A Study"

OBJECTIVES OF THE STUDY

The objectives are as follows:

1. To determine the level of information literacy of the members of the faculty of Arts and Science Colleges and Engineering Colleges.
2. To identify the academic productivity of the members of the faculty of Arts and Science Colleges and Engineering Colleges.
3. To compare the information literacy among the faculty members based on gender, locality, age, type of colleges, experience, discipline, familiarity of computer and familiarity of internet.

HYPOTHESES OF THE STUDY

Hypotheses framed for the study are as follows

1. Faculty members have high information literacy skills.
2. Academic productivity is more among the members of the faculty.
3. There is no significant difference in information literacy among the faculty members based on gender, locality, age, type of colleges, experience, discipline, familiarity of computer and familiarity of internet.

LIMITATIONS OF THE STUDY

1. The present study is confined only to the members of the faculty of Engineering Colleges and Arts and Science Colleges under the jurisdiction of Manonmaniam Sundaranar University.
2. Faculty members of other higher educational institutions are excluded from the study.
3. Academic productivity of the members of the faculty are measured in terms number of articles published in journals, number of book published and number of projects they have.

METHODOLOGY

The present study intends to analyse the information literacy skills of the members of the faculty in the higher education system and to determine the academic productivity of the members of the faculty. Therefore, the method adopted for the present study is survey method. The population for the present study is the faculty members in the higher education system. It includes faculty members from Arts and Science Colleges, and Engineering Colleges in the jurisdiction of Manonmaniam Sundaranar University, which cover the geographical area namely three southern districts of Tamilnadu viz. Kanyakumari district, Tirunelveli and Thoothokudi district. The structured questionnaire prepared by the investigator in consultation with experts in the field is used for the collection of data from the respondents. The structured questionnaire is administered among 650 respondents using stratified random sampling techniques by giving due weightage to various categories of the variable under the study. Out of 650 respondents, 562 respondents had returned the questionnaire completed in every aspect, response rate is 86 per cent. Thus, the sample for the study is 562 members consisting of 281 faculty members each from Arts and Science College and Engineering Colleges. Information literacy assessment tool consists of 12 statements on various facets of information identification to evaluation of information from different sources. The academic productivity is assessed based on their contribution to the scholarly communication. The respondents are asked to respond the statements on five-point scale ranging from 'do not know' to 'excellent'. The collected data are analysed using SPSS ver17.0.

ANALYSIS AND INTERPRETATION OF DATA

Data collected from the respondents are analysed by using the descriptive and inferential statistics and interpreted accordingly.

Distribution of Respondents Based on Demographic Variables

The respondents for the present study consist of 562 respondents, of which equal numbers of respondents are from both Engineering and Arts & Science Colleges. The gender wise distribution of samples reveals that 256 respondents (45.5 per cent) belong to male category and the remaining (54.5 per cent) belongs to female; and the locality wise distribution of the faculty members reveals that 56.2 per cent from rural and the remaining 43.8 per cent belong to urban. The age wise classification reveals that 29.5 per cent of the respondents have age less than years, 51 per cent have age between 30-40 years, 15.5 per cent have age between 40-50 years and 3.9 per cent have age greater than 50 years.

Similarly, discipline wise classification reveals that 37.2 per cent respondents belong to Engineering discipline, 35.5 per cent belongs to Science, 15.8 per cent belong to Arts and 11.4 per cent belongs to Social Science.

Academic Productivity in Terms of Papers Published in Journals

The efficiency of the faculty members are generally assessed based on their contribution in academic environment such as teaching-learning process, research and contribution in scholarly communication and so on. The academic productivity of the faculty members are measured based on their contributions in articles published in refereed and reputed journals. Academic productivity in terms of papers published in journals by the faculty members with respect to type of college is given in Table 1.

Table-1 informs that 63.7 per cent (358 out of 562) of the respondents have published a maximum of 1-2 publications in journals, only 6.7 per cent of the respondents have maximum of 3-4 articles in journals, and 6.4 per cent have maximum of 5-6 articles in journals. It also reveals that 20.6 per cent of the respondents (116 out of 562) have no publication in journals so far.

When comparing to Engineering Colleges, faculty members from Arts and Science Colleges have published more articles namely maximum 1-2 publication by 65.8 per cent (185 out of 261), maximum of 3-4 articles by 9 per cent, maximum of 5-6 articles by 10 per cent. Faculty members from Engineering College have maximum of 1-2 articles by 61.6 per cent, a maximum of 3-4 articles by 4.1 per cent and a maximum of 5-6 articles by 2.8 per cent (8 out of 281).

Academic productivity of faculty members in terms publications in journals are more among Arts and Science Colleges than Engineering Colleges.

Academic Productivity in Terms of Papers Published in Seminars/Conferences

Academic productivity of the faculty members are also evaluated through the number of articles published in seminars and conferences. They are simply a scholarly communication of their research findings in the forum of intellectuals. Number of papers published by the faculty members with respect to type of colleges is given in Table 2.

Table 2 indicates that 333 out of 562 respondents (59.4 per cent) have published at least 1-2 papers in seminars/conferences, 13.7 per cent of the respondents have 3-4 publications, 7.7 per cent have 5-6 publications, and 1.8 per cent have more than 8 publications in seminars/conferences. It also indicates that 18.3 per cent of the respondents (103 out of 562) have no publications in seminar and conferences.

When comparing college wise, 201 out of 281 (71.5 per cent) respondents from Arts and Science College have 1-2 publications and 132 out of 281 respondents (47.2 per cent) from Engineering Colleges have 1-2 publications. Similarly, 27 out of 281 respondents (9.6 per cent) from Arts and Science Colleges have 3-4 publications and 50 out of 281 respondents (17.8 per cent) from Engineering Colleges have 3-4 publications in seminars; and 8 out of 281 respondents (3 per cent) from Arts and Science Colleges

Table 1. Academic productivity of faculty: Papers published in journals

S.No	Academic Productivity – Papers in Journals	Type of College		Total
		Arts & Science College	Engineering College	
1	No Publication	35(12.5)	81(28.8)	116(20.6)
2	Between 1-2 Papers	185(65.8)	173(61.6)	358(63.7)
3	Between 3-4 Papers	25(8.9)	12(4.2)	37(6.7)
4	Between 5-6 papers	28(10.0)	8(2.8)	36(6.4)
5	Between 7-8 Papers	1(0.3)	0	1(0.2)
6	More than 8 papers	7(2.5)	7(2.5)	14(5.0)
Total		281(100.0)	281(100.0)	562(100)

Source: Primary Data Items in parentheses are percentage

Table 2. Academic Productivity of faculty in terms of papers published in seminars/ conferences

S.No	Academic Productivity – Papers published in Seminars/Conferences	Type of College		Total
		Arts & Science College	Engineering College	
1	No Publication	43(15.3)	60(21.3)	103(18.3)
2	Between 1-2 Papers	201(71.5)	132(47.0)	333(59.4)
3	Between 3-4 Papers	27(9.6)	50(17.8)	77(13.7)
4	Between 5-6 papers	8(2.8)	29(10.3)	37(7.7)
5	Between 7-8 Papers	1(0.2)	0	1(0.2)
6	More than 8 papers	1(0.2)	9(3.2)	10(1.8)
Total		281(100.0)	281(100.0)	562(100)

Source: Primary Data Items in parentheses are percentage

have 5-6 publications and 29 out of 281 respondents (12.4 per cent) from Engineering Colleges have 5-6 publications in seminars.

Academic Productivity in Terms of Number of Books Published

Another criterion to assess the academic productivity of the faculty members in higher education system is in terms of number of books published by them. Number of books published by the respondents with respect to type of college is in Table 3.

Table 3 shows that 500 out 562 respondents (88.9 per cent) have no book publications. Only 6.7 per cent of the respondents have published at least 1 to 2 books, 3.8 per cent have published at least 3-4 books and 0.4 per cent has published more than 4 books.

Moreover, books published between 1-2 category, 22 out of 281 respondents (7.8 per cent) from Arts and Science College and 16 out of 281 (5.6 per cent) from Engineering Colleges. Also, books published between 3-4 category, 15 out of 281 (5.2 per cent) are from Arts and Science Colleges and 7 out of 281 (2.4 per cent) are from Engineering Colleges.

Table 3. Academic productivity of faculty in terms of books published

S.No	Academic Productivity – Book Published	Type of College		Total
		Arts & Science College	Engineering College	
1	Nil	242(43.0)	258(45.9)	500(88.9)
2	Between 1-2 Books	22(3.9)	16(2.8)	38(6.7)
3	Between 3-4 Books	15(2.6)	7(1.2)	22(3.8)
4	More than 4 Books	2(0.4)	0	2(0.4)
Total		281(50.0)	281(50.0)	562(100)

Source: Primary Data Items in parentheses are percentage

Academic Productivity in Terms of Number of Project

Academic productivity of the faculty members are further analysed based on the number of projects completed or ongoing in their service. Academic productivity of the faculty members with respect to type of college is given in Table 4.

Table 4 discloses that 92.3 per cent of the respondents have not undertaken any projects. Only 39 respondents out of 562 (6.9 per cent) have undertaken projects at a maximum of 1-2 project, and 0.5 per cent have a maximum of 3-4 projects.

Moreover, 21 out of 281 faculty members from Arts and Science College have completed 1-2 projects and 18 out of 281 have 1-2 projects from Engineering Colleges; and 2 out of 281 from Arts and Science Colleges have completed 3-4 projects and 1 out of 281 from Engineering Colleges.

Thus, only a minimum number of faculty members have served as a principal investigator of projects in both type of college.

Information Literacy Score of the Faculty Members

Information literacy of the faculty members are determined based on the responses provided by the respondents on the predetermined statements on five-point scale ranging from 'do not know'; to 'excellent.' The Information literacy score of the respondents with mean score, standard deviation, minimum value, maximum value and percentile value along with other statistic are provided in Table 5.

Table 5 reveals that the mean information literacy score of the respondents is 47.10 and standard deviation 7.39. The minimum and maximum information literacy score are respectively 12 and 60. Moreover, 33.33 percentiles and 66.667 percentiles are respectively 46 and 51. The information literacy score less than 46 is categorized as low, the score between 46 and 51 as medium and score above 515 as high. Deviation in information literacy scores shows that faculty members differ in their information literacy level.

The average score 47.10 indicates that faculty members have information literacy at the medium level

Table 4. Academic Productivity of faculty in terms of projects/completed

S.No	Academic Productivity – Project Completed/Ongoing	Type of College		Total
		Arts & Science College	Engineering College	
1	Nil	258(91.8)	261(46.4)	519(92.9)
2	Between 1-2 Projects	21(7.4)	18(6.3)	39(6.9)
3	Between 3-4 Projects	2(0.8)	1(0.4)	33(0.5)
6	More than 4 projects	0	0	0
Total		281(100.0)	281(100.0)	562(100)

Source: Primary Data Items in parentheses are percentage

Table 5. Information literacy score of the respondents

S.No	Statistic – Information Literacy	Value
1	N	562
2	Mean	47.10
3	Standard Deviation	7.39
4	Minimum	12
5	Maximum	60
6	Percentile 33.33	46
7	Percentile 66.66	51
8	Low - Information Literacy	Less than 46
9	Medium - Information Literacy	Between 46 and 51
10	High – Information Literacy	Greater than 51

Source: Computed Value

Level of Information Literacy

Information literacy is vital for the effective retrieval of information from various e-resources. Respondents of the study are classified in to three group namely low, medium and high based on information literacy score and percentiles values of 33.33 and 66.66. Distribution of respondents based on their level of information literacy is given in Table 6.

Table 6 reveals that 38.8 per cent of the members of the faculty have medium level, 31.1 per cent have low level and 29.1 per cent have high level information literacy.

Comparison of Information Literacy Score Based on Gender

Information literacy scores of the members of the faculty are further compared based on gender using 't' test. The details of analysis are given in Table 7.

- **Hypothesis:** There is no significant difference in information literacy among the faculty members based on gender.

Table 6. Level of Information literacy and type of college

S.No	Level of Information Literacy	Total
1	Low	181(31.1)
2	Medium	218(38.8)
3	High	163(29.1)
Total		562

Source: Primary Data/Items in parentheses are percentage

Table 7 reveals that the mean information literacy scores of the male and female faculty members are respectively 46.88 and 47.28; standard deviation are respectively 7.83 and 7.02. The calculated value of 't' is not significant at any level ($t = -0.647, p > 0.05$). Therefore the hypothesis is accepted. Thus, it is interpreted that the members of the faculty do not differ significantly on information literacy score based on gender.

Comparison of Information Literacy Based on Locality

Information literacy scores of the respondents are further compared based on their locality. Details of 't' test are given in Table 8.

- **Hypothesis:** There is no significant difference in information literacy among the faculty members based on locality.

Table 8 discloses that the calculated value of 't' is not significant at any level ($t = -1.428, p > 0.05$). Therefore the null hypothesis is accepted. This shows that the faculty members do not differ significantly on information literacy score based on locality.

Comparison of Information Literacy Based on Age

Information literacy scores of the respondents are further compared based on age. For comparison, one way analysis is carried out to determine the significant difference exists among the subgroups based on age. The details of 'F' statistics are given in Table 9.

- **Hypothesis:** There is no significant difference in information literacy among the faculty members based on age.

Table 7. Comparison of information literacy score based on gender

Gender	N	Mean	Std. Dev.	t Value	Remarks
Male	256	46.88	7.83	- 0.647 $p = 0.518$	Not Significant
Female	306	47.28	7.02		
Total	562	47.1	7.39		

Table 8. Comparison of information literacy based on locality

Locality	N	Mean	Std. Dev.	t Value	Remarks
Rural	316	46.71	7.79	-1.428 $p = 0.156$	Not Significant
Urban	246	47.1	6.8		
Total		47.1	7.39		

Table 9. Comparison of information literacy based on age

<table>
<tr><th>Age</th><th>N</th><th>Mean</th><th>S. D</th><th>Source of Variation</th><th>df</th><th>Mean Square</th><th>F Value</th></tr>
<tr><td>Below 30 Yrs</td><td>166</td><td>46.82</td><td>7.93</td><td rowspan="2">Between Groups</td><td rowspan="4">3
558</td><td rowspan="4">202.02
53.93</td><td rowspan="4">3.746*
p=0.011</td></tr>
<tr><td>30-40 Yrs</td><td>287</td><td>47.90</td><td>7.05</td></tr>
<tr><td>40-50 Yrs</td><td>87</td><td>44.93</td><td>7.65</td><td rowspan="2">Within Groups</td></tr>
<tr><td>Above 50 Yrs</td><td>22</td><td>47.41</td><td>4.38</td></tr>
</table>

* Significant at 0.05 level

The Table 9 reveals that the calculated value of 'F' among the faculty members is significant at 0.05 level ($F = 3.746; p < 0.05$). Thus, the null hypothesis is rejected at 0.05 level. Hence, it is interpreted that the members of faculty differ significantly on information literacy based on age.

Comparison of Information Literacy Based on Experience

Information literacy score of the respondents are compared based on their experience using one way analysis of variance. The details of ANOVA are provided in Table 10.

- **Hypothesis:** There is no significant difference in information literacy among the faculty members based on experience.

Table 10 clearly indicates that the calculated value of F among the faculty members of based on experience is not significant at any level ($F = 0.195, p > 0.05$). Thus, the null hypothesis is accepted. Hence, it is interpreted that the faculty members do not differ significantly on information literacy score based on experience.

Comparison of Information Literacy Based on Type of College

Information literacy score of the respondents are further compared based on type of College using 't' test. The details are provided in Table 11.

- **Hypothesis:** There is no significant difference in information literacy among the faculty members based on type of college.

Table 10. Comparison of information literacy based on experience

<table>
<tr><th>Age</th><th>N</th><th>Mean</th><th>S. D</th><th>Source of Variation</th><th>df</th><th>Mean Square</th><th>F Value</th></tr>
<tr><td>Below 5 Yrs</td><td>192</td><td>46.86</td><td>7.87</td><td rowspan="2">Between Groups</td><td rowspan="4">3
558</td><td rowspan="4">10.73
54.9</td><td rowspan="4">0.195
p=.900</td></tr>
<tr><td>5-10 Yrs</td><td>265</td><td>47.29</td><td>7.61</td></tr>
<tr><td>10-15 Yrs</td><td>73</td><td>47.27</td><td>6.63</td><td rowspan="2">Within Groups</td></tr>
<tr><td>Above 15 Yrs</td><td>32</td><td>46.56</td><td>5.57</td></tr>
</table>

Table 11. Comparison of information literacy based on type of college

<table>
<tr><th>S.No</th><th>Institution</th><th>N</th><th>Mean</th><th>Std. Dev.</th><th>t value</th></tr>
<tr><td>1</td><td>Arts & Science College</td><td>281</td><td>47.75</td><td>6.8</td><td rowspan="3">2.08*
$p = 0.038$</td></tr>
<tr><td>2</td><td>Engineering Colleges</td><td>281</td><td>46.45</td><td>7.9</td></tr>
<tr><td colspan="2">Total</td><td>562</td><td>47.1</td><td>7.39</td></tr>
</table>

* Significant at 0.05 level

Table 11 indicates that the mean information literacy score of the members of the faculty of Arts and Science and Engineering colleges are respectively 47.75 and 46.45; standard deviation are respectively 6.8 and 7.9 and the calculated value of t is 2.08, which is significant at 0.05 level ($t = 2.08$; $p < 0.05$). Therefore the null hypothesis is rejected at 0.05 level. Hence, there is a significant difference exists among the members of the faculty on information literacy competency based on type college. Thus, members of the faculty differ in information literacy and members of Arts and Science colleges have more information literacy than the faculty of Engineering colleges.

Comparison of Information Literacy Based on Discipline

Respondents for the present study consist of faculty members belong to various disciplines namely Arts, Social Science, Science and Engineering. The significance difference in information literacy based on discipline is analyzed using ANOVA. The details are given in Table 12.

- **Hypothesis:** There is no significant difference in information literacy among the faculty members based on discipline.

Table 12 clearly indicates that the calculated value of 'F' among the faculty members is not significant at any level ($F = 2.39$, $p > 0.05$). Thus, the null hypothesis is accepted. Hence, it is interpreted that the faculty members do not differ significantly on information literacy score based on discipline.

Comparison of Information Literacy Based on Familiarity of Computer

Information literacy score of the respondents are compared based on familiarity of computer using ANOVA. The details of ANOVA statistic are given in Table 13.

Table 12. Comparison of information literacy based on discipline

<table>
<tr><th>Discipline</th><th>N</th><th>Mean</th><th>S. D</th><th>Source of Variation</th><th>df</th><th>Mean Square</th><th>F Value</th></tr>
<tr><td>Arts</td><td>89</td><td>48.09</td><td>6.6</td><td rowspan="2">Between Groups</td><td rowspan="4">3
5.58</td><td rowspan="4">130.92
54.5</td><td rowspan="4">2.39
p=.067</td></tr>
<tr><td>Social Science</td><td>64</td><td>48.88</td><td>5.9</td></tr>
<tr><td>Science</td><td>200</td><td>46.68</td><td>7.2</td><td rowspan="2">Within Groups</td></tr>
<tr><td>Engineering</td><td>209</td><td>46.54</td><td>8.12</td></tr>
</table>

Table 13. Comparison of information literacy based on familiarity of computer

User of Computer	N	Mean	S. D	Source of Variation	df	Mean Square	F Value
Advanced	236	47.88	7.38	Between Groups	2	226.99	4.195*
Intermediate	268	46.90	7.07		559	54.11	*p*=.016
Beginner	58	44.83	8.30	Within Groups			

* Significant at 0.05 level

- **Hypothesis:** There is no significant difference in information literacy among the faculty members based on familiarity of computer.

Table 13 reveals that the calculated value of F among the faculty members based on familiarity of computer is significant at 0.05 level ($F=4.195$; $p <0.05$). Therefore, the null hypothesis is rejected. Hence, the faculty members differ significantly on information literacy based on familiarity of computer. Computer knowledge influences their information literacy specifically on electronic environment.

Comparison of Information Literacy Based on Familiarity of Internet

Information literacy score of the respondents are compared based on familiarity of Internet using ANOVA. The details of one way analysis are given in Table 14.

- **Hypothesis:** There is no significant difference in information literacy among the faculty members based on familiarity of internet.

Table 14 clearly indicates that the calculated value of F among the faculty members is not significant at any level ($F = 2.79$, $p > 0.05$). Thus, the null hypothesis is accepted.

Hence, it is interpreted that the faculty members do not differ significantly on information literacy score based on familiarity of internet and type of college.

FINDINGS OF THE STUDY

Findings emerged from the study are:

Table 14. Comparison of information literacy based on familiarity of internet

User of Internet	N	Mean	S. D	Source of Variation	df	Mean Square	F Value
Advanced	227	47.89	6.86	Between Groups	2	151.94	2.79
Intermediate	265	46.79	7.56		559	54.38	*p*=.062
Beginner	70	45.70	8.2	Within Groups			

1. Faculty members have information literacy only at average level.
2. Academic productivity of faculty members in terms publications in journals are more among Arts and Science Colleges than Engineering Colleges.
3. Majority of the faculty members (59.4 per cent) have published at least 1-2 papers in seminars/ conferences, and 13.7 per cent of the respondents have 3-4 publications.
4. Only 6.7 per cent of the faculty members have published at least 1 to 2 books, 3.8 per cent have published at least 3-4 books and 0.4 per cent have published more than 4 books.
5. Among the Faculty Members, 38.8 per cent of the members of the faculty have medium level, 31.1 per cent have low level and 29.1 per cent have high level information literacy.
6. Faculty Members do not differ significantly on information literacy based on gender, locality, experience, discipline, and familiarity of Internet.
7. Faculty Members differ significantly on information literacy based on age, type of college and familiarity of computer.

CONCLUSION

No doubt, faculty members in the higher education system have information literacy above the average level and the academic productivity is prevalent among them in terms number of publications in seminars, in journals, publication of books and supervising various projects. This clearly shows the effect of information literacy on growth of literature. More training and awareness are to be provided to the faculty to acquaint with latest technology for retrieval of information and its effect which will reflect in terms of publication output in scholarly communication.

REFERENCES

Adeleke, D. S., & Emeahara, E. N. (2016). Relationship between information literacy and use of electronic information resources by postgraduate students of the university of Ibadan. *Library Philosophy and Practice*. Retrieved from http://digitalcommons.unl.edu/libphilprac/1381

Bibina, C. B., & Kabir, S. H. (2016). Understanding the factors related to digital literacy: A survey among science-research scholars of university of Kerala. *Journal of Knowledge & Communication Management*, *6*(2), 93–104. doi:10.5958/2277-7946.2016.00008.5

Catalano, A., & Philips, S. R. (2016). Information literacy and retention: A case study of the value of the library. *Evidence Based Library and Information Practice*, *11*(4), 2–13. doi:10.18438/B82K7W

Khan, J. (2015). Use of information sources and need of information literacy among students in Aligarh Muslim university, Aligarh. *International Journal of Library and Information Science*, *7*(1), 10–13. doi:10.5897/IJLIS2014.0447

Prince, G. & Saravanan, P. (2015). A study on awareness and perception towards open access resources among the users in the higher educational institutions in Kanyakumari District. *International Journal of Next Generation Library and Technologies, 1*(3), 01-09.

Salauddin, N. (2015). Usage of information communication technology application in national physical laboratory library, Delhi. *International Journal of Next Generation Library and Technologies*, *1*(1), 1–8.

Saravanan, P., & Mary, A. L. (2007). Teachers approach to internet and online information resources. *IASLIC Bulletin*, *52*(3), 150–158.

Chapter 2
Open Educational Resources and Scientific Literature:
Indian Perspective

G. Prince
Mar Eprahem College of Engineering and Technology, India

ABSTRACT

This chapter is intended to present briefly the open educational resources initiatives in India especially related to growth of scientific literature. Scholarly communication can be defined as the system through which research and other scholarly writings are created, evaluated for quality, disseminated to the scholarly community, and preserved for future use. An old tradition and a new technology have converged to make it possible as an unprecedented public good. The old tradition is the willingness of scientists and scholars to publish the fruits of their research in scholarly journals without payment, for the sake of inquiry and knowledge. The new technology is the internet and web 2.0 tools. Thus, the scholarly communications are available in the public domain in the open access platform. The system includes both formal means of communication, such as publication in peer-reviewed journals, and informal channels in electronic media of open access. Also presented is an overview of all available open source resources in Indian context.

INTRODUCTION

Scholarly communication can be defined as the system through which research information and other scholarly writings are created, evaluated for quality, disseminated to the scholarly community, and preserved for future use. An old tradition and a new technology have converged to make it possible as an unprecedented public good. The old tradition is the willingness of scientists and scholars to publish the fruits of their research in scholarly journals without accepting any payment, for the sake of inquiry and knowledge. The new technology is the internet and web 2.0 tools. Thus, scholarly communications are available in the public domain in the open access platform. The system includes both formal means of communication, such as publication in peer-reviewed journals and informal channels in electronic media of open access (UNESCO, 2015).

DOI: 10.4018/978-1-5225-3457-0.ch002

OPEN ACCESS RESOURCES

Open access resources are electronic resources which are freely available to the end user at free of cost. There is no need to pay either by the institution or the individual for accessing, and downloading articles. Users can read, download, copy, distribute, print, search, or link to full texts of these articles (Prince and Saravanan, 2015). Open access resources are not limited only to open access journal but it also may vary in forms such as open e-book, videos, audios, personal websites, discussion forums, open access databases, institutional websites, personal blogs, open access repositories, and institutional repositories and open source software.

Suber (2012) defined Open Access as "Open Access literature is digital, online, free of charge, and free of most copyright and licensing restrictions". The concept of open access has emerged in response to the restrictive access to knowledge in scholarly and scientific journals imposed by commercial publishing houses via subscription fees, license fees or pay-per-view fees (Christian, 2008).

Open access to knowledge is a generic term used for knowledge resources made available in the public domain for public access or public consumption at large scale, without any hindrance of subscription fee or access charges.

HISTORY AND DEVELOPMENT OF OPEN ACCESS

The concept of open access has emerged due to the conventional method of dissemination scholarly content through restricted access and against payment system of research and scholarly articles publications. It has been discussed in various forums consisting of educationists, publishers and policy makers.

Three OA declarations which form the basis for evolution of open access are commonly known as BBB declarations (The Budapest Open Access Initiative -2002, Berlin Declaration -2003 and the Bethesda Statement -2003) in the beginning of the 21st century have shaped OA publishing environment in the successive decades. These declarations have also hinted strong philosophical foundations for supporting the ideas and principles of OA. These three pioneering declarations got wide supports from the academicians, and renowned global thinkers. The regional, national and similar committees constituted for the purpose follows the BBB declarations, all of them endorse the principles of the OA model for maximizing the access and benefit to scientists, scholars and the public throughout the world. Thereafter, many organizations developed their open access policies.

EVOLUTION OF OPEN ACCESS IN INDIA

During the year 1999, Indian Academy of Science hosted a meeting on geographical information and virtually every one of the speakers focused on public access to geographical information. The evolution of open access policy in India began on a two day conference on Advances in Information Access and Science Communication held at M.S. Swaminathan Research Foundation, Chennai, (MSSRF) on 16-17 September 2000, as a tribute to Dr. Eugene Garfield on his 75th birthday. At this conference Prof. Stevan Harnad, open access arch evangelist spoke about "Scholarly Skywriting" and the need for every research performing institution to adopt open access self-archiving of preprints.

The Indian Academy of Science convened a meeting in April 2001, the second ICSU-UNESCO International conference on electronic publishing in science decided to encourage Indian Science and technology journal publishers to adopt electronic publishing of Indian journals by open access. The open access movement in India is acknowledged worldwide. In India it started modestly from few institutions and now spread all over. The Indian Institute of Science was the first institution in the country to set up an interoperable institutional repository eprints@IISc, and followed by Indian Institute of Management, Kozhikode; Indian Statistical Institute, Bangalore; Indian Institute of Technology, Delhi; National Institute of Technology, Rourkela; National Aerospace Laboratories, Bangalore; National Chemical Laboratory, Pune; INFLIBNET, Ahmadabad; National Institute of Oceanography, Goa, and Raman Research Institute, Bangalore. India has launched many new open access journals and also converted some reputed subscription-based journals to open access (Sahu and Parmar, 2006).

The National Knowledge Commissions (NKC) working group on open access recommended that public-funded research literature is made available to public through open courseware repositories for countrywide dissemination of quality courseware to many cross-sections of people. The scholarly literature and life-long learning materials produced by state-sponsored institutions would then be made accessible through open access channels such as national and institutional repositories. It also recommended that research papers resulting from public-funded research should be peer-reviewed before it is made available through open access channels. In 2005, the University Grants Commission of India (UGC) drafted a national policy framework entitled "UGC (Submission of Metadata and Full-text of Doctoral Theses in Electronic format) Regulations, 2005". These regulations proposed two sets of planned actions, such as Creation of Indian National Theses and Database and Submission of Ph.D Theses in electronic form (Das, 2008). Consequently, the UGC has established Shodhganga, an e-thesis database and now e-PG Pathshala @ Inflibnet, INFOPORT@ INFLIBNET and INFLIBNET IR.

OPEN EDUCATIONAL RESOURCES

The term open educational resources first came into use at a conference hosted by UNESCO in 2002, defined as "the open provision of educational resources, enabled by information and communication technologies, for consultation, use and adaptation by a community of users for non-commercial purposes" (Johnstone, 2005). The most common definition of Open Educational Resources (OER) now used is "Open educational resources are digitized materials offered freely and openly for educators, students and self-learners to use and reuse for teaching, learning and research".

OPEN ACCESS INSTITUTIONAL REPOSITORY

Institutional repositories (IRs) are also known as digital repositories, or open access repositories. There are four types of repository publications: the subject-based repository, the research repository, the national repository system and the institutional repository (Jain, 2012).

An institutional repository is an online archive for collecting, preserving, and disseminating in digital form of the intellectual output of an institution. Usually, it may be a research institution which would include materials such as research journal articles, digital versions of theses and dissertations, digital documents, curriculum, and course materials.

OPEN ACCESS PUBLISHERS AND EDUCATIONAL REPOSITORIES IN INDIA

A large number of the journals are published from India belong to learned societies and associations, and published by the association or the editor themselves without the involvement of any commercial publisher. The members of these learned societies receive the print copies of the journals without paying an annual or recurring fee. For continuing their publication activities the associations depends on non-member subscriptions, which are limited in number and restricted by and large to the Indian universities and colleges. Most of the Indian journals suffer from low circulation, low visibility, and low impact syndrome. With many fewer paid regional or international subscriptions, these journals have limited visibility and restricted mainly to the members of the association. The Indian open access journals will be able to reach to a wider audience. At the same time, loss, if any of paid non-member subscription is less likely to have a major effect on the economics of these journals. Impact of open access publishing has certainly helped the Indian journals to reach an international audience, as could be seen by the number of distribution of article downloads. The present chapter intended to present briefly the open educational resources initiatives in India especially for Science literature growth. Major open access platform for access of open educational resources from India include NISCAIR, Shodhganga@INFLIBNET, e-PG Pathshala, Infoport, INFLIBNET IR, NPTEL, DRTC, IETE, National Digital Library of India, DOAR, ROAR, Indian Academy of Sciences, Indian National Science Academy, IndMED, DOAJ, DRDO Journals, OA Library, High Wire Press, Indian Agricultural Research Institute, Bioline International, Indian Journals.Com, Bentham Open Publishes, MDPI, PDFDRIVE, PLOS and Medknow Publishing. The contributions of these institutions through open educational resources for the scientific literature growth are critically presented in this chapter.

NISCAIR

National Institute of Science Communication and Information Resources (NISCAIR) came into existence on 30th September 2002 with the merger of National Institute of Science Communication (NISCOM) and Indian National Scientific Documentation Centre (INSDOC). Both NISCOM and INSDOC, the two premier institutes of the Council of Scientific and Industrial Research (CSIR), were devoted to dissemination and documentation of S&T information. The core activity of NISCAIR is to collect/store, publish and disseminate S&T information through a mix of traditional and modern means, which will benefit different segments of society.

NISCAIR offers 18 journals in open access platform. They are Journal of Scientific and Industrial Research, Indian Journal of Biotechnology, Indian Journal of Chemistry, Sec A, Indian Journal of Biochemistry and Biophysics, Indian Journal of Chemistry, Sec B, Indian Journal of Chemical Technology, Indian Journal of Experimental Biology, Indian Journal of Engineering & Materials Sciences, Indian Journal of Fibre & Textile Research, Indian Journal of Geo-Marine Sciences, Indian Journal of Pure and Applied Physics, Indian Journal of Radio and Space Physics, Indian Journal of Traditional Knowledge, Journal of Intellectual Property Rights, Indian Journal of Natural Products and Resources, Annals of Library and Information Studies, Journal of Scientific Temper and Bharatiya Vaigyanik evam Audyogik Anusandhan Patrika. The users can download full text of articles without any restrictions.

Figure 1.
Source: http://www.niscair.res.in/

SHODHGANGA

The Shodhganga offered by INFLIBNET Centre provides a platform for research students to deposit their Ph.D. theses and make it available to the entire scholarly community in open access. The repository has the ability to capture, index, and store, disseminate and preserve ETDs submitted by the researchers. The full texts of all the documents submitted to Shodhganga are available to read and to download in open access to the academic community worldwide.

Shodhganga is an important Indian initiative and will facilitate open access to Indian theses and dissertations too the world academic community. Online availability of electronic theses through centrally maintained digital repositories will not only ensure easy access and archiving of these but will also help in raising the quality and standard of research. Presently 281 universities are contributing to the shodhganga database and more than 136718 full text theses are available.

E-PG PATHSHALA

e-PG Pathshala is an initiative of the MHRD under its National Mission on Education through ICT (NME-ICT) being executed by the UGC. The content and its quality being the key component of education system, high quality, curriculum-based, interactive e-content in 77 subjects across all disciplines of social sciences, arts, fine arts and humanities, natural & mathematical sciences, linguistics and languages are being developed by the subject experts and other R&D institutes across the country. Every subject has a team of principal investigator, paper coordinators, content writers, content reviewers and multimedia team.

Figure 2.
Source: http://www.shodhganga.inflibnet.ac.in

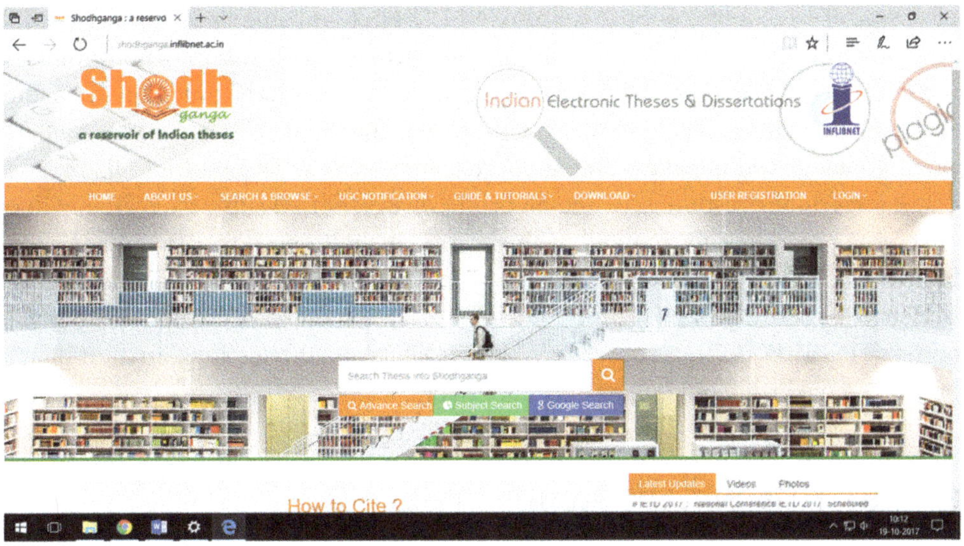

Figure 3.
Source: http://www.epgp.inflibnet.ac.in

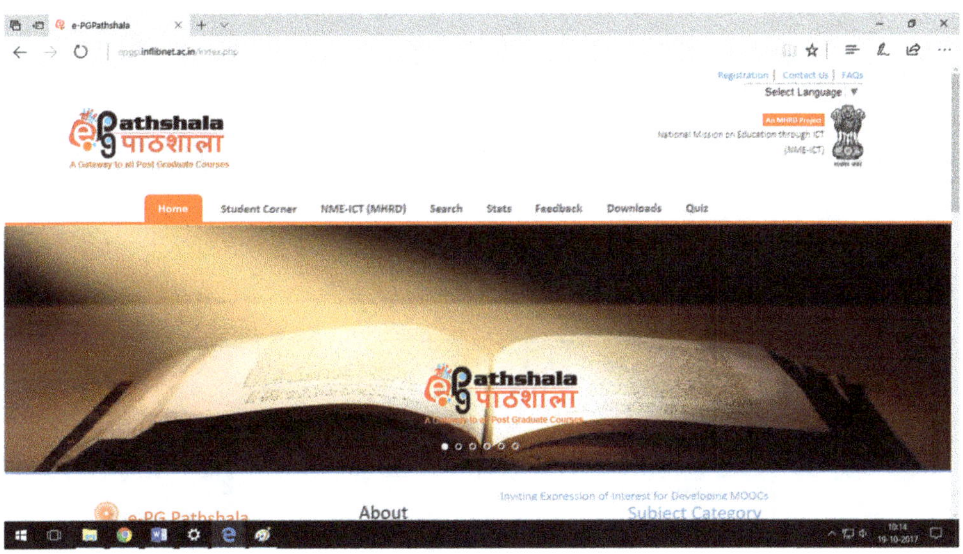

e-PG Pathshala offers 12800 e-textbooks and 10000 videos are in open access mode. Make e-content available to students and peers using different delivery modes to impart formal and informal education and for supplementing and complementing the process of teaching and learning in higher education and promote usage of e-content amongst students and peers. The materials available for PG courses in different universities shall benefit from those students studying at remote places.

INFOPORT SUBJECT GATEWAY FOR INDIAN ELECTRONIC RESOURCES

Infoport is a gateway of e-resources published from INFLIBNET and other institutions. The INFLIBNET centre promotes open access to Indian scholarly content through the InfoPort. The InfoPort is designed and developed to serve as a comprehensive gateway to all Indian scholarly content and a subject gateway for Indian electronic-resources. The gateway open-ups the Indian scholarly content scattered over the internet through an integrated interface that support search, browse and multiple listing. The InfoPort selectively catalogues online resources of Indian origin on diversified subjects available in open access through an elaborate process of testing and evaluation. The centre proposes to collaborate with librarians and scholars in college and universities in the process of identification and selection of resources. InfoPort is classified according to DDC, indexed subject wise and arranged alphabetically.

InfoPort offered INFLIBNET e-resources are Union Database, Shodhganga, Institutional Repository, N-List Resources, e-consortium and Open Journal system.

INFLIBNET INSTITUTION REPOSITORY

Institutional repository of INFLIBNET Centre provides in which the articles published in all conventional proceedings of INFLIBNET Centre. More over various training material, press clippings, newsletters etc. are available. DSpace open source software is used for this repository creation and retrieval system. The repository offered three main community collections that are freely available to the users. They are

1. INFLIBNET in Press and Media (News clipping about INFLIBNET)
2. INFLIBNET's Convention Proceedings (Caliber, Planner and Seminar proceeding)
3. INFLIBNET's Publications (INFLIBNET Annual Reports, INFLIBNET News letter, IRTPLA

Figure 4.
Source: http://www.infoport.inflibnet.ac.in

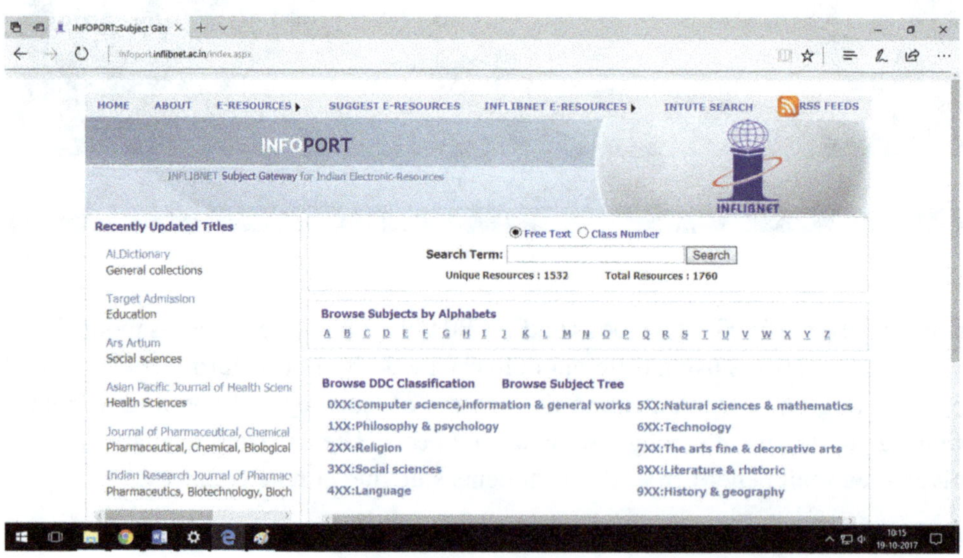

Figure 5.
Source: http://www.ir.inflibnet.ac.in

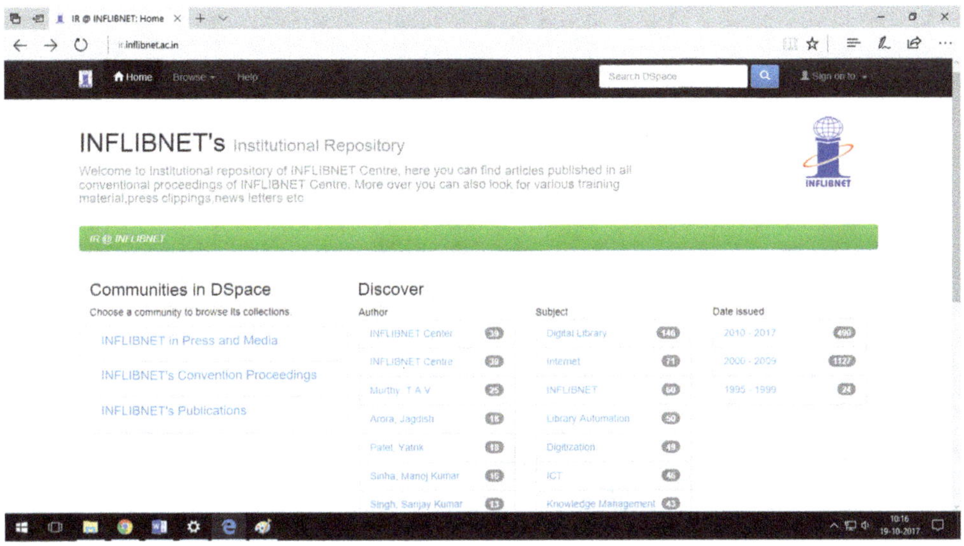

Course materials, Workshop and Publications by INFLIBNET Staff).

NPTEL

NPTEL is an acronym for National Programme on Technology Enhanced Learning which is an initiative by seven Indian Institutes of Technology's (IIT- Bombay, Delhi, Guwahati, Kanpur, Kharagpur, Madras and Roorkee) and Indian Institute of Science (IISc-Bangalore) for creating course contents in engineering and science.

Five engineering branches (Civil, Computer Science, Electrical, Electronics and Communication and Mechanical) and core science programmes that all engineering students are required to take in their undergraduate engineering programme in India were chosen initially. Contents for the above courses were based on the model curriculum suggested by All India Council for Technical Education (AICTE) and the syllabi of major affiliating universities in India. Course contents will be useful for teacher training and through them improve the quality of students. In addition, the course materials (both web and video) are freely accessible by everyone independent of their geographic location.

DRTC

The DRTC (Documentation Research and Training Centre) digital library offered the collection of library and information science resources like theses and dissertations, conference proceedings, open source library software and etc. are in open platform. DSpace open source software is used for this repository creation and retrieval system.

It includes the following communities for Library an Information Science,

Figure 6.
Source: http://www.nptel.ac.in

Figure 7.
Source: http://www.drtc.isibang.ac.in

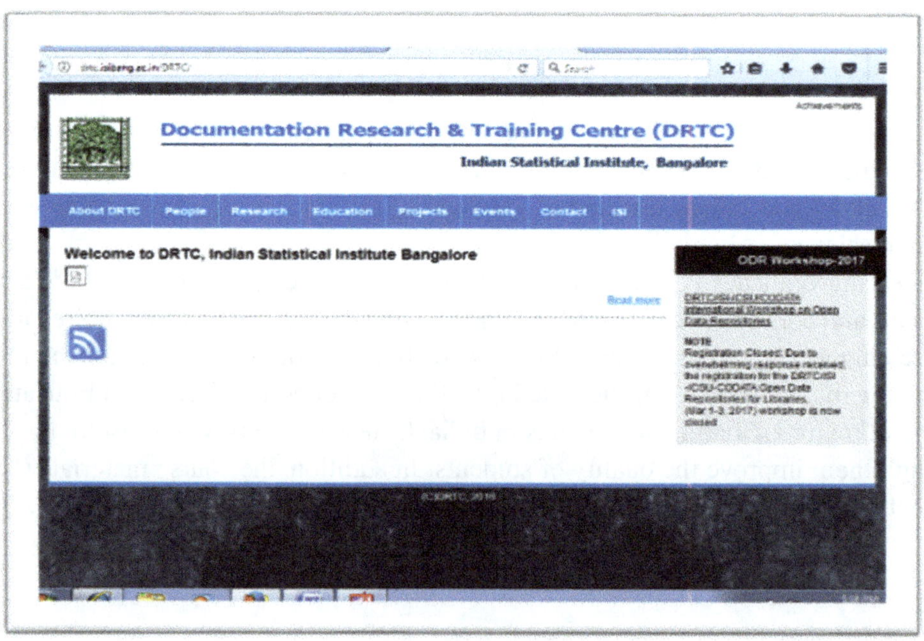

1. Theses and Dissertations
2. Presentations
3. Photographs of LIS Activities
4. Photographs of S.R. Ranganathan

5. Demo of Multi Lingual Documents
6. Publications and Articles
7. From Dspace.org
8. LIS Related Software.

IETE

The Institution of Electronics and Telecommunication Engineers (IETE) is India's leading recognized professional society devoted to the advancement of Science and Technology of Electronics, Telecommunication & IT, founded in 1953. The e-learning portal is open to all.

Content for the subjects is developed by IETE experts. User can directly access the content by clicking on each topic in the syllabus. It has been searched topic wise, so as to save user time. The reference links for the subjects in Section-A & Section-B of AMIETE and DIPIETE are being updated. The latest news about the events, activities of e-learning will be displayed time to time.

NATIONAL DIGITAL LIBRARY OF INDIA

NDL is a new initiative from MHRD and IIT Karapur. The objective of the project is to integrate all the existing digitized and digital contents across educational institutions of the nation to provide a single-

Figure 8.
Source: http://www.iete-elan.ac.in

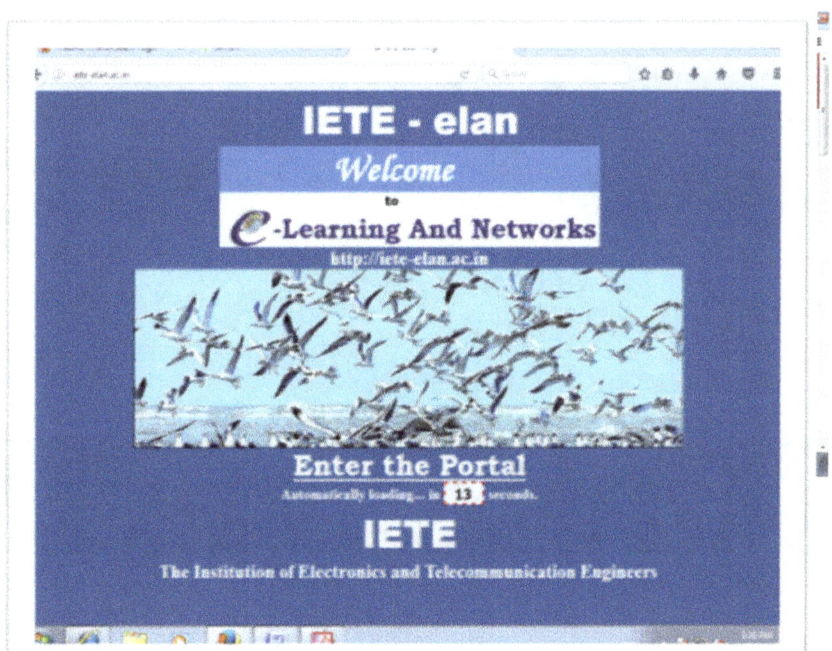

window access with e-learning facility to different groups of users ranging from primary level to higher education level of our county. NDL will harvest metadata and contents from all the Institutional Digital Repositories (IDR) of Universities and institutions, all other digital library initiatives, and NMEICT projects and index in the national digital library server so that all the e-contents can be searched and accessed in the full-text by the users through a single window.

NDL offered more than 15 lakhs of educational materials from primary to postgraduate level in several languages and repository host contents form multiple subject domains like technology, science, humanities, agriculture and others in the form of books, audio and video contents. Registered users can access national digital library at free of cost.

DOAR

DOAR (Directory of Open Access Repositories) provides a quality-assured listing of open access repositories around the world. Open DOAR is maintained by SHERPA Services, based at the centre for research communications at the University of Nottingham. Open DOAR is primarily a service to enhance and support the academic and research activities of the global community. It provides a quality-assured listing of open access repositories around the world and maintains a comprehensive and authoritative list of institutional and subject-based repositories. It also encompasses archives supported by funding agencies like the National Institutes for Health in the USA and the Wellcome Trust in the UK. (http://www.opendoar.org).

Figure 9.
Source: https://www.ndl.iitkgp.ac.in

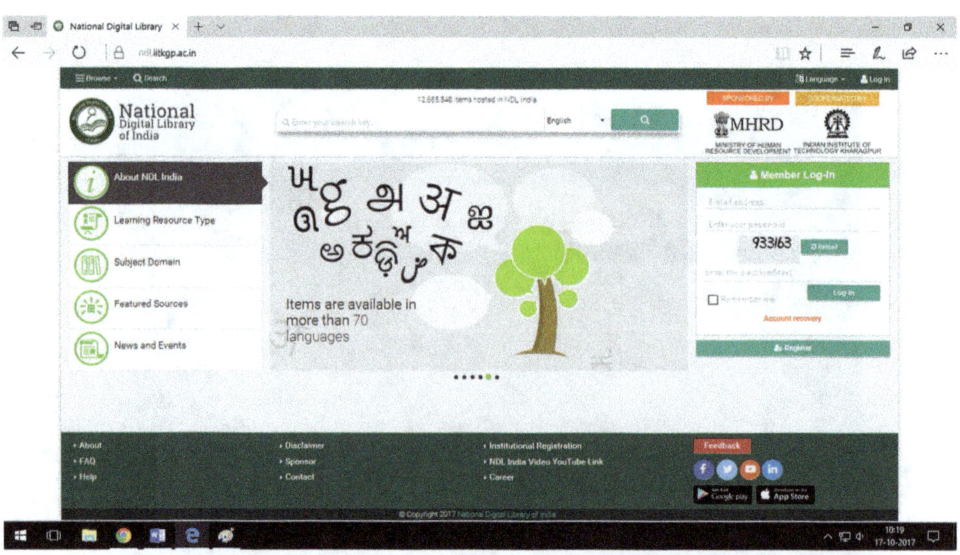

ROAR

The aim of Registry of Open Access Repository (ROAR) is to promote the development of open access by providing timely information about the growth and status of repositories throughout the world. ROAR is a searchable international registry charting the growth of open access mandates adopted by universities, research institutions and research funders that require their researchers to provide open access to their peer reviewed research article output by depositing it in an open access repository. ROAR is hosted at the University of Southampton and is made possible by funding from the JISC. (http://www.roar.eprints.org).

INDIAN ACADEMY OF SCIENCES

Founded in 1934 by Sir C V Raman, the Indian Academy of Sciences, Bangalore was registered as a society with the main objective is promoting the progress and upholding the cause of science. The academy from its very beginnings has taken great interest and initiative in publishing science research, and today publishes 10 science journals in different disciplines, uses an online submission and review management system that's tractable, papers are peer reviewed and published content is made available online and open access.

The academy have published 10 journals, several of which grew out of the original proceedings consists of Proceedings – Mathematical Sciences, Sadhana – Proceedings in Engineering Sciences, Journal of Chemical Sciences, Journal of Earth System Science, Journal of Genetics, Pramana–Journal of Physics, Bulletin of Materials Science, Journal of Astrophysics and Astronomy, Journal of Biosciences, Resonance – Journal of science education and Current Science. Since 2007, all our journals are co-published with Springer, and since 2015, with Springer Nature.

Figure 10.
Source: http://www.ias.ac.in

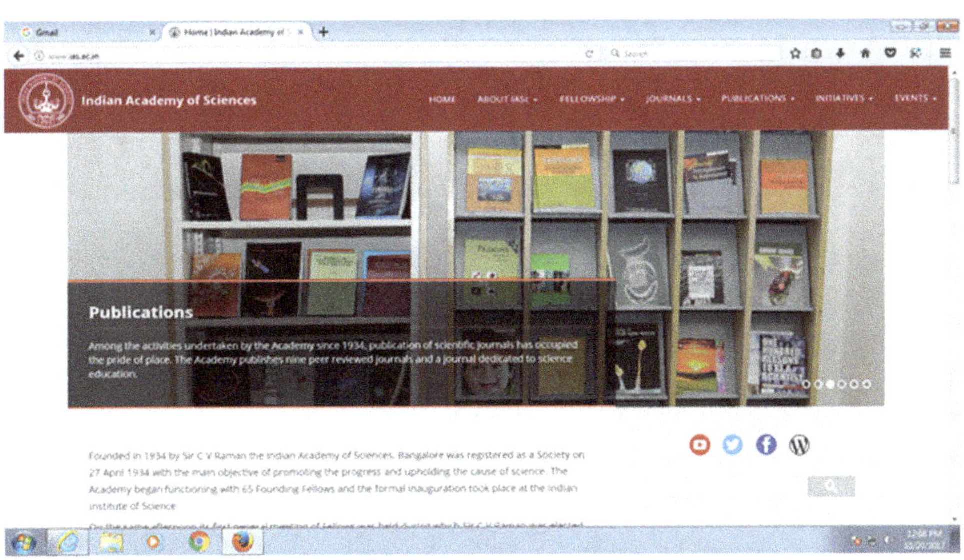

INDIAN NATIONAL SCIENCE ACADEMY

Indian National Science Academy is a premier science academy in the country, plays crucial role in promoting, recognizing and rewarding excellence. Another important task is to publish journals, organize scientific discussions and bring out proceedings and monographs. The academy promotes public awareness and understanding of science. Acting as links between the scientific community and the planners, they advise the governments on critical issues. Importantly they serve as forums for interaction among scientists within and outside the country. It recognizes many top young scientists, engineers and technologists by awarding them medals, providing modest support for pursuing research activities to needy young researchers. Senior scientists are honored with different awards/medals and the prestigious research professorship.

The Academy publishes three leading journals in open access system, they are

1. Proceedings of the Indian National Science Academy
2. The Indian Journal of Pure and Applied Mathematics (IJPAM)
3. The Indian journal of History Science (IJHS)

In addition, the academy also publishes its year book, annual report, compendium of fellows and biographical memoirs of deceased fellows.

INDMED (INDEXING OF INDIAN MEDICAL JOURNALS)

National Informatics Centre (NIC) and Indian Council of Medical Research (ICMR) had collaborated to setup ICMR-NIC centre for biomedical information. The centre, known as Indian Medlars Centre,

Figure 11.
Source: http://www.insa.nic.in

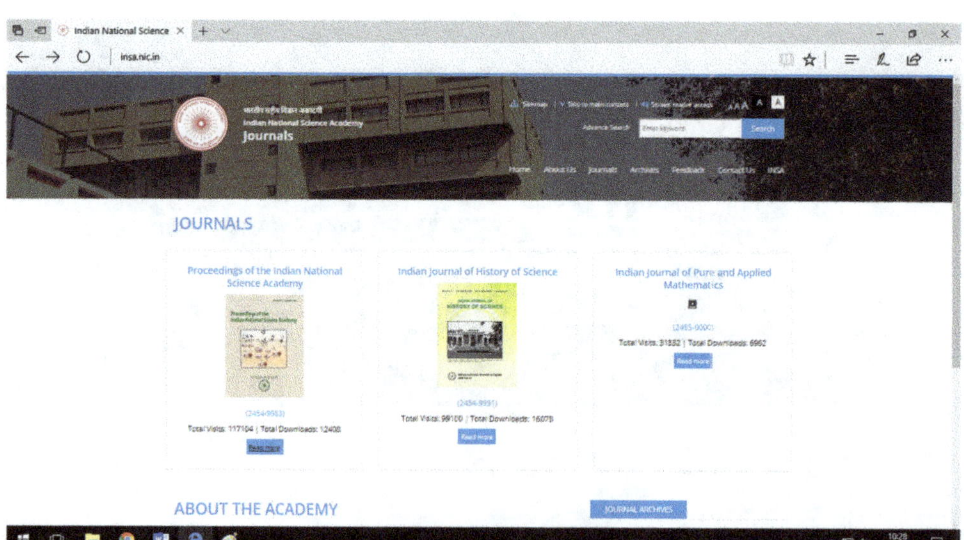

provided information support services to medical research community. It also produced two important resources, a bibliographic database IndMED - by indexing about 100 Indian medical journals and a portal medIND providing free full text access to 62 Indian medical journals.

IndMed is a bibliographic database for biomedical information seekers. It is available free of charge to the individuals through internet.

DOAJ

The Directory of Open Access Journals (DOAJ) was initially started in 2003 with financial support from the open society institute, since then has been developed and maintained by Lund University Libraries. It is one of the largest open access directories. The prime concern of the DOAJ is to provide the users with a quality controlled peer-reviewed scientific open access journals in full text. The DOAJ is defining OA journals as journals that are using financial model that doesn't charge readers or their institutions for access. DOAJ is taking the right of users to read, download, copy, distribute, print, search, or links to the full text of these articles.

Currently DOAJ database offered 9000 open access journals covering in all areas of science, technology, medicine, social science and humanities.

DRDO PUBLICATIONS

DRDO have published three journals in open access mode, they are

Figure 12.
Source: http://www.indmed.nic.in

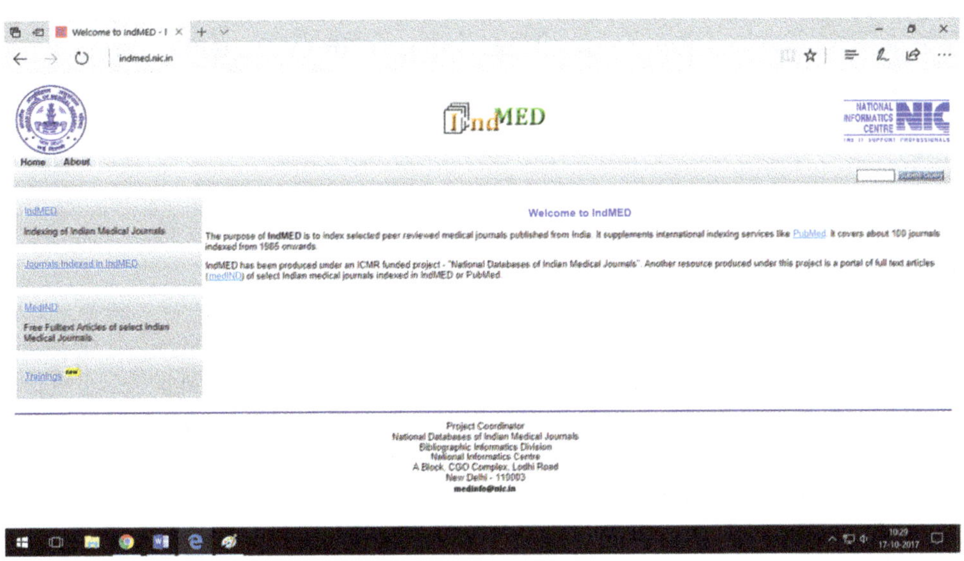

Figure 13.
Source: http://www.doaj.org

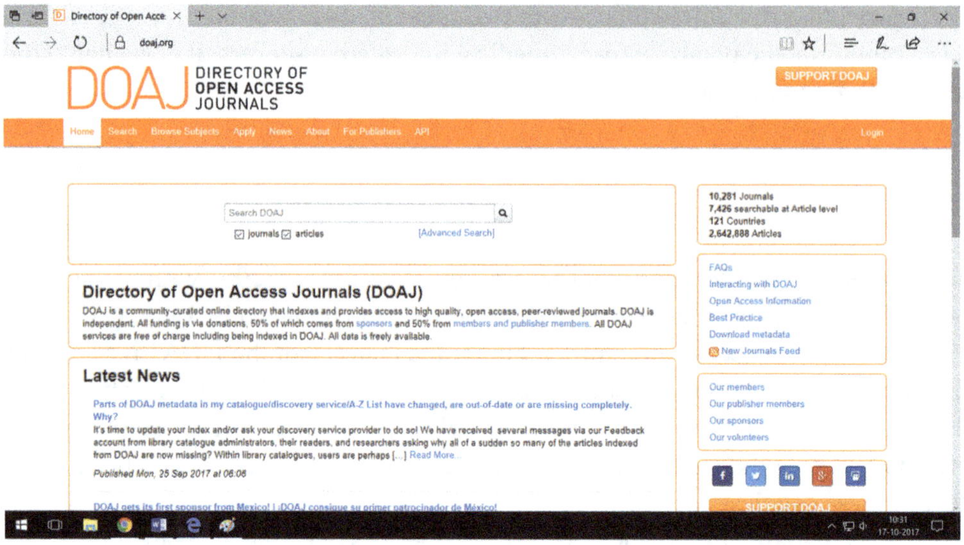

1. **Defence Life Science Journal:** The Journal publishes original research to feature recent progress in the field of defence/military life support system. It is a quarterly research journal. The Journal has been conceptualized to cater to the needs of scientists, researchers, academicians of life sciences and allied disciplines like; biotechnology, bio-medicine, bio-engineering, bio-electronics, non-invasive life imaging, pharmacology and toxicology, physiology, NBC warfare, food technology, and psychology.
2. **Defence Science Journal:** Defence Science Journal (DSJ) is a peer-reviewed, multidisciplinary research journal in the area of defence science and technology. Journal feature recent progresses made in the field of defence/military support system and new findings/breakthroughs, etc.
3. **DESIDOC Journal of Library & Information Technology:** DESIDOC Journal of Library & Information Technology (DJLIT*)* is a peer-reviewed, open access, bi-monthly journal that publishes original research and review papers related to IT applied to library activities, services, and products. Topics covered include automation, digitization, user interfaces, networks, hardware and software development, and technology.

OA LIBRARY

OA Library is an open source platform. More number of e-books and e-journals in all fields are available online. Users can freely download and printout for their academic needs. Open access library consists of the search engine, based on a database with metadata of open access papers. It is currently hosting links and metadata to more than 4,230,329 open access articles covering a wide range of academic disciplines. All full text articles from the search results are free to download. Besides searching, it also provides the publishing, indexing and depositing services.

Figure 14.
Source: http://www.publications.drdo.gov.in

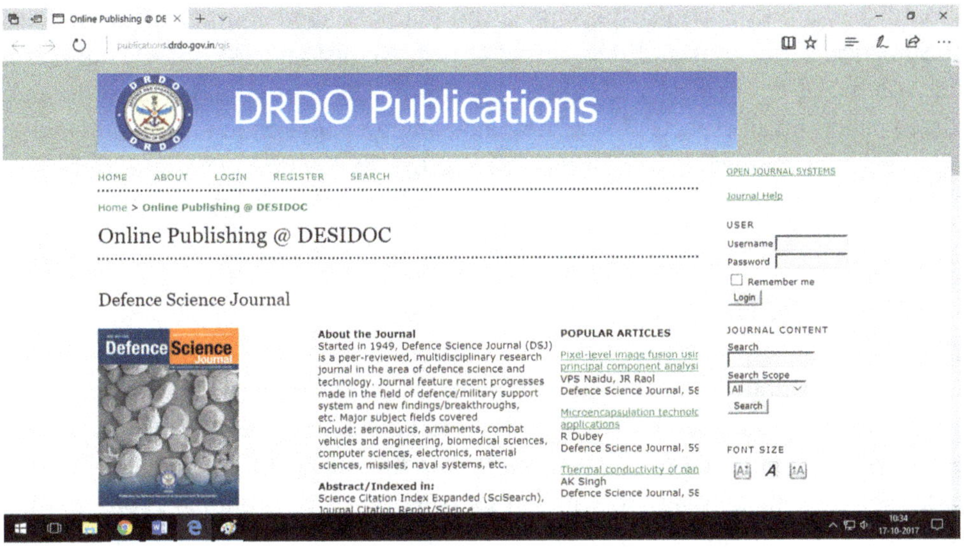

Figure 15.
Source: http://www.oalib.com

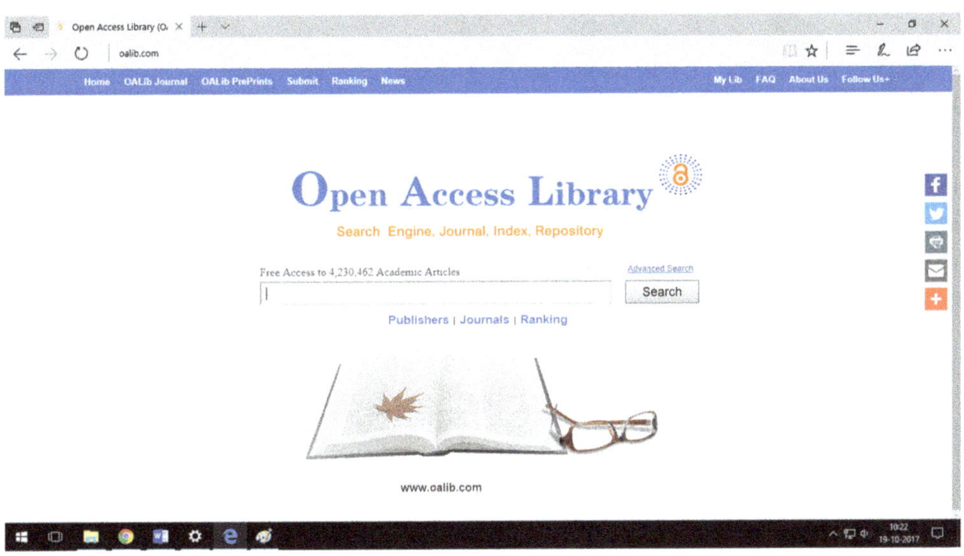

HIGHWIRE PRESS

Highwire press is a division of Stanford University libraries. Highwire press is a pioneer e-publisher in the area of scientific and medical journals. It offers 2,434,604 free full-text articles and 7,659,003 total articles (http://www.highwire.org).

INDIAN AGRICULTURAL RESEARCH INSTITUTE

The society published an international peer reviewed journal called Journal of the Indian Society of Agricultural Statistics. Three issues of the journal (April, August and December) are published annually. The first volume of the journal was released in 1948. The journal devoted to the publication of original research papers on all aspects of statistics and computer applications preferably with innovative applications in agricultural sciences or that have a potential application in agricultural sciences (http://www.isas.org.in).

BIOLINE INTERNATIONAL

Bioline International is a non-for-profit scholarly publishing cooperative committed to providing open access to quality research journals published in developing countries. BI's goal of reducing the south to north knowledge gap is crucial to a global understanding of health (tropical medicine, infectious diseases, epidemiology, emerging new diseases), biodiversity, the environment, conservation and international development. By providing a platform for the distribution of peer-reviewed journals, currently from Bangladesh, Brazil, Chile, China, Colombia, Egypt, Ghana, India, Iran, Kenya, Malaysia, Nigeria, Tanzania, Turkey, Uganda and Venezuela. Bioline is a non-profit electronic publishing service provider, committed to providing open access to quality bioscience research journals and medical journals (http://www.bioline.org.br).

INDIAN JOURNALS.COM

Indian journals.com provides a vast collection of interdisciplinary Indian journals and research publications with open access and subscription mode. It includes 334 research journals in 35 subjects are which in open access platform (http://www.indianjournals.com).

BENTHAM OPEN PUBLISHES

A Bentham Open publisher offers number of peer-reviewed open access journals. These free-to-view online journals cover all major disciplines of science, technology, medicine and social sciences (http://www.benthamopen.com).

MDPI

MDPI (Multidisciplinary Digital Publishing Institute) is an academic open-access publisher with headquarters in Basel, Switzerland. MDPI publishes 172 diverse peer-reviewed scientific open access electronic journals. All articles published by MDPI are immediately available worldwide under an open access license. Everyone has free and unlimited access to the full-text of all articles published in MDPI journals and free to re-use the published material. (http://www.mdpi.com).

PDFDRIVE

PDFDRIVE is a free search engine which permits the user to search, preview and download millions of PDF files. When PDF files are withdrawn from the web, these files are also withdrawn from PDF drive search results instantly. In this way, PDF Drive library constantly grows, stays up to date and offers you a huge database. It has more than 297,679,804 e Books are available freely as on date 13 October 2017. They are frequently updating by adding PDF files in the database (https://www.pdfdrive.net).

PLOS

PLOS was founded in 2001 as a nonprofit open access publisher, innovator and advocacy organization with a mission to accelerate progress in science and medicine by leading a transformation in research communication. PLoS published peer-reviewed open access journals. The journals vary in their selectivity and contain differing amounts of commentary articles from opinion leaders in a variety of scientific disciplines (https://www.plos.org).

MEDKNOW PUBLISHING

Medknow provides publishing services for peer-reviewed, online and print-plus-online journals in medicine on behalf of learned societies and associations with a focus on emerging markets. Medknow operates the open access model of publishing services, providing unrestricted online access to peer-reviewed scholarly research. Now Medknow provides publishing services to over 350 medical society journals in over 40 specialties (http://www.medknow.com).

Figure 16.
Source: https://www.pdfdrive.net/

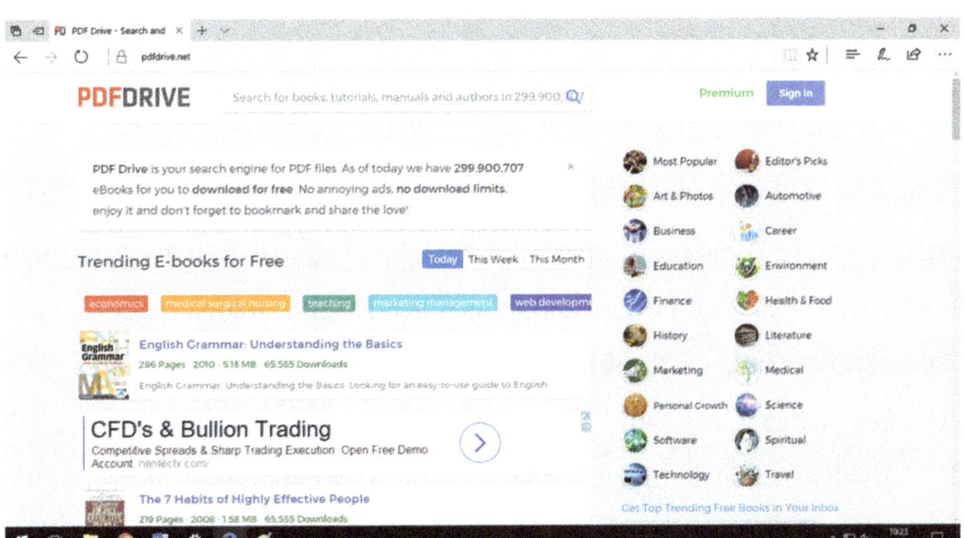

CONCLUSION

Access to open educational resources has emerged as one of the latest phenomenon in the process of making scholarly information free to all. Parallel to electronic resources on payment mode, the open access resources have emerged as gateway for accessing scientific literature. The users can download qualitative research articles in science and related to science through open access platform such as open access journals, open access repositories, institutional and personal websites. Thus the contribution of open educational resources for scientific literature growth is remarkable one and its contribution goes beyond our expectations.

REFERENCE

Bentham Open Publisher. (2017). Retrieved March 27, 2017 from http://www.benthamopen.com

Bioline International. (2017). Retrieved March 27, 2017 from http://www.bioline.org.br

Christian, G. E. (2008). *Open access initiative and developing world.* Retrieved March 25, 2017 from https://ssrn.com/abstract=1304665

Das, A. K. (2008). *Open Access to Knowledge and Information: Scholarly Literature and Digital Library Initiatives -The South Asian Scenario.* Retrieved March 25, 2017 from http://www.portal.unesco.org/ci/en/files/.../12075628443open_access.../open_access_book_en.pd

DOAJ. (2017). Retrieved March 27, 2017 from http://www.doaj.org

DOAR. (2017). Retrieved March 25, 2017 from http://www.opendoar.org

DRTC. (2017). Retrieved March 25, 2017 from http://www.drtc.isibang.ac.in

e-PG Pathshala@Inflibnet. (2017). Retrieved March 25, 2017 from http://www.epgp.inflibnet.ac.in

High Wire Press. (2017). Retrieved March 27, 2017 from High Wire Press: http://www.highwire.org

IETE. (2017). Retrieved March 25, 2017 from http://www. iete-elan.ac.in

Indian Academy of Sciences. (2017). Retrieved March 27, 2017 from Indian Academy of Science: http://www.ias.ac.in

Indian Agricultural Research Institute. (2017). Retrieved March 27, 2017 from http://www. isas.org.in

Indian Journals. *Com.* (2017). Retrieved March 27, 2017 from http://www.indianjournals.com

Indian National Science Academy. (2017). Retrieved March 27, 2017 from http://www.insa.nic.in

IndMED. (2017). Retrieved March 27, 2017 from http://www.indmed.nic.in

INFLIBNET Institution Repository. (2017). Retrieved March 25, 2017 from http://www.ir.inflibnet.ac.in

Infoport@ INFLIBNET. (2017). Retrieved March 25, 2017 from http://www.infoport.inflibnet.ac.in/

Jain, P. (2012). Promoting open access to research in academic libraries. Retrieved March 30, 2017, from http://unllib.unl.edu/LPP/

Johnstone, S. (2005). *International institute for educational planning,Open educational resources and open content for higher education background note*. UNESCO International institute for educational planning. Retrieved March 25, 2017 from http://www.unesco.org/iiep/virtualuniversity/media/forum/oer_forum_session_1_note.pdf

O.A. Library. (2017). Retrieved March 30, 2017 from http://www.oalib.com

MDPI. (2017). Retrieved March 30, 2017 from MDPI: http://www.mdpi.com

Medknow Publishing. (2017). Retrieved March 30, 2017 from http://www.medknow.com

National Digital Library of India (NDL). (2017). Retrieved April 20, 2017 from https://www.ndl.iitkpg.ac.in/

NISCAIR. (2017). Retrieved March 25, 2017 from http:// www.niscair.res.in

NPTEL. (2017). Retrieved March 25, 2017 from http://www.nptel.ac.in

PDFDRIVE. (2017). Retrieved October 14, 2017 from https://www.pdfdrive.net/

PLOS. (2017). Retrieved March 30, 2017 from PLOS: https://www.plos.org

Prince, G. & Saravanan, P. (2015). A study on awareness and perception towards open access resources among the users in the higher educational institutions in Kanyakumari district. *International Journal of Next Generation Library and Technologies, 1*(3), 01-09.

DRDO Publications. (2017). Retrieved April 20, 2017 from http://www.publications.drdo.gov.in

ROAR. (2017). Retrieved March 25, 2017 from http://www.roar.eprints.org

Sahu, D. K., & Parmar, R. C. (2006). Open access in the developing world: Regaining the lost impact In *Workshop on electronic publishing and open access: Developing country perspectives*, Bangalore, India, *November 2-3*.

Shodhganga@INFLIBNET. (2017). Retrieved March 25, 2017 from http://www.shodhganga.inflibnet.ac.in

Suber, P. (2012). Open Access. In Open Access (pp. 4-10). Cambridge: MIT Press.

UNESCO. (2015). *Introduction to open access*. Retrieved March 25, 2017 from http://www.unesdoc.unesco.org/images/0023/002319/231920e.pdf

Chapter 3
Health Information Systems and Their Usefulness Among the Women Folk of Kanyakumari District, India

R. Gnanabai
Scott Christian College, India

ABSTRACT

This chapter describes how health is the most precious component for the happiness and all-round development of human being in the society. Alongside this, information is an important resource for individual growth and survival. Therefore, a Health Information System (HIS) is a system for collecting/processing of data from various sources, and using the information for policy-making and management of health services. This chapter discusses health literacy and its association with health information needs and health information literacy, the status of women in India and their influence in the well-being of the family, and the impact of HIS. This chapter also proves that with enough data obtained from the women of Kanyakumari District in India, the government's policy needs to be directed towards women for the success of its health-care programmes.

INTRODUCTION

Health is a relative state in which one is able to function well physically, morally, socially, and spiritually in-order-to express the full range of one's unique potentialities within the environment in which one is living. In the words of Dubos (2007), "…health is primarily a measure of each person's ability to do and becomes what he wants to become…"

According to Buddha, of all the gains, the gains of health are the highest and the best. Health is not only essential for leading a happy life for an individual but it is also necessary for all productive activities in the society. Thus the concept of health is actively discussed in various period from the early stage of the society due to its social relevance and importance.

DOI: 10.4018/978-1-5225-3457-0.ch003

Since independence, India has built up a vast health infrastructure and health personnel at primary, secondary and tertiary care in public, voluntary, and private sectors. For producing skilled human resources, a number of medical and paramedical institutions including Ayurveda, Yoga and Naturopathy, Unani, Siddha, and Homeopathy (AYUSH) institutions have been set up.

HEALTH

The word 'health' is derived from the old English word 'hoelth', which means a state of being sound and was generally used to infer a soundness of the body (Dolfman, 1973). Health as a state of being can range from very good to very poor. Perhaps health can be more exactly described as an index, or a measurement of function rather than of status.

The term 'health' is viewed differently by different people and authorities all over the world. India's First Five Year Plan (1951-1956) stated health as a "positive state of well being in which harmonious development of mental and physical capacities of the individual lead to the enjoyment of a rich and full life".

HEALTH AND SOCIAL DEVELOPMENT

Health is the most precious component for the happiness and all round development of human being in the society. An individual's health and the health of a society are considered complementary to each other. It is a fact that the individual's health contributes to higher productivity and economic development, which in turn provides them with higher wages and better prospects for good health. Thus, societal development largely depends on the health of its community. It has been rightly pointed out in the report on the Census of India, 1971 that the expenditure on improving the health of a nation is regarded as good investment yielding indirect return in increased efficiency and productivity. Health is a major instrument of social and economic development and it can play a very important role in the creation of a new world. The level of development achieved by a society is often determined on the basis of the level of health and the system of health care and health services prevalent in the society.

The Universal Declaration of human Rights, states "Everyone has the right to a standard of living, adequate for the well-being of himself or herself and his or her family" (Sedamkar, 2011).

HEALTH INFORMATION

Information is an important resource for individual growth and survival. The progress of modern societies as well as individuals depends a great deal upon the provision of the right kind of information, in the right form and at the right time. Information is needed to take a right decision and also to reduce uncertainty.

The term 'health information' has been defined in different ways. In many countries, the term is interpreted to mean health statistical, epidemiological and other health oriented data used in the planning and management of health services by the decision-makers and administrators. It has been interpreted to mean health science literature covering health information at various levels. Some authorities safely consider health information just as a health literature.

According to World Health Organization (WHO, 1980), health information is information that contributes to knowledge and understandings that, in turn, provides part of the basis for making decisions and developing and managing services to improve health and health care. It also recognizes the three components of health information as (a) management and operational information; (b) health statistics and (c) health literature. Rolison (1998) defined health information as "...information on a continuum between health education and health promotion. Therefore, access to health information may contribute to health education and promote healthy lifestyle choices..."

Information is the first step to every healthy choice. Improvements in our health depend on us taking control over, and responsibility for, health as an important component of our everyday lives. This active participation requires full and continuing access to information: information about our bodies, their workings in health and illness, and the services available to us in treatment and care, support and cooperation... (Gann, 1986).

Ugboma (2010) has rightly mentioned the need for health information as "Information is a major resource that is needed in every sphere of life endeavour especially in health matters".

HEALTH INFORMATION SYSTEM

Health information systems refer to any system that captures, stores, manages or transmits information related to the health of individuals or the activities of organizations that work with the health sector. Health is an integrated effort to collect, process, report and use health information and knowledge to influence policy and decision-making programme action, individual and public health outcomes, and research. Sound decision-making at all levels of a health system requires reliable health statistics that are disaggregated by sex, age and socioeconomic characteristics. At the policy level, decisions informed by evidence contribute to more efficient resource allocation and at the delivery level; information about the quality and effectiveness of services can contribute to better outcomes.

Health Information System (HIS) is a system for collecting/processing of data from various sources, and using the information for policy-making and management of health services. HIS integrates data from civil/vital registrations, censuses, population surveys, facility surveys, individual records, service records and resource records for policy making and efficient management of health services.

HIS can be defined as "a set of components and procedures organized with the objective of generating information which will improve health care management decisions at all levels of the health system" (Lippeveld, Sauerborn & Bodart, 2000).

The health information system provides the underpinning for decision-making and has four key functions: data generation, compilation, analysis and synthesis, and communication and use. The health information system collects data from the health sector and other relevant sectors, analyses the data and ensures their overall quality, relevance and timeliness, and converts data into information for health-related decision-making (WHO, 2008).

The health information system is sometimes equated with monitoring and evaluation but this is too reductionist perspective. In addition to being essential for monitoring and evaluation, the information system also serves broader ends, providing and alert and early warning capability, supporting pattern and health facility management, enabling planning, supporting and stimulating research, permitting health situation and trends analysis, supporting global reporting, and underpinning communication of health challenges to diverse users. Information is of little value if it is not available in formats that meet the needs of multiple users-policy-makers, planners, manager's health care providers, communities, individuals. Therefore, dissemination and communication are essential attributes-of the health information system.

Health Information System (HIS) in most of the countries is meant denote as a system consisting of health statistics, epidemiological and other health related information useful for health planning and management. According to WHO, a national HIS is constituted to assemble, analyze, interpret and transmit information related to health to the right users.

WHO's conferences on Health Information System, held in Copenhagen in 1973 defined as, "A mechanism for the collection, processing, analysis and transmission of information required for organising and operating health services and also for research and training".

Health information system is a process whereby health data (input) are recorded, stored, retrieved and processed for decision-making (output). Decision making broadly includes managerial aspects such as planning, organising, and control of health care facilities at the national, state and institution levels and clinical aspects which can be subdivided into (i) providing optimal patient care, (ii) training of medical personnel to generate appropriate human resources, and (iii) facilitating research and developmental activities in various fields of medicine (Devi, 2010).

HEALTH OF WOMEN

Women's health issues have attained higher international visibility and renewed political commitment in recent decades. While targeted policies and programs have enabled women to lead healthier lives, significant gender-based health disparities remain in many countries. With limited access to education or employment, higher illiteracy rates and increasing poverty levels are making health improvements for women exceedingly difficult (Radha, 2012).

The slogan, "Healthy Women Healthy World" embodies the fact that as custodians of family health, women play a crucial role in maintaining the health and well-being of their communities (State of the World Population, 2005).

A woman's health is her total well-being not determined solely by biological factors and reproduction, but also by effects of workload, nutrition, stress, war and migration, among others.

Basic health care, family planning and obstetric services are essential for women yet these remain unavailable to millions. Gender-equitable approaches to health are needed to enable women's full participation in the planning and delivery of health services. The health of families and communities are tied to the health of women. The illness or death of a woman has serious and far reaching consequences for the health of her children, family and community (WHO, 2000).

Health information is necessary to women for their healthy life in the society. The following facts on women health rationalized the need for effective health information system exclusively for women.

- Maternal conditions are leading causes of death and disability among women (WHO, 1946). More than 99 percent of the estimated 5, 36,000 maternal deaths each year occur in the developing world (WHO, 2000).
- Every year, about 10 million women endure life- threatening complications during pregnancy and child birth, sometimes leading to long term disability (Nanda, 2005).
- According to World Health Organisation (2007), globally, women comprise half of the adults living with HIV/AIDS – in sub-Saharan Africa, the proportion rises to 61 per cent. A woman affected by HIV/AIDS is plunged further into poverty, losing the ability to provide for herself and her children.
- Report of the International Conference on Population and Development states that early and unwanted child-bearing, HIV and other sexually transmitted infections, and pregnancy related illnesses and deaths account for a significant proportion of the burden of illness experienced by women-especially in low-income countries.
- Nearly all maternal deaths are preventable through timely prenatal and postnatal care, skilled birth attendance during delivery and the availability of emergency care to deal with complications. The health benefits of spacing and limiting births for mothers and children with family planning services are well known.
- United Nations Economic and Social Council (2008) International funding from public and private donors and other non-governmental organizations accounts for only 15 per cent of the expenditures on reproductive health and family planning activities in developing countries. The majority of spending comes from within the countries themselves, with more than half coming from consumers' pockets. Women health has become a branch of medicine based on emerging trends. It refers to the branch of medicine that focuses on the diagnosis and treatment of diseases and conditions that affect a woman's physical and emotional well-being.

According to the Global Health Council, "the health of families and communities are tied to the health of women, the illness or death of woman has serious and far reaching consequences on the health of her children, family and community." But in the society, the socio-cultural factors that prevent women to benefit from quality health services and attaining the best possible level of health include:

- Unequal power relationships between men and women;
- Social norms that decrease education and paid employment opportunities;
- An exclusive focus on women's reproductive roles; and
- Potential or actual experiences of physical, sexual and emotional violence (Radha, 2012).

HEALTH INFORMATION NEED

The term, 'information need' has been a troublesome one, which has debated the difference between 'needs' and 'wants', between 'perceived needs' and 'expressed needs', and so on. According to Line (1974) defines, Information need as "what an individual ought to have for his work, his research, his education and his recreations."

The terms 'information needs' and 'information requirements' are often used for the same concept and represent the one and the same. Information requirements are however user specifications that applies when information is requested or exchanged and include aspects such as accessibility and relevance of information (Gericke, 2001).

According to Dervin (1983), "an information need is an impediment preventing an individual from moving forward in cognitive time and space. The person is faced with a gap that must be bridged by 'asking questions, creating ideas, and/or obtaining resources'. Such gaps do not occur in the abstract but arise out of particular events and situations".

The Librarian's Thesaurus defines 'Information Need' as "that needs which library services or materials are intended to satisfy" (Soper, 1990). The objectives of studying information needs are:

1. The explanation of observed phenomena of information use or expressed need;
2. The prediction of instances of information uses;
3. The control and thereby improvement of the utilization of information manipulation of essentials conditions.

Human needs are of two types namely Physical and Psychological: Physical needs are requirements for physical health and Psychological needs are requirements for mental health. However, Maslow (1970) classified human needs into five namely:

1. **Self-Actualizing Needs:** Formal education, leisure activities, ethics and values.
2. **Esteem Needs:** Multicultural awareness, emotional awareness, social system knowledge, sex education, ethics and values.
3. **Love and Belonging Needs:** Multicultural awareness, emotional awareness, leisure activities, interpersonal skills, ethics and values and sex education.
4. **Safety Needs:** Crime avoidance, traffic rules, emergency procedures, basic literacy and sex education.
5. **Psychological Needs:** Personal hygiene, nutrition, general health issues, AIDS prevention, drug, tobacco, alcohol abuse and child abuse.

For the need of information by the user certain conditions are to be followed.

These conditions may be a) whether the awareness of the need for information is either necessary or sufficient for concluding that the need exists. b) Secondly, whether the presence of a desire for information is necessary condition for saying that there exists a need for information. c) Lastly, is the lack of information a necessary or sufficient condition for concluding whether the information need exists.

The term 'information need' therefore does not necessarily imply that people are "in need of" information as such but that the use of information can lead to the satisfaction of a more basic need (Wilson, 1981).

The need is concluded when the purpose for the use of information is present. According to Derr (1983), the presence of "information purpose" as a necessary condition of information need has stressed two necessary conditions for information need as follows:

1. The presence of an information purpose.
2. The information in question, contributes to the achievement of an information purpose.

HEALTH INFORMATION LITERACY

Before defining health literacy and health information literacy, it is necessary to understand the meaning of information literacy because the later concepts are developed from the information literacy. Information literacy is a modern concept that enables the user to identify, use and evaluate the required information from the modern complex information sources.

According to the definition of American Library Association (1989) information literacy is a set of abilities requiring individuals to recognize when information is needed and have the ability to locate, evaluate, and use effectively the needed information. The information literacy is also recognized as the basis for lifelong learning. It is common to all disciplines, to all learning environments, and to all levels of education. It enables learners to master content and extend their investigations, become more self-directed and assume greater control over their own learning.

Broadly defined, information literacy is a set of abilities that allow a person to recognize when information is needed and to effectively and efficiently act on that need. An information literate individual is able to:

- Determine the extent of information needed.
- Access the needed information effectively and efficiently.
- Evaluate information and its sources critically.
- Incorporate selected information into his or her knowledge base.
- Use information effectively to accomplish a specific purpose.
- Understand the economic, legal, and social issues surrounding the use of information and access and use information ethically and legally (Rockman, 2003).

Information literacy is no longer just a library issue. It is a common phenomenon and required for everyone to be self-reliant on information and skills for life-long learning. Library professionals have to play an effective role inculcating the spirit of information literacy among the different category of users to fulfill their information needs.

Association Between Health Literacy and Health Information Literacy

The two term 'health literacy' (HL) and 'health information literacy' are the two widely used and related concepts. While the former term tends to dominate medical and nursing research, the latter is more directed towards issues relevant to Information Studies. Health literacy is defined as "the degree to which individuals have the capacity to obtain, process, and understand basic health information and services needed to make appropriate health decisions" (Healthy People, 2010). Based on a review of the definitions found in the literature and consultation with an expert panel, Berkman, Davis & McCormack (2010) offer a modified and further elaborated definition of HL as "the degree to which individuals can obtain process, understand, and communicate about health-related information needed to make informed health decisions."

The concept of health literacy represents a constellation of skills, which includes print literacy as understanding, locating, and interpreting documents; numeracy as mastering quantitative information; and oral literacy as abilities to speak and listen effectively. Alongside these three aspects of HL, also broader definitions have been brought in, including working knowledge of disease processes, ability to use technology, to network and interact with others socially, motivation for political action regarding health issues, and self-efficacy (Berkman Davis & McCormack, 2011). When an emphasis is put on preventive health, HL may become an everyday resource outside health care settings (Peerson & Saunders, 2009).

In 2003, the Medical Library Association (MLA) formed Health Information Literacy (HIL) task force to assist the association in addressing health information literacy issues (Rombo, 2004). The task force considered the Department of Health and Human Services (DHHS) definition of health literacy and American Library Association's (ALA) definition of information literacy to develop MLA's definition 'health information literacy'. According to this, HIL is

...the set of abilities needed to recognize a health information need, identify likely information sources and se them to retrieve relevant information, assess the quality of the information and its applicability to a specific situation, and analyse, understand, and use the information to make good health decisions...

This definition pre-supposes that a person has reading and writing skills, but it does not include the social dimension of information into health information literacy. However, despite the narrower approach, MLA's definition of HIL has two advantages:

1. It is specific enough and applicable to empirical research and
2. It is appropriate for assessing general everyday health information literacy among the literate population.

The concept of (HIL) pre-supposes that an individual recognizes the need for information and how to find, evaluate, use and subsequently communicate information effectively to solves particular problems or make decisions. More importantly, whether information comes from the internet or world wide web, online databases, books, films, conversations, posters, pictures or other images or any other possible sources, inherent in the concept of information literacy is the ability to understand and critically evaluate the information.

IMPACT OF WOMEN ON HEALTH-CARE

It is vital for a woman's health and life that she has access to medical and health care needs for preventive, as well as curative health, so as to be able to play her full part in the development of this country. Indian society has always recognized the need for special consideration for women in its traditional health sciences. In ancient India, it was the rishis or sages who composed the medical treatises. Women have easy access to them among all sections of the population except in the case of very aristocratic.

The World Health Organization has defined Community Health workers as workers who live in the community they serve, are selected by that community, are accountable to the community they work within, receive a short, defined training, and are not necessarily attached to any formal institution (WHO Study Group, 1989).

Good health is an invaluable asset for better economic productivity, both at the individual and national level, but it is valued by those who own it as a prerequisite for a better quality of life and better standards of living. Those who are the highest risk of poor health and its effects on longevity and morbidity are the poor, women and children, especially those in the lowest strata of society.

The inability to access healthcare and to spend on healthcare are the main reasons among these marginalized groups for their poor health status. Public healthcare systems do not have enough outreach to them; it is a known fact that rural India is facing an acute shortage of trained medical personnel. Across states, only 6.3 per cent posts are vacant whereas 67 per cent of doctors enrolled remain absent from duty.

Community Health Cell's empowering the poor rural women to be health activists community health workers and its impact on primary health care is narrated. The Ministry of Health and Family Welfare initiated a project towards empowering women for health with financial assistance from WHO-SEARO in 1998 to address the physical and emotional health problems of vulnerable groups of people in 15 states (Sudhamani, 2008).

Women account for 50% of all people living with HIV/AIDS globally. In the year 2000, there were 80 million unwanted pregnancies. 20 million unsafe abortions, 5lakhs maternal deaths, 99% of these cases were reported in developing countries.

WOMEN IN INDIA

The Eleventh Five Year Plan will provide an opportunity to restructure policies to achieve a New Vision based on faster, broad-based, and inclusive growth. One objective of the Eleventh Five Year Plan in to achieve good health for people, especially the poor and the underprivileged. In order to do this, a comprehensive approach is needed that encompasses individual health care, public health, sanitation, clean drinking water, access to food, and knowledge of hygiene, and feeding practices. The Plan will facilitate convergence and development of public health systems and services that are responsive to health needs and aspirations of people. Importance will be given to reducing disparities in health across regions and communities by ensuring access to affordable health care. The health status of women has improved over the last few decades; however, it remains a major development task. Long standing challenges like reducing unwanted fertility still exist in some countries while other countries have moved on to new and different challenges.

Close to 245 million Indian women lack the basic capability to read and write. Adult literacy rates for ages 15 and above for the year 2000 were female 46.4% male rate of 69%. The average nutritional intake of women is 1400 calories daily. The necessary requirement is approximately 2200 calories. 38% of all HIV positive people in India are women yet only 25% o beds in AIDS care centers in India are occupied by them. 92% of women in India suffer from gynaecological problems. 300 women die every day due to childbirth and pregnancy related causes. The maternal mortality ratio per 10,00,000 live births in the year 1995 was 440.

Maternal conditions are the leading cause of death and disability among women. More than 99 per cent of the estimated 5, 36,000 maternal deaths each year occur in the developing world. Every year, about 10 million women endure life-threatening complications during pregnancy and childbirth, sometimes leading to long-term disability. Early and unwanted childbearing, HIV and other sexually transmitted infections, and pregnancy-related illness and deaths account for a significant proportion of the burden of illness experienced by women, especially in low-income countries. Nearly all maternal deaths are preventable through timely prenatal and postnatal care, skilled birth attendance during delivery and the availability of emergency care to deal with complications. Millennium Development Goal 5 focuses on reducing the maternal mortality ration (MMR) by 75 per cent between 1990 and 2015 and ensuring universal access to reproductive health by 2015.

DETERMINANTS OF WOMEN'S HEALTH

Health status is influenced by complex biological, social, and cultural factors that are highly interrelated. These factors affect men and women differently. Women's reproductive biology, combined with their lower socio-economic status, results in women bearing the greater burden from unsafe sex which includes both infections and the complications of unwanted pregnancy. Biological and social factors affect women's health throughout their lives and have cumulative effects. Therefore, it is important to consider the entire lifecycle when examining the causes and consequences of women's poor health (Mukherjee, 2012).

Biological Determinants

Unlike men, women are subject to risks related to pregnancy and childbearing. Where fertility is high and basic maternity care is not available, women are particularly vulnerable. Certain conditions including hepatitis, anemia, malaria, and tuberculosis, can be exacerbated by pregnancy. Biological factors, women have a higher risk per sexual exposure of contracting sexually transmitted infections (STIs), including the human immunodeficiency virus (HIV) than do men. In addition, because women with STIs are less likely to have recognizable symptoms, they may delay treatment until an advanced stage, with more severe consequences. Human papilloma virus infection results in genital cancer much more frequently in women that in men, and it is the single most important risk factor for cancer of the cervix. Gynecological cancers including breast, cervical, uterine, and ovarian account for 27 per cent of all malignancies occurring to women kin developing countries.

Socio-Economic Determinants

Poverty underlies the poor health status of developing country population, and women represent a disproportionate share of poor. Furthermore, the cultural and socioeconomic environment affects women's exposure to disease and injury, their diet, their access to and use of health services, and the manifestations and consequences of disease. In all regions reproductive health continues to be worst among the poor. Women in the poorest households have much higher fertility rates than those in the wealthiest and far fewer births in the presence of skilled health professionals, contributing to higher maternal mortality ratios.

Women's disadvantaged social position, which is often related to the economic value placed on familial roles, helps perpetuate poor health, inadequate diet, early and frequent pregnancy, and a continued cycle of poverty. In countries where women are less educated and have less control over decision making and family resources, they are also less apt to recognize health problems or to seek care. Women's low socioeconomic status makes them more vulnerable to physical and sexual abuse and mental depression. Unequal power in sexual relationships exposes women to unwanted pregnancy as well as STIs. Their low social status has also led to more and more women in forced prostitution.

STATUS OF WOMEN HEALTH IN THE DEVELOPING WORLD

When women are sick, there is no one to look after them. When men are sick, they can be looked after by women. In developing countries, women's health status is changing in response to several emerging trends. On the positive side, more girls are attending school, delaying marriage and childbearing, and having smaller families. However, the rate of HIV/AIDS infections is accelerating among women, with young women particularly at risk.

The world has witnessed an increase in life expectancy at birth, primarily because of the improved survival of infants and young children. Developing countries are now faced with an unfinished health agenda of problems such as continuing high maternal mortality ratios and malnutrition, and the new challenge of an increasing prevalence of chronic diseases such as cardiovascular disease resulting from an aging population. Socio-medical problems, such as gender-based violence, are also an increasing source of concern. Many health concerns merit attention to improve women's health.

THE STUDY AND ITS FINDINGS

In this context a survey was conducted among the women in Kanyakumari district with intention to assess the health information needs of women. The survey on health awareness among the women that majority (50.4%) of the women folk have only medium level awareness towards health. Moreover, their health awareness is influenced by age, locality, educational qualification, occupation, religion, marital status and size of the family. They have aware on various health parameters such as vaccination (84.2%), child care (84.0%), basic reasons for fever (78.6%), age of puberty (75.7%), preventive measures (74 per cent) and 71.9 per cent on head ache. They have poor awareness on cholesterol level (34.3 per cent), sugar level (58.9 per cent), blood pressure level (56.5 per cent, and reason for diabetes (13 per cent). Among those who have high level awareness towards various health parameters, 37.6 per cent of the respondents are aware of vaccination, child care (37 per cent), first aid of various health issues (34.7 per cent), basic reasons of fever (34.7 per cent), age of puberty (34.5 per cent), reproduction (30.6 per cent), basic reasons of headache (30.2 per cent), blood pressure level (24.7 per cent) and sugar level (14.6 per cent).

The study revealed that majority of the women (83.1 per cent) are aware of Allopathic Medicine, and 71 per cent are aware of Herbal Medicine and followed by Naturopathy Medicine by 57.6 per cent, Siddha by 55.7 per cent, Ayurvedic medicine by 51.6 per cent and Chiropractic Medicine by 57.4 per cent. In-spite-of the fact that majority of the women aware of both the medical insurance scheme and free ambulance services, only 46.9 per cent of the women have availed medical insurance scheme. Awareness of the women on various media sources for seeking health information. Television is the first preference and followed by magazines, Books, Newspapers, Library, Internet, Posters and radio. Women are more aware of 'Television' as a source in dissemination of health information than other media sources. Awareness of women towards various personal resources majority of the women preferred 'primary health centres', health departments, NGOs, local chemists, and birth attendants. Thus, women are aware of personal sources for health information from primary health centres. Majority of the women (65.4 per cent) are aware of the availability of health information on the web. Most preferred libraries for seeking health information by the women are academic libraries followed by public libraries and medical libraries.

Health information need of women reveals that top most priority is to information related 'puberty and reproduction', followed by 'sickness', 'child health', 'physical exercise' and 'symptoms and preventive measures'. The least priority areas are 'first aid', 'hygienic and sanitation' and 'vaccination'. Frequency of health information needed by women is as follows: 33 per cent needs the health information occasionally, 23.8 per cent rarely need health information, and 12.6 per cent needs health information often and 2.7 per cent needs health information always. Majority of the women (69 per cent) needed health information before visiting the medical practitioner and during the treatment. Availability of some information on the symptoms of the disease helps in early detection and also in its prevention.

Health seeking behavior of women among the personal sources, 85.2 per cent of the women reports that they collect health information from general hospitals, it is followed by from medical professionals by 81.6 per cent, from public health centers by 80.1 per cent, from friends and relatives by 64.4 per cent, by attending class or course by 62.6 per cent, and NGOs by 62.1 per cent. Among the impersonal sources, books are preferred by 90.1 per cent of the respondents to collect health information; public libraries are used by 78.9 per cent and academic libraries by 63.7 per cent. Women collect health information through discussions with medical professionals, discussions with friends, relatives and colleagues, participating in health awareness programmes, and through individual effort such as reading books, magazines and other documentary sources.

Health Information literacy majority of the respondents (68.6 per cent) are familiar with the term health information literacy. Women have health information literacy at medium level only. It also shows that 31.4 per cent have high health information literacy, 31.1 per cent have medium health information literacy and 37.5 per cent have low level health information literacy. Women do differ significantly on health information literacy based on age, educational qualification and occupation.

Barriers in accessing health related information location, language and lack of time are the major individual barriers among the women in accessing relevant health information. Lack of support from the library professionals, lack of up-to-date information on health, lack of libraries and lack internet access are the major barriers perceived by the women in accessing health information. Most of the women perceived the problems in locating health information is only at minimum and the intensity of the perceived problems is only at low level.

CONCLUSION

The present study reveals that women in Kanyakumari District have health information literacy at moderate level and it is significantly associated with their health awareness. Moreover, their health awareness on various health indicators is at medium level. Health awareness of the women is influenced by the demographic variables such as age, educational qualifications, occupation, health information literacy and monthly income. Moreover, women use modern information and communication technologies for accessing health information. The study also revealed that majority of the women depends on academic libraries for seeking health information in Kanyakumari District. Public libraries and information centers, and public health centers are also used for accessing health information. It is concluded from the study that, the existing information system must give importance and priority to fulfill their health information needs which helps to develop a health society. Continuous education programme, awareness programme and training are required to improve health information access, and to be self-reliant on their health information for health decisions making and helps to empower decision taking ability.

SUGGESTIONS

Improving women's health requires a strong and sustained commitment by governments and other stakeholders, a favorable policy environment, and well-targeted resources. Long-term improvement in education and employment opportunities for women will have a positive impact on the health of women and their families. In the short term, significant progress can be achieved by strengthening and expanding essential health services for women, improving policies, and promoting more positive attitudes and behavior towards women's health.

The women are to be motivated to participate in health camp and health education programmes. The authorities concerned should take efforts to establish an effective health information system to disseminate health information to the public. Effective community health care mechanism and information centers for creating health awareness among the women are the need of the hour. The primary health information system should take the health tips to the door steps of the public. The contents of the mass media such as Television and Newspaper should be improved to provide health information and preventive steps periodically to the public. The district administration as well as voluntary agencies should organize health camps and health awareness programmes on various health issues and on seasonal diseases.

One of the most reliable measures of prevention is education and training. The Council on Graduate Medical Education has been providing funds for numerous centers to research women health issues, which has to be increased. More individual-level preventive and personal habits are vital for good health. Most physicians believe that a baseline physical examination is a reliable comparative tool. Counseling for special issues concerning cigarette smoking, exercise, diet, primary disease prevention, safe sexual practices, alcohol abuse, psychological disorders and violence should organized.

Knowledge of family history is important since many diseases have strong propensity among first-degree relatives. Blood pressure should normally be measured every other year. Screening tests for breast, cervical, and colorectal cancer is recommended. Pap smears taken during routine pelvic examinations can screen for disease processes in the reproductive tract. Serum cholesterol monitoring and reduction are advised. Proper documentation of indigenous knowledge on treatment system, and prevention is to be carried out for the benefits of coming generation.

REFERENCES

American Library Association (ALA). (1989). ALA Presidential Committee on Information Literacy Final Report. Chicago: American Library Association.

American Medical Association Foundation Health Literacy. (2003). *Help Your Patients Understand.* CME Programme.

Berkman, N. D., Davis, T. C., & McCormack, L. (2010). Health Literacy: What Is It? *Journal of Health Communication, 15*(2), 9–19. doi:10.1080/10810730.2010.499985 PMID:20845189

Derr, D. L. (1983). A conceptual analysis of information needs. *Information Processing & Management, 19*(5), 273–278. doi:10.1016/0306-4573(83)90001-8

Dervin, B. (1983). An overview of sense-making research: Concepts, methods and results to date. In *Proceedings of the International Communication Association Annual Meeting,* Dallas, TX.

Devi, S. (2010). Medlars: Health information system in library. *University News, 48*(40), 14–22.

Dolfman, M. (1973). The concept of health: An historic and analytic examination. *The Journal of School Health, 43*(8), 491–497. doi:10.1111/j.1746-1561.1973.tb03740.x PMID:4583082

Dubos, R. (2007). *Saunders Comprehensive Veterinary Dictionary* (3rd ed.). Elsevier.

Gann, R. (1986). *The Health Information Hand Book: Resources for self care*. England: Gower.

Health Metrics Network Framework and Standards for Country Health Information Systems. (2008). WHO.

Line, M. B. (1974). Draft definitions: Information and library needs wants, demands and users. *Aslib Proceedings*, *26*(2), 87. doi:10.1108/eb050451

Lippeveld, T., Sauerborn, R., Bodart, C., & World Health Organization. (2000). Design and Implementation of Health Information Systems. Geneva: WHO.

Maslow. (1970). *Motivation and Personality* (2nd ed.). New York: Harper and Row.

Mbagaya, G. M., & Odhiambo, M. O. (2005). Mother's health seeking behaviour during child illness in a rural western Kenya community. *African Health Sciences*, *5*(4), 322–327. PMID:16615844

Medical Library Association (MLA). (2003). The Medical Library Association Task Force on Health Information Literacy. July 23, 2003. Retrieved January 7, 2011 from http://www.mlanet.org/resources/healthlit/ define.html

Mukhereje, M. (2012). Improving Women's Health. *HealthAction*, *25*(4), 4–7.

Radha, P. (2012). Empowering women for healthy families. *HealthAction*, *25*(4), 8–10.

Rockman, I. F. (2003). Integrating information literacy into the learning outcomes of academic disciplines, A critical 21st century issue. *College & Research Libraries News*, *64*(9), 612–615.

Rolison, J. (1998). Health information for the teenage years, what do they want to know? *Information Research, 3*(2). Retrieved from http://information.net/lr/3 3/papaer42html

Soper, M. E. (1990). *The librarian's thesaurus: A concise guide to library and information terms*. Chicago: American Library Association.

Sudhamani, N. (2008). Revitalizing Primary Health Care through Women's Empowerment. *HealthAction*, *21*(12), 48.

Ugboma, M.U. (2010). Provision and use of legal information among civil rights groups in the oil producing communities of Nigeria. *Library Philosophy and Practice.*

The United Nations. (1948Universal Declaration of Human Rights (UDHR). *General Assembly resolution 217A*.

Wilson, T. D. (1981). On user studies and Information Needs. *The Journal of Documentation*, *37*(1), 5–6. doi:10.1108/eb026702

Wilson. (1981). On user studies and information needs, *Journal of Documentation, 37*(1), 3-15.

World Health Organization. (1946). *Preamble to the constitution of World Health Organization* as *adapted by the International Health Conferences.*

Chapter 4
How to Design and Frame a Questionnaire

Rayees Farooq
Lovely Professional University, India

ABSTRACT

The present chapter explores the essential stages in designing and developing the questionnaires and offers a basic introduction to beginners who are at the preliminary stages of their research. The study was approached purely from a theoretical perspective and various issues and challenges have been addressed including the length of questionnaires, how to distribute a questionnaire and how to analyze the data. The study concluded that researchers often rely on the previously developed questionnaire rather than developing the new one because it requires the knowledge and expertise. For novice researchers, developing entirely a new questionnaire requires a basic understanding of the format of the questionnaire, length of the questionnaire and how to distribute a questionnaire. Therefore, the chapter presents a systematic way to develop a questionnaire keeping in view all the issues and problems faced by researchers who are engaged in study for a thesis or in other research projects.

INTRODUCTION

The questionnaire is the research instrument based on different set of questions for the purpose of gathering information from the respondents across various fields. According to Rowley (2014), questionnaires are one of the most widely used means of collecting data and therefore many novice researchers in business and management and other areas of the social sciences associate research with questionnaires. It is not easy to design a questionnaire; it requires a lot of effort and dedication in creating the questionnaire that answers your questions and fulfills your purpose. Moreover, the questionnaire is a sort of document which includes a various set of questions including open and closed-ended, dichotomous questions, ranking and Likert scale and the respondents are asked to respond the questionnaire. Developing a questionnaire is critical to the success of the survey because ambiguous questions, inappropriate wording, the length of the questionnaire can affect the response rate and makes the survey inconsequential. According to Beri (2008), a marketing researcher intending to collect primary data has to be extremely

DOI: 10.4018/978-1-5225-3457-0.ch004

careful in deciding what information is to be collected, how many questions are to be formulated, what should be their sequence, what should be the wording of each question and what should be the layout of the questionnaire. The issues of response and non-response bias are very common in evidence-based research leading to the biased results and findings.

If you are a novice to research and want to design a questionnaire for your dissertation or thesis then this study will help you in designing, distributing and collecting the data of the questionnaire in a very efficient and effective way.

The present study is an attempt to highlight the some of the core issues and challenges in designing the questionnaire viz. length of the questionnaire, distribution of questionnaire, response and non-response bias.

The remainder of the study is organized as follows: the first section discusses the introduction of the study. The second section discusses the types of questionnaires. The third section discusses why questionnaires are being used? The fourth section discusses the rationale of using the different types of instruments in designing the questionnaire. The fifth section discusses what kind of research can be conducted through questionnaires.

Section sixth discusses the type of questions asked through questionnaires. The seventh section discusses the length of the questionnaire. Section eighth highlights the selection of respondents. Section ninth discusses the distribution of questionnaires. Section tenth highlights the role of response and non-response bias. Section eleventh discusses the analysis of data and preparation of data for analysis.

TYPES OF QUESTIONNAIRE

According to Saunders et al. (2005), the design of a questionnaire differs to how it is administered, and in particular, the amount of contact you have with the respondents. Generally, there are two types of questionnaires such as self-administered questionnaire and interviewer administered. The self-administered questionnaires are usually completed by respondents themselves. These questionnaires are usually delivered and returned through email. The questionnaires which are delivered by email or the internet are called online questionnaires. Google docs play a major role in framing the online questionnaires and reduce the time and effort. On the other hand, the questionnaires which are delivered by post to the respondents are called postal questionnaires, usually, respondents return the questionnaire by post also. The questionnaires which are delivered to respondents by hand and are collected later are referred to as delivery and collection questionnaire (Saunders et al., 2005). The second one is interviewer administered, where interviewer physically meets the respondents and grabs the useful information by interviewing the respondent. The widely used interview schedules are a telephonic questionnaire or telephonic interviews where investigator asks the questions from the respondents by telephonic conversation. Whereas, the structured interview is a quantitative method where interviewer exactly asks the same questions in the same order.

Self-administered questionnaire:

1. Online questionnaire
2. Postal questionnaire
3. Delivery and collection questionnaire

Interviewer administered:

1. Telephone questionnaire
2. Structured interview

WHY QUESTIONNAIRES?

Questionnaires are mostly used in conducting quantitative research, where the researcher wants to profile the sample in terms of numbers (e.g. the proportion of the sample in different age groups) or to be able to count the frequency of occurrence of opinions, attitudes, experiences, processes, behavior's, or predictions (Rowley, 2014). Questionnaires are typically designed to collect the data from a large number of respondents, however; interviews are limited to the smaller number of respondents. For example, if we want to study the impact of market orientation on the business performance, we can generate a bunch of items or questions from the literature to measure the market orientation and business performance. For measuring a market orientation and business performance we need to generate a set of items and then decide who will be the respondents for the same. If the researcher wants to adopt the different approach other than questionnaire like interviews, which can be time-consuming.

COULD YOU USE AN EXISTING INSTRUMENT

A researcher can use three types of instruments including the one which has been previously developed. Researchers can develop a new instrument and can modify the previously developed questionnaire. However, the standardized questionnaire can save time and resources as the researcher has no need to generate the items for developing the questionnaire. The researcher should give outline details of the standardized questionnaire which he/she has adopted from the different studies. If the standardized questionnaire is not available, you will have to develop your own questionnaire. The best way is to modify the previously developed questionnaire which can save the need to develop a new construct. According to Farooq (2016a)

...there are various approaches or methods for developing a scale. The first approach deals with the development of a new scale. The second approach deals with the selection and use of a scale that has been developed previously. The third is concerned with the modification of an existing scale or the introduction of a set of new items to an existing scale.

It is not necessary to validate the contents of a standardized questionnaire, however developing entirely a new scale requires the validation of contents by subject matter experts.

TYPE OF RESEARCH CONDUCTED THROUGH QUESTIONNAIRES

Different type of research can be conducted through questionnaires including descriptive research and analytical research. Descriptive research involves describing situations and does not determine cause and effect relationship. The most commonly used descriptive methods are the survey, observation, and case study method. Observation method deals with the direct observation of phenomena, while as in survey method respondents answer the questions administered through questionnaire or schedule. However, case study method is more prominent and involves in-depth study of an individual or group of individuals.

According to Kothari & Garg (2014), descriptive research includes surveys and fact-finding enquiries of different kinds and the notion of descriptive research is a description of the state of affairs as it exists at present. Analytical research is a specific type of research which includes critical thinking skills and the measurement of facts about the research being conducted. The purpose of analytical research is to study the relationship between different variables. Analytical research involves examining the multifaceted relationship between variables. The most commonly used analytical research includes regression and structural equation modeling. Regression can be used to study the impact of the independent variable on dependent variable for instance, if we have to study the impact of market orientation on business performance regression can be the best possible option available to researchers and academicians. On the other hand, structural equation modeling can be used to study the multiple dependence relationships. For example, image creates satisfaction and satisfaction creates loyalty, satisfaction is both independent and dependent variable.

TYPE OF QUESTIONS ASKED THROUGH QUESTIONNAIRE

The researcher should use the language which respondents can easily understand so that he can grab the relevant information about a construct or a scale under study. If the researchers are interested in studying the relationship between market orientation and business performance; it is unlikely that he will ask the questions separately for both the constructs. What type of question should a researcher ask? depends upon the nature of the study as well. For instance, if a researcher wants to study the subjective measures of performance, he can use 5 point or 7 point likert scale.

When designing or developing a questionnaire, a researcher is often confronted with what type of questions to ask the respondents such as open and closed-ended questions, dichotomous questions, likert type and questions based on nominal or ordinal scale.

Over the years, various rules of thumbs have been formulated pertaining to-dos and donts of asking questions. Below are some rules which can guide a researcher for avoiding mistakes in designing the questions.

1. Always bear in mind research questions
2. What do you want to know
3. How do you answer it
4. Avoid ambiguous terms in questions
5. Avoid long questions
6. Avoid double-barreled questions
7. Avoid technical terms

These rules can be very fruitful for the researchers who want to design questions without any ambiguity and want to ensure that information gathered actually represents the construct (Bryman and Bell, 2010). Respondents can ask a different type of questions including open and closed-ended questions, dichotomous, Likert type and ranking questions. Open-ended questions allow respondents to give responses in their own way, while as closed-ended questions limits the respondents with a grid of questions from which they must choose to answer the question. However, when a question has two possible options such as yes/no, true/false there are different ways of posing these questions to the respondents. For instance, do you believe that corporal punishments are justified? A) Yes B) No.

Likert scales are frequently used to collect the opinion data. When responding to a Likert scale, respondents specify their level of agreement or disagreement on a five or seven point scale for a particular set of items. The researcher should include negative statements to ensure that respondents read each statement carefully however, the researcher should bear in mind that negative statements are reverse coded. The final category deals with ranking questions, where the researcher asks the respondents to rank a group of items or statements depending upon the importance of each statement or item. Respondents compare the items to each other in order of their preference.

HOW DO I MAKE SURE THAT RESPONDENTS REPLY MY QUESTIONS ACCURATELY?

There are various questions which arise in the minds of the researchers before designing a questionnaire. Am I targeting the right respondents? Are they knowledgeable enough to respond my questions? Are they taking the enough time to respond the questions? How to tackle the problems if the respondents are biased? Well, a researcher should target the right respondents to grab the accurate information or data. For instance, if a researcher is studying the relationship between knowledge sharing and business performance and he is targeting the wrong population then there is a likelihood of response bias. Respondents should be knowledgeable enough to answer the given set of statements.

The researchers or academicians should not pose any question which is equivocal rather he should consult the subject matter experts for validating the items he wishes to measure. Choosing the right subject matter expert for measuring the content validity can mitigate the chances of response bias. For example, if you want to measure the business performance of an organization you could select a battery of items measuring business performance including sales growth, return on investment, market share and asset growth and the subject matter experts selected should be from the same field so that items selected for measuring business performance actually reflects the same.

LENGTH OF THE QUESTIONNAIRE

Don't make questionnaires too long, failure to observe this rule will produce a decline in response rate not because of the fact that respondents become reluctant to respond but also because the response from each respondent will decline (Burchell & Marsh, 1992). According to Herzog & Bachman (1981), many researchers are convinced that survey instruments have a maximum length beyond which there is an increasing probability of premature termination, random responding, or other behavior patterns which result in data of lower quality. Childers & Ferrell (1979) opined that the issue of questionnaire length has been the subject of considerable investigation but with often confusing and conflicting results. There has been a growing perception about increasing the length of the questionnaire which may lead to poor response rates. However, this perception may change depending upon the nature/type of study.

Questionnaire length is an important factor in determining the response rate in various studies. For instance, if a researcher wants to measure knowledge sharing of an organization and he/she selects 20 items from different studies, the response rate will be much lower than 10 item scale of knowledge sharing. A study conducted by Lund & Gram (1998), on a random sample of one thousand Norwegian women between 35-49 years old; five questionnaires were developed according to the length and title. A total of 3106 questionnaires were returned with response rate 62.1% and the highest response rate 70.2% was achieved with a questionnaire having two pages as compared to the questionnaire having four and six pages respectively with a response rate of 62.8% and 63.3%.

HOW DO I CHOOSE RESPONDENTS?

Respondents should be selected on the basis of nature of the study, for instance if you are conducting a study on knowledge management, your population will comprise chief knowledge officers or the respondents who are having the decision-making power including top management. The first step in identifying and selecting respondents is to deliberate the population of your study. Respondents should be selected carefully otherwise it will change the purpose of the study, for instance, if you are conducting a research on information technology orientation, your respondents will comprise of chief information officers, IT experts, information science researchers and data scientists. However, LinkedIn, CII (Confederation of Indian industry), and American technology association can be very supportive in selecting the respondents depending upon the membership.

Respondents should be selected on the basis of prior experience and contact details also called sampling frame. The sampling frame is the list of people forming a population and may include individuals, households, and institutions. There are various approaches for selecting a sample including probability and non-probability sampling. According to Lathman (2007), the choice to use probability or non-probability sampling depends on the goal of the research.

A detailed summary of probability and non-probability sampling methods available to the researcher as shown in Table 1. Probability sampling is also called random sampling where every item has equal chances of selection whereas; non-probability sampling does not ensure the equal chances to every item.

DISTRIBUTION OF A QUESTIONNAIRE

The distribution of a questionnaire depends upon the mode of delivery available to the researcher such as self-administered mail, email, administered by telephone and face-to-face (Bird, 2009). Once your questionnaire is designed, pilot tested and amended and your sample selected, the questionnaire can be used to collect data. This final stage is called administering the questionnaire. In administering your questionnaire, it is important that you abide by your university's or professional body's code of ethics (Saunders et al., 2011).

The best way to distribute a questionnaire is to use an online survey such as Survey Monkey. This site allows the researcher to develop his own questionnaire and provides extra options of a target audience. Survey Monkey helps in creating online surveys and results can be viewed graphically.

Mail survey can be very supportive and researcher should make sure to include a self-addressed envelope so that respondents can easily mail their responses. Conducting face-to-face interviews can

Table 1. Sampling techniques

	Sampling Technique	Description
Probability	Simple random sampling	Every item of the universe has equal chances of inclusion in the sample.
	Cluster sampling	It is a modified version of simple random sampling, where population is infinite and sampling frame does not exist
	Systematic sampling	Every ith item is selected from the list randomly from the first k units of the list and then every kth unit is selected. In this type of sampling, the population is finite.
	Stratified sampling	If a sample is drawn from a population which does not constitute a homogenous group, stratified sampling can be used for the same. Therefore we divide the whole population into different strata's that are individually homogenous and items are selected from each stratum
Non-probability	Quota sampling	Suppose a researcher wants to study the impact of income on education, he may select a quota of 100 students and collects the data from the same. The sample of 100 students selected is called quota sampling
	Convenience sampling	If a researcher collects the data from the respondents on the basis of his own convenience and respondents are willing to cooperate for providing information then such a sampling is called convenience sampling
	Judgmental sampling	If a researcher is to select a sample of 200 IT experts, he may exercise his own pre-conception and judgment for including IT experts in the sample such a sampling is called judgmental sampling.
	Snowball sampling	A group of individuals are selected and asked to recommend others in the same or different field such a sampling is called snowball sampling

reduce response bias and it is more challenging for respondents to avoid answering a question when the researcher asks it directly. However, creating online questionnaire through Google docs will reduce time and effort and can be very helpful when researchers find it difficult to edit the questionnaire.

RESPONSE AND NON-RESPONSE BIAS

A paper published by Kanuk and Beerenson in 1975 titled, *"Mail Surveys and Response Rates: A Literature Review"*, published in Journal of marketing research, raised many questions viz., How reliable are mail questionnaire surveys today? What empirical techniques have been developed to improve the response rates, to reduce or disarm the problems of response and nonresponse bias? In brief, what is the state of the art today? The unwillingness to participate in surveys may be due to several reasons: The length of the questionnaire and the respondents are chosen from different fields which do not represent the study leading to the lower response rate.

According to Kanuk and Beerenson (1975), empirical studies designed to improve the validity and reliability of mail surveys can be divided into two categories: Those designed to improve response rates and those designed to reduce biases due to nonresponse. The respondents are generally divided into early respondents and late respondents and the responses may or may not differ in both the cases. For example, if we want to study the relationship between knowledge management and business performance and the respondents selected are from a different background than knowledge management, leading to the response and nonresponse bias. The respondents selected should be from knowledge management background which may reduce the response and nonresponse bias.

Sheikh and Mattingly (1981) proposed two methods of identifying the nonresponse bias: Firstly, the difference in outcome between early and late respondents may indicate non-response bias, and its extent may be proportional to the delay in response. The second method makes use of population characteristics. For example, the difference between respondents and non-respondents in terms of age, sex, and social class may indicate non-response bias.

HOW TO ANALYZE THE DATA?

The researcher should decide at the very outset which technique to use for analyzing the data prior to the questionnaire design. If a researcher wants to study an impact of the independent variable on the dependent variable and both variables are based on metric data then he/she could use regression and if a researcher comes across with a situation where he/she has multiple independent variables and single dependent variable he/she could apply multiple regression. However, if a researcher wants to study the association between income and education, he should apply chi-square test. However, data can be analyzed on the basis of the nature of study for instance, if a researcher wants to measure the multiple dependence relationships between the different variables, structural equation modeling can be used for the same. Structural equation modeling is the only tool which can specify a series of multiple dependence relationships, for example, image creates satisfaction and satisfaction creates loyalty. In this example, satisfaction is both independent as well as an dependent variable which can only be tested using structural equation modeling.

With the advent of different statistical packages including SPSS, SAS, R and AMOS analysis can be conducted very quickly. There is no need to calculate regression manually as it can be done using Microsoft excel or SPSS. According to Rowley (2014), data analysis software can be divided into three categories including web survey software (Survey monkey), office software such as Microsoft excel and specialist data analysis software such as SPSS, AMOS, LISREL, EQS, MINITAB, and SmartPLS. Selection of particular statistical software depends on the availability and expertise of that particular software package. For instance, excel is widely available and user-friendly and is not based on the assumptions like univariate or multivariate normality. However, if you are dealing with the SPSS it requires certain assumptions like normality of data. One of the limitations of SPSS is that it does not calculate convergent validity and discriminant validity, while as statistical packages like AMOS, SmartPLS and LISREL can be very supportive for calculating the construct validity, convergent validity, and discriminant validity.

PREPARATION OF DATA FOR ANALYSIS

When the researcher completes the data collection, the first step in data preparation is to check whether the questionnaires fit for the analysis or not. If the questionnaire checking was not undertaken at the time of collection of data, then it can be taken into consideration after data collection is completed.

The data collected through questionnaire will not be acceptable if it contains the missing frequency, incomplete information; information provided by respondents who are not knowledgeable enough or does not qualify for the participation (Kothari & Garg, 2014).

Apart from questionnaire checking, editing should be taken into consideration by checking whether information collected through questionnaire is complete and uniformly entered. The editing should be followed with the coding of data. Coding is defined as the process of assigning numerical to the responses. Coding decisions should be taken into account at the time of designing of questionnaires. Data can be classified on the basis of metric and non-metric, former deals with the ratio and interval scale while as later deals with nominal and ordinal data. Data having similar attributes can be placed into one group and in this way, entire data can be categorized into various groups or clusters.

Graphical representation of the data helps in understanding the data easily. Every statistical package including Microsoft excel, SPSS and R offers a wide range of graphs in the form of Bar chart, pie-chart, histogram, and surface. Data cleaning involves the checking of data in the form of outliers and missing values. Outliers can be detected and removed using boxplots and histogram, however, missing values can be identified using descriptive statistics.

DISCUSSION AND CONCLUSION

The present study has explored the issues and challenges faced by researchers in designing and developing the questionnaires. It can be concluded that researchers often rely on the previously developed questionnaires rather than developing the new one due to lack of understanding and expertise. For novice researchers, developing entirely a new questionnaire requires a basic understanding of the format of the questionnaire, length of the questionnaire and how to distribute a questionnaire. Therefore, this article presents a systematic way to develop a questionnaire keeping in view all the issues and challenges faced by researchers who are engaged in study for a thesis or in other research projects. The article was approached from a theoretical perspective and does not cover all the issues in designing a questionnaire. However, this article will be very supportive for beginners (M.Phil and Ph.D. scholars) who wish to design the questionnaire and want to learn how to deal with challenges like the length of the questionnaire, analyzing the data, data preparation and how to choose respondents. The designing of questionnaire has become very easy with the advent of some online outsourcing agencies including survey monkey. This site allows the researcher to develop his/her own questionnaire and provides extra options of a target audience.

Questionnaire involves gathering a large number of data in a short period of time and the results can be easily quantified using different statistical packages including SPSS, Minitab, AMOS, and SmartPLS.

According to Beri (2008, p. 108), a marketing researcher intending to collect primary data has to be extremely careful in deciding what information is to be collected, how many questions are to be formulated, what should be their sequence, what should be the wording of each question and what should be the layout of the questionnaire. However, the length of the questionnaire has been a challenge for researchers, questionnaires should be as short as possible because making questionnaires too long reduces the response rate and also creates the boredom among the respondents. According to Herzog & Bachman (1981, p.549), many researchers are convinced that survey instruments have a maximum length beyond which there is an increasing probability of premature termination, random responding, or other behavior patterns which result in data of lower quality. The length of the questionnaire is not the only issue but there are other aspects which should be taken into consideration while designing a questionnaire including how to choose respondents, distribution of questionnaire, how to analyze the data and how to make sure that respondents respond my questions accurately. Rowley (2014) suggests that when respondents read your questions they will each do this from their view of the world, including their understandings, interpretations, values, views, and attitudes.

The issues of response and non-response should be carefully estimated so that the findings can be properly generalized. According to Kanuk and Beerenson (1975), despite a large number of research studies reporting techniques designed to improve response rates, there is no strong empirical evidence favoring any techniques other than the followup and the use of monetary incentives. Dividing the respondents into early respondents and late respondents and then measuring the differences between the two groups can be one of the best strategies to identify the non-response bias in evidence-based research.

The study reviewed a limited number of previous studies and excluded other journals of high quality which are often considered as significant publication means among social science researchers. Future research needs to discuss the other aspects which were beyond the scope of this study including content validity and its different formats and subject matter experts.

REFERENCES

Beri, G. C. (1993). *Marketing Research*. Tata: McGraw-Hill.

Bird, D. K. (2009). The use of questionnaires for acquiring information on public perception of natural hazards and risk mitigation –a review of current knowledge and practice. *Natural Hazards and Earth System Sciences*, *9*(4), 1307–1325. doi:10.5194/nhess-9-1307-2009

Bryman, A., & Bell, E. (2010). *Business research methods*. USA: Oxford University Press.

Burchell, B., & Marsh, C. (1992). The effect of questionnaire length on survey response. *Quality & Quantity*, *26*(3), 233–244. doi:10.1007/BF00172427

Childers, T. L., & Ferrell, O. C. (1979). Response rates and perceived questionnaire length in mail surveys. *JMR, Journal of Marketing Research*, *16*(3), 429–431. doi:10.2307/3150720

Farooq, R. (2016a). Role of structural equation modeling in scale development. *Journal of Advances in Management Research*, *13*(1), 75–91. doi:10.1108/JAMR-05-2015-0037

Herzog, A. R., & Bachman, J. G. (1981). Effects of questionnaire length on response quality. *Public Opinion Quarterly*, *45*(4), 549–559. doi:10.1086/268687

Kanuk, L., & Berenson, C. (1975). Mail surveys and response rates: A literature review. *JMR, Journal of Marketing Research*, *12*(4), 440–453. doi:10.2307/3151093

Kothari, C. R., & Garg, G. (2014). *Research methodology: Methods and techniques* (3rd ed.). New Delhi: New age international publishers.

Lathman, B. (2007) Sampling: What is it? *Quantitative Research Methods,* Retrieved 15 October, 2014 from http://webpages.acs.ttu.edu/rlatham/Coursework/5377(Quant))/Sampling_Methodology_Paper.pdf

Leung, W. C. (2001). How to design a questionnaire. *Student BMJ*, *9*(11), 187–189. PMID:11407060

Lund, E., & Gram, I. T. (1998). Response rate according to title and length of questionnaire. *Scandinavian Journal of Public Health*, *26*(2), 154–160. doi:10.1080/14034949850153680 PMID:9658516

Rowley, J. (2014). Designing and using research questionnaires. *Management Research Review*, *37*(3), 308–330. doi:10.1108/MRR-02-2013-0027

Saunders, M. N. (2005). *Research Methods for Business Students, 3/e*. Pearson Education India.

Saunders, M. N. (2011). *Research Methods for Business Students, 5/e*. Pearson Education India.

Sheikh, K., & Mattingly, S. (1981). Investigating non-response bias in mail surveys. *Journal of Epidemiology and Community Health*, *35*(4), 293–296. doi:10.1136/jech.35.4.293 PMID:6461711

Chapter 5
Role of Bibliographical Databases in Measuring Information:
A Conceptual View

S. Hariharan
Marthandam College of Engineering and Technology, India

V. S. Prasanth
Lekshmipuram College of Arts and Science, India

P. Saravanan
Lekshmipuram College of Arts and Science, India

ABSTRACT

The quantum of information being produced has made the information scientists to devise various ways and means to effectively organize and disseminate them when needed. Bibliographical, citation and abstracting databases are the some of the tools. This chapter intends to describe the conceptual base of bibliographical databases, their genesis and impact in library science perspective and their importance in measuring information. The chapter also presents the science of measuring information using various quantitative and qualitative approaches like bibliometrics and scientometrics.

INTRODUCTION

Information can simply be defined as data that have been collected, collated and communicated in a meaningful form. The advancements in science and technology Post-World War created several by-products and one such is the production of information in huge quantity. Along with this the advent of modern fifth generation computers and communication technology gave birth to phrases and terms such as 'information explosion,' 'information overload,' 'bandwidth,' 'channel capacity,' 'information redundancy,' 'noise,' etc.

DOI: 10.4018/978-1-5225-3457-0.ch005

Information explosion, which is also known as exponential growth of information leads to formulate a mechanism to retrieve the information easily and effectively without waste of time. Presently information explosion has become a common phenomenon due to the fact new information are generated more frequently and disseminated through ICT based communication technology. Price (1963) states that "once in ten years the number of articles in a field (particularly in science) doubles". The development of science with a rapid increase in the number of scientific researches, scientists and researchers, and the advent of novel scientific disciplines have led to an exponential rise in the number of journals, books, conference proceedings, dissertations, patents, technical reports and other publications bringing research results and updating science literature. These publications are referred to as primary publications or primary sources of information, pointing to the original character of the information presented. Convenient search of such a huge number of publications issued in different countries, in different languages and stored in different media would be impossible to access without special aids, i.e. secondary publications or secondary information sources, which process, analyze and summarize primary publications and help in their target search. As such, a great amount of publications and their authors require proper evaluation, these secondary sources of information are also used for scientific validation, based on the high criteria they employ in the selection of primary publications to be systematically followed and processed.

Bibliographies served this purpose to some extent before the invention of computers. The modern day bibliographical databases play the role of secondary sources of information leads the readers to access primary sources of information. The concept of bibliography is more relevant and important to the publishers and library professionals for the management of collections in the library in earlier days while researchers too are concerned about it though for different reasons. The word 'bibliography' is a systematic description of books, their authorship, printing, publication, editions and so on. It is synonymously used with terms like, list of reference, book list, lists of books, catalogue, and record.

BIBLIOGRAPHY

The term bibliography was first used by Louis Jacob de Saint Charles in his bibliographic *parsiana* (1645 – 50) which gained popularity in the eighteenth century. The *termis* coined from the two Greek words, viz., 'Biblion' and 'graphein'. The meaning of the word *biblion* is Books and *graphein* is to write. Thus, etymologically bibliography means 'writing of books'. Bibliography as defined by Shores (1954), is a "…list of written, printed or otherwise produced record of Civilization, which may include books, serials, pictures, films, maps, records, manuscripts and any other media of communication." According to Ranganathan (1963), 'bibliography' "…is a list of documents listed together for some purpose. The purpose is to bring to the attention of the reader an exhaustive or selective list of documents relevant to his pursuit of study or enquiry."

Bibliography is a complex structure of lists which extends from the local library to the region, nation and the world. A bibliography records not only what is available, but also what had been available in the past and what will be available in the immediate future.

Bibliography or bibliographic databases as a secondary source of information arise from the need to make the follow up, search and access to most relevant literature ever more convenient of the users. Current bibliographic databases were preceded by printed publications, their best known and oldest representatives being *Index Medicus* and *Chemical Abstracts*. These secondary publications consisted of several types of indexes such as listing data on author names, topic, journal title, citations, etc. in alphabetical order, thus pointing to original articles. That is why this type of publications are occasionally referred to as index publications. In the 1960s, when computers were initially introduced in the production of these publications, their computer-read equivalent – bibliographic databases appeared first as an accompanying phenomenon. With time, however, the traditional printed publications gave way to electronic databases in most cases.

Nowadays, advancement of Science and Technology in publication leads to tremendous growth of published documents and in print form consequently to the alternate medium of electronic and digital form. Information scientists have predicted that literature in science and technology doubles at least once in three years and in social sciences it happens in five years. The proliferation of information through scholarly publication gains momentum in the digital era. The electronic publication has reduced the time lag between knowledge generation and dissemination in scholarly platform. In this context, database management system on bibliographic databases helps the academic community to retrieve the scholarly communication. This chapter intends to describe the role of bibliographic databases in assessing the evaluation of scientific literature.

BIBLIOGRAPHIC DATABASE

A bibliographic database is a database of bibliographic records, an organized digital collection of references to published literature, including journal and newspaper articles, conference proceeding, reports, government and legal publications, patents, books, etc. Primarily these databases are intended as a tool to aid the information-seeker in locating the information need without wastage of time. Nevertheless, these databases which are similar to abstracting and indexing services, citation databases become the basis and major sources for quantification of literature growth or output, evaluate the growth of specific subject or discipline, and in a specific filed. In the present competitive world, measuring the research output of an institution or individual scientists and also that of the particular knowledge domain or geographical entity are essential for the policy makers and planners.

An electronic record of an academic database contains information on article title, names of authors, their affiliation, institutional address, journal title, pagination, issue number, volume number, year of publication, abstract, DOI (Digital Object Identifier), and other relevant metadata. Five types of academic databases are usually available to academic community, namely:

- Bibliographic Databases
- Citation Databases
- Full-text Databases
- E-Journal Gateways
- Online Directories of Journals

Bibliographic databases contain bibliographic records of papers, published in different peer-reviewed scholarly journals. Many indexing & abstracting (I&A) periodicals covering contents of published literature in different disciplines are available in print format. These periodicals systematically obtain and disseminate bibliographic records of recently published literature in their respective academic disciplines.

Later, many of these Indexing and Abstracting periodicals have discontinued publishing in print format. Instead, they started offering online indexing and abstracting services in machine readable formats. Some of these indexing services are available in dual print and online formats. These online databases are searchable using metadata. As these databases contain abstracts of scholarly literature, free text search is also made possible. These databases also provide external full-text links to journal contents available in publishers' portal, so that users can easily obtain full-texts of relevant literature.

Each bibliographic database follows and processes a great number of carefully selected publications; most of them are articles from (scientific) journals. It is otherwise called indexing of journal articles. The selection and processing are performed by renowned experts from the scientific fields covered by the given database. Each paper is represented by bibliographic record that contains data such as author name(s), title of the article, name of the publication where the article has appeared, year of publication, key words, abstract, author affiliation(s), original language of the article, type of the article, digital object identifier (doi), etc. These data are structured and classified into special fields such as authors, title, source, key words, etc. Where the fields in a database are more, it is more convenient for a more precise target search and more accurate search result. Ensuring an adequate amount of data on each paper included and its classification in particular boxes enables the user to search and survey a great number of papers, and to find those of interest for his/her query. Initially, bibliographic databases did not provide an insight into the extensive text of the articles, however, in some bibliographic databases links have lately been attached by use of special tools to the extensive text of the papers found in e-journals.

CITATION DATABASES

Citation databases make a separate entity within bibliographic databases. The procedure of citation is a usual practice in scientific communication. In simpler terms it is an author giving the credit to another for an earlier work. Therefore, the authors of articles appearing in scientific journals add bibliography or the list of references at the end of the paper. Citation databases are specific for presenting each article included in the base also by the respective list of references in addition to bibliographic record. These lists of references are called cited references or citations. The search according to cited references is more complete because it enables target follow up of a particular topic through all articles on the topic which are included in the database. Namely, citations are presumed to be related to the topic of the current paper by their contents, irrespective of the reasons for their citing i.e. favourable, such as paying credit to, or for criticism and correction. In addition to allowing literature searching according to topics, citation databases provide data on the number of citations received by a particular journal, author, or paper.

SPECIFIC BIBLIOGRAPHIC DATABASES

At present there are more than 100,000 scientific, scientific-professional and professional journals are currently disseminating scholarly communication in the world. Preparation of an absolutely comprehensive bibliographic and citation databases including all scholarly communication with global coverage is neither possible nor feasible and economically impractical. Hence, based on specific information needs, there are number of databases on various discipline and sub-discipline of knowledge. Thus, one of the main characteristics of bibliographic and citation databases is their selectivity of content coverage. Analyses of scientific publications have revealed that in each particular field, the majority of relevant scientific results appear in a relatively small number of journals, referred to as "core journals". Between 30% and 40% of the overall number of journals are being selected and processed in the relevant discipline-oriented bibliographic databases. There are a great number of databases (about 900) worldwide, the most widely known in the field of science and biomedicine being Medline and PubMed (biomedicine and related fields), Chemical Abstracts (chemistry and related fields), EMBASE (biomedicine, pharmacology), International Pharmaceutical Abstracts (pharmacy), Biological Abstracts and Biological Sciences (biology and biological sciences).

The Current Contents (CC) bibliographic database, and Science Citation Index (SCI), Social Sciences Citation Index (SSCI) and Arts and Humanities Citation Index (AHCI) citation databases are international databases of highest selectivity. Until 2004, these were published by the Institute for Scientific Information (ISI) from Philadelphia. In 2004, Thomson Corporation, now Thomson-ISI took-over their publication. Since 1997, the above-mentioned citation databases have been unified into a unique base, Web of Science (WoS). Users in a particular field may not use all bibliographic databases if all are available in a combined form. Thus, to satisfy the specific information needs of target groups there are number of specific bibliographical databases such as *Scopus, Web of Science, PubMed, Embase, MEDLINE, BIOSIS, AGRIS, AGRICOLA, PsychARTICLS, PsychINFO, ProQuest* and so on. Each of these databases is designed to meet the information requirements specific user groups.

BIBLIOGRAPHIC DATABASES FOR MEASURING INFORMATION

The primary purpose of bibliographical databases at the initial stage was searching the literature, but due to effective retrieval mechanism of database management system, features like links to full text articles and citation details and complete list of references were added. These features make the databases attractive to develop several quantitative indicators leading to evaluate the growth and other aspects of knowledge both qualitatively and quantitatively. The criteria used for indexing the articles of journals become the criteria to measure scientific contribution. Moreover, the trend in choosing those journals which are having more visibility, impact, indexing agencies, are some of the selected criteria for publishing their scholarly articles by the scholars.

Indexing or representation of the journal in relevant databases contributes to its better visibility and availability, and is found to be influencing the impact of a publication over the international scientific production. Representation in ISI databases are considered highly relevant for their high quality filtering, and there certainly is no scientific periodical editor who would not want his journal to become part of the "core of the world knowledge". Scientists tend to publish their papers in renowned journals, thus to ensure the best possible visibility in the international scientific community, and due respect and career promotion for them personally. The criteria for indexing a journal in renowned databases play a key role in measuring information or research output or productivity of an individual or an organisation. Now-a-days, higher educational institution insists the scholars to possess at least a particular minimum number of papers in indexed or impact factor journal also shows the role of bibliographical databases in measuring information. The more number of papers published in journals indexed by these databases is considered as equivalent of high prestige.

Another criterion of good bibliographic database is complete list of references, which leads the user to identify the frequently used sources, frequently cited author, subject domain, interdisciplinary nature of the subjects, citations per year, etc. Thus, it aids to develop several quantitative indicators of information. Thus, it is concluded that bibliographic databases are a fundamental requirement for the measurement of information. Scientific productivity, research output or academic productivity is calculated based on well structured, criterion based and comprehensive bibliographic databases.

APPROACHES TO MEASURE INFORMATION

Measuring information is not an easy task, several methods and techniques are adapted time to time to measure the growth of information. Several attempts made earlier are all based on quantitative indicators. The approach for measuring information based on bibliometric record of recorded knowledge is generally classified into two namely bibliometric and scientometric approaches. Apart from these some more metrics are developed by the information scientists namely informetrics, cybermetrics and webometrics are based on the quantitative indicators only. For all type of metrics bibliographic databases provide a platform to conduct an exhaustive study.

Librametry proposed by Ranganathan in 1948 is the application of mathematical and statistical techniques to library problems (Ranganathan, 1969). Librametrics is the quantitative analysis of various facets of library activities and library documents by the application of mathematical and statistical calculus to seek solution to library problems (Sengupta, 1985). However, librametrics has not gained much currency among information professionals.

Though Henri Paul Otlet used the French term *bibliometrie* in the year 1934, *Alan* Pritchard coined the term '*bibliometrics*' in the year 1969 to replace the term 'statistical bibliography'. Pritchard (1969) defines bibliometrics as

...to shed light on the processes of written communication and of the nature and course of development of a discipline, by means of counting and analyzing the various facets of written communication ... the application of mathematics and statistical methods to books and other media of communication...

White and McCain (1989) defined bibliometrics as the quantitative study of literatures as they are reflected in bibliographies.

According to Tague-Sutcliffe (1992),

Scientometrics is the study of the quantitative aspects of science as a discipline or economic activity. It is part of the sociology of science and has application to science policy-making. It involves quantitative studies of scientific activities, including, among others, publication, and so overlaps bibliometrics to some extent.

Scientometrics is the science of measuring the "quality "of science. It includes all quantitative aspects of the science of science, communication in science, and science policy. It deals with analysis, evaluation and graphic representation of science and technology information. Science mapping, visualization and research trend analysis are the some of the scientometric techniques which are in their nascent stage.

Informetrics was used as a generic term to mean the use and development of a variety of measures to study and analyze several properties of information in general and documents in particular. Tague-Sutcliffe (1992) defines

Informetrics is the study of the quantitative aspects of information in any form, not just records or bibliographies, and in any social group, not just scientists. Thus it looks at the quantitative aspects of informal or spoken communication, as well as recorded, and of information needs and uses of the disadvantaged, not just the intellectual elite. It can incorporate, utilise, and extend the many studies of the measurement of information that lie outside the boundaries of both bibliometrics and scientometrics... Two phenomena that have not, in the past, been seen as a part of bibliometrics or scientometrics, but fit comfortably within the scope of Informetrics are definition and measurement of information, and types and characteristics of retrieval performance measures.

According to Ingwersen and Christensen (1997) "the term Informetrics designates a recent extension of the traditional bibliometric analyses also to cover non-scholarly communities in which information is produced, communicated, and used". Wilson (1999) defines "Informetrics is the quantitative study of collections of moderate-sized units of potentially informative text, directed to the scientific understanding of informing processes at the social level".

In the mid-1990s attempts were made to study the nature and properties of the World Wide Web by applying modern informetric methods to its contents, links and search engines. In the year 1997, Almind and Ingwersen named these studies as webometrics. Webometrics have several similarities to bibliometrics, informetrics and scientometrics. In the digital era webometrics and its sister terms like 'netometrics' and 'cybermetrics' are significant. It can be stated that they are the application of bibliometric and scientometric methods to web resources. Webometrics and Cybermetrics are the study of the quantitative aspects of the construction and use of information resources, structure and technologies on the www drawing on bibliometrics and informetric approaches.

According to Brookes (1990), the techniques of scientometrics and bibliometrics are closely similar but their roles are distinguished by their different contexts. Bibliometrics stresses the material aspects of the analysed unit such as a paper, citations or any other information irrespective of the subject orientation. Scientometrics on the other hand emphasizes the measurement of specific information related to its scientific value. Scientometrics includes all quantitative aspects and maps related to the production and dissemination of scientific and technological knowledge.

BIBLIOMETRIC INDICATORS

Evaluation of quality and quantity of publications can be done using a set of statistical and mathematical indices called bibliometric indicators. There are two major categories of indicators namely, (i) quantitative indicators that measure the research productivity of a researcher and (ii) performance indicators that evaluate the quality of publications. Bibliometric indicators are important for both the individual researcher and organizations. They are widely used to compare the performance of the individual researchers, journals and universities. Research and academic appointments, promotions and allocation of research funds are predominantly based on these indicators.

The paradigm shift in the approaches to measuring scientific information using sciencetometric / bibliometric tools has provided science policy-makers a basis for evaluating and directing research and development (Nederhof, 2008; Junquera & Mitre, 2007; Jansen, et al., 2007). Impact factor (IF) is the most popular one among the scientometric indicators. IF is a figure stating how many times on an average a scientific paper from a journal has been cited during a given period of time. IF for the current year is calculated by dividing the number of citations received in the current year for the papers published over the preceding two years by the number of papers published during this two-year period. Analysis of citations at ISI citation databases provides numerical indicators on the basis of which the journal's echo in a particular field is estimated.

H-Index was introduced by Hirsch for quantifying a researcher's publication output while balancing it with its citations. It attempts to measure both the scientific productivity and the apparent scientific impact of a scientist. This index is based on the set of the scientist's most cited papers and the number of citations that they have received in other researcher's publications. The impact of citations can be understood by the statement "I was able see farther because I was standing on the shoulders of giants" made by Sir. Isaac Newton.

Some of the other popular qualitative and quantitative indicators are as follows:

- **The Number of Papers:** Reflects scientific output, as measured by articles, letters and reviews count
- **The Number of Article Citations:** Measures the impact on the scientific community.
- **The Number of Co-Signers:** Indicates co-operation at national or international level.
- **The Number of Publications in the Top 10%:** Allows to characterize international visibility
- **Position Index:** Measures the implication of the author among the signers.
- **Co-Publications:** Measures interactions and scientific relationships between networks, teams, institutions and countries.

BIBLIOGRAPHIC DATABASES

The information age has made it possible to have number of bibliographic and citation databases that are specific to the subject or a specific branch of a subject. The complete enumeration of available databases is beyond the scope of the chapter. But commonly used bibliographical databases are given as follows.

Web of Science (WoS)

Web of Science (WoS) was actually launched by Eugene Garfield, father of citation indexing. It is a bibliographic database system developed by the *Institute for Scientific Information (ISI)* and maintained by *Thomson Reuters*. It includes three citation databases, i.e. Science Citation Index (SCI), Social Sciences Citation Index (SSCI) and Arts & Humanities Citation Index (AHCI), which taken together index more than 8,700 scientific periodicals from all scientific fields. There is 90%-100% overlapping between Current Contents and citation databases. Currently, they cover 8,700 journals. When ISI launched the development of Science Citation Index as the first citation index in the 1960s, it was based on some 600 titles in the fields of natural and applied sciences. As the number of periodicals grew in the world during the decades to come, the journal "core" increased accordingly. Until the recent advent of Scopus in 2004, Web of Science was the only world citation database. Access: http://wos.irb.hr

Scopus

Scopus is the latest product launched by Elsevier, the biggest scientific publisher in the world. It is a bibliographic and citation database like Web of Science but with a wider content coverage than Web of Science. More than 14,000 reviewed journals from natural, technical and social sciences and biomedicine journals issued by more than 4000 publishers are being indexed. In addition to periodicals, Scopus covers 250 million of quality and relevant web sites including 13 million patents. Also, Scopus completely covers two separate biomedicine databases, Medline and EMBASE. The specificity of this database is that more than 60% of the journals included come from non-American countries. The conditions that a journal has to meet to be covered by Scopus are as follows: paper title and abstract in English language (whereas extensive text of the paper may be in any other language), regular appearance, some form of quality control (e.g., reviewing procedure), and high overall quality (which is assessed by the number of citations the journal receives *via* Scopus, reputation of the publisher, authors and editorial board, and some other parameters). In contrast to ISI citation databases that tend to exclusivism, Scopus plans to include all valuable sources of information and to develop further depending on the rise in the number of new information. Access: http://www.scopus.com

PubMed

PubMed services started in 1996 piloted in the era of free computerized searching of the *MEDLINE* database of references and abstracts on life sciences and biomedical topics. The *PubMed* provides over 21 million references to biomedical and life sciences journal articles back to 1946. *MEDLINE* includes citations from over 5,600 scholarly journals published around the world. *PubMed* is a free search engine which provides access to The United States National Library of Medicine (NLM).

Google Scholar

Google Scholar (GS) provides a simple way to search and access scholarly literature across many disciplines. The source coverage includes articles, theses, books, abstracts and court opinions, online repositories of universities and other web sites. Though GS is not planned as a direct challenge to other citation databases it is a colossal database and Google proposes to enlarge its coverage. The most important advantages of GS are its speed and free availability. Bibliographic databases like *Scopus* and *WoS* are affordable only to those institutions that are able pay the high subscription cost. Nevertheless, GS comprehensively covers journals and web resources that are not included in any other bibliographic database.

CONCLUSION

In the information exploration era, bibliographic databases play a dominant role in information management arena not only for the selection of information sources but also for determining various metrics related to quantitative and qualitative measurement of information. The specific subject oriented bibliographic databases and its features like access to full text, references, and citations make it is a reliable source for accessing latest information. The bibliometrics, scientometrics or any other metric analysis on information are made based on the bibliographic databases. The subject-specific bibliographic database plays crucial role in measuring relevant specific areas of knowledge. Providers of bibliographic databases are also continuously adding value added metrics in their databases, which helps to evaluate the information accurately. Research output of the institutions, scientific productivity of authors and progress of research agencies are assessed based on indexing their scholarly communication in prominent bibliographic databases.

REFERENCES

Almind, T., & Ingwersen, P. (1997). Informetric analyses on the world wide web: Methodological approaches to 'webometrics'. *The Journal of Documentation*, *53*(4), 60–64. doi:10.1108/EUM0000000007205

Brookes, B. C. (1990). Biblio-, sciento-, infor-metrics??? What are we talking about. In L. Egghe & R. Rousseau (Eds.), *Informetrics 89/90* (pp. 31–43). Amsterdam: Elsevier Science Publishers B.V.

Gašparac, P. (2006). The role and relevance of bibliographic citation databases. *Biochemia Medica*, *16*(2), 93–102. doi:10.11613/BM.2006.009

Ingwersen, P., & Christensen, F. H. (1997). Data set isolation for bibliometric online analyses of research publications: Fundamental methodological issues. *Journal of the American Society for Information Science*, *48*(3), 205–217. doi:10.1002/(SICI)1097-4571(199703)48:3<205::AID-ASI3>3.0.CO;2-0

Jansen, D., Wald, A., Franke, K., Schmoch, U., & Schubert, T. (2007). Third party research funding and performance in research. On the effects of institutional conditions on research performance of teams. *Kolner Zeitschrift Fur Soziloogie Und Sozialpsychoogie*, *59*(1), 125–149. doi:10.1007/s11577-007-0006-1

Junquera, B., & Mitre, M. (2007). Value of bibliometric analysis for research policy: A case study of Spanish research into innovation and technology management. *Scientometrics*, *71*(3), 443–454. doi:10.1007/s11192-007-1689-9

Nederhof, A. J. (2008). Policy impact of bibliometric rankings of research performance of departments and individuals in economics. *Scientometrics*, *74*(1), 163–174. doi:10.1007/s11192-008-0109-0

Price, D. (1963). *Little Science, Big Science*. New York: Columbia University Press.

Pritchard, A. (1969). Statistical bibliography or bibliometrics. *The Journal of Documentation*, *25*(4), 348–349.

Ranganathan, S. R. (1963). *Documentation and its facts*. Bombay: Asia Publishing House.

Ranganathan, S. R. (1969). Librametry and its scope. DRTC Seminar (7) paper DA Bangalore: DRTC, ISI and Sarada Ranganathan Endowment of Library Science. Reprinted in: *JISSI: The International Journal of Scientometrics and Informetrics*, *1*(1), 15-21.

Ravichandra Rao, I. K. (1985). Bibliometric Model: An entropy approach and their application. *IASLIC Special Publication*, *25*, 1–8.

Roy, P. M. (1983). Towards a theory of citing in citation analysis studies. Developing horizons in Library and Information Science, 1.

Sengupta, I. N. (1985). Ranganathan's Philosophy and Librametrics. In T. S. Rajagopalan (Ed.), *Ranganathan's Philosophy: Assessment, impact, and relevance* (pp. 447–458). New Delhi: Vikas Publishing House.

Shores, L. (1954). *Basic reference sources.* Chicago: American Library Association.

Tague-Sutcliffe, J. (1992). An Introduction to Informetrics. *Information Processing & Management, 28*(1), 13. doi:10.1016/0306-4573(92)90087-G

White, H. D., & McCain, K. W. (1989). Bibliometrics. In M. E. Williams (Ed.), *Annual Review of Information Science and Technology* (Vol. 24, pp. 119–186). Amsterdam: Elsevier Science Publishers B.V.

Wilson, C. S. (1999). Informetrics. *Annual Review of Information Science & Technology, 34*, 107–247.

Chapter 6
Information Needs and Use Pattern of Educationists:
A Citation Study Based on Masters Dissertations in Education

P. Sheela
N.V.K.S.D College of Education, India

ABSTRACT

The chapter attempts to identify the information needs of educationists based on the citations in Master's degree dissertations in Education. The source for the study is the Master's degree dissertations submitted to the Manonmanium Sundaranar University during the period 1996 - 2006 from N.V.K.S.D. College of Education. The citation-based approach is followed in the study. Number of citations, cited articles, authors, frequency and percentage distributions were the tools used to analyze the data. Findings showed that journals were the most utilized reference materials in the dissertations. Indian periodicals are ranked first, followed by that from the USA. Mostly cited periodicals are published in the year 1993 with 181 citations (14.23%) and mostly-cited books were published in the year 1996 with 71 (13.95%) citations. Moreover, highest number of citations was in the year 1997 with 702(10.19%) citations and the lowest number of citations was during the year 2001 with 337(4.89%) citations.

INTRODUCTION

Information is produced every now and then across the globe, as a result of scientific, academic and industrial research. The new information is communicated to the users through scientific publications. This has led to the growth of literature. Bibliometrics includes the study of the relationship among the documents. Typically, these descriptions focus on constituent patterns involving authors, monographs, journals, subject language etc. Bibliometrics have established itself as a viable and distinctive technique for evaluation of science based on bibliographical and citation data.

DOI: 10.4018/978-1-5225-3457-0.ch006

Citation analysis is the examination of pattern, frequency of citations in books and articles. It was citation in scholarly works to establish links to another work or other researchers. The study of the relationship between cited and citing documents come under the purview of citation analysis. A cited document is one which is taken as a source of reference in this study and a citing document is the document which makes references of other documents. Citation analysis is thus the analyses of citation to and from documents, that is appended with the research communication it count citation given at the end of each scientific article counting citation is often called citation analysis.

Citation analysis has been used in the assessment of scientific activity for decades. Librarians have used citation count to study the adequacy of collection of information materials. The quality of such publications can be known by analyzing citations made by the paper itself and citations got by other authors. This can be done through citation analysis.

According to RavichandraRao (1983):

The main objective of the citation analysis are to evaluate and interpret citations received by articles, authors, institutions and other aggregates of scientific activities. It is also used as a tool for measuring communication links in the sociology of the science.

Citation Analysis has emerged as a useful technique for deriving trends in a discipline. Studies based on citations are generally termed as Citation Analysis. The primary function of citation is to provide a connection between two documents, on that cites and other, which is cited. Citation Analysis has been increasingly used in the study of Science and Technology theses to provide qualitative data on their utility and relationships that could be used in several ways.

According to Garfield "the term Citation Analysis is one which deals with works cited as having actually been used in the preparation of or having otherwise contributed to the source of paper".

Citation Analysis is a technique that shows how many citations are there for a given document, author and journal over a period of time. In the present study citations from 158 M.Ed. Dissertations has taken as sample during the year 1996-2006 and collected on the data sheet. The data includes number of reference types of documents and number of most cited Journals.

Citation Analysis is one of the popular methods employed in recent days for the investigation of core documents in various subject fields or for a particular scientific community in a geographical proximity. In fact, citation study is one of the effective and indirect methods to understand the information requirements of users. Citation method is used in the present study to understand the Information needs and Use patterns.

Citation Analysis has emerged as a useful technique for deriving trends in a discipline. Studies based on citation are generally termed as citation analysis. The primary function of citation is to provide a connection between two documents, one that cites and the other which is cited. Citation Analysis has been increasingly used in the study of scientific field to provide qualitative data. The utility and relationships of that data could be used in several ways.

CITATION ANALYSIS: A THEORETICAL OVERVIEW

The concept of citation Analysis has existed under different names. Shepard's citation is the oldest major citation index in existence. It was started in 1873 to provide the legal profession with a tool for searching legal decisions. It was done by listing the citations to precedents used in the cases decided by Federal and State Courts and various Federal Administrative Agencies. The legal 'citator' system provided a model of how a citation index could be organized to function as an effective search tool.

APPLICATION OF CITATION ANALYSIS

Citation analysis has been applied extensively to study various aspects of documents in their subject context and use environment. As an emerging technique, it provides a new approach to the study of information phenomena. The uses of the technique (Smith, 1981; Narin and Moll, 1977) broadly fall into the following categories.

1. **Collection Development and User Studies:** Citation analysis has been employed to formulate journal subscription policy by evaluating journals on the basis of the number of times a journal is cited. Another use is the formulation of retirement policies based on citations. The relative values of different types of documents to various categories of users have also been studied.
2. **Information Retrieval:** Citation analysis has been used to develop document surrogates, user-document-keyword relationship and search strategy, computer identification of citing statements, and access to interdisciplinary literature.
3. **Development and Growth of Subjects and Their Literature:** Productivity of authors and their influence on others have been measured through citations. Citation coupling and co-citation have been employed to study the structure of scientific growth and to map boundaries of subjects.
4. **Historical and Research Standing Studies:** These includes tracing the development of subjects through time, density and context of the citation and using a citation network as a measure to access inter relationships and influence of different authors and their works.
5. **Pattern of Research Communication:** Another area of application has been the impact of isolation due to barriers of language, distance, non-availability of research literature and scientific papers and also identification of problem areas in communication.

USEFULNESS OF CITATION ANALYSIS

1. Citation analysis is used to study the citation links between Scientific Papers, technical notes and reviews; for example, it may be used in the periodicals librarian for study of the structure of literature and to identify core journals.
2. Citation analysis provides relevant measures of utility and relationships of journals whose primary function is to communicate research results.
3. Citation analysis helps in identification of key documents and creation of core lists of journals.
4. It helps in clustering of documents according to common references and citations.

CITATION INDEX

According to Eugene Garfield, a citation index is an ordered list of cited articles, each of which is accompanied by a list of citing articles. The citing article is identified as a source, the cited article as a reference citation. The Index is arranged by reference citation.

Citation Indexing originated first in U.S.A. The first citation Index was prepared by Frank Shepard in 1873. English common law operated on the principle of stand by decisions and not to disturb settled matters. This is the doctrine or policy of following rules and principles laid down in previous judicial decisions. Since later decisions may reverse or modify earlier decisions, it is legally important to find all later decisions that effect on earlier decisions. Shepard's citations enable searches to find all later decisions.

Moreover, a number of studies were conducted by Eugene Garfield on citation indexing it can be applied to scientific literature during 1950. In 1952 Shepard's technique was applied for indexing the scientific literature and also a new project was started for indexing of medical literature known as "Welch Project", was to investigate the role of automation in the organization and retrieval of medical literature.

In the early 1960's Eugene Garfield and Associates developed two pilot projects that would test the viability and efficiency of citation indexing. The first project involved the creation of a database that would index the citations of 5,000 Chemical patents held by the private pharmaceutical companies. The referenced citations in this instance were to prior patents, the documentation sources that the Govt. Patents' Examiners were using to support a decision to grant or deny a patent. The connections that the patent citation Index made were then analyzed with two comparable classifications and indexing systems that were currently being used by the participants. Based on this investigation and analysis, the project sponsors determined that citation indexing permitted the retrieval of relevant literature across arbitrary classifications in a way that subject oriented indexing could not.

A second pilot project in 1962 involved Garfield's recently incorporated enterprise, the Institute for Scientific Information, with the United States National Institutes of Health in building an index based the published literature on genetics. This project was more complexes in nature than the patents index. Three databases were built to cover the literature over 1 year, 5 year and 14 years with a varying number of source publications indexed in each. While this project was to test the feasibility and utility of a narrow, discipline-oriented citation index at most comprehensive and useful guide to the published literature in the field of genetics. The database for the single-year term had drawn not just on journals that were primarily devoted to the field of genetics research but had drawn as well as from a large pool of journals that published genetics papers as a more frequent or occasional basis. Additionally, while the automated system required a certain level of effort in standardizing the entries from a wide variety of published materials, the project demonstrated the cost-effectiveness of citation indexing as opposite to the expensive traditional subject indexing processes.

SCIENCE CITATION INDEX

Science citation Index is the best known and most commonly used citation index. Shepard's citations serve the legal field and Garfield's Science Citation Index serves the scientific filed. It is publishing quarterly.

The online version of SCI is known as Science Citation Index Search. It contains data from 1974. It is updating bi-weekly. The SCI published by the Institute for Scientific Information is one of the best known and most of the 2000 journals are searched for references to be used in the SCI. The author's last name and initials, the name and the volume number and beginning page number of each article or other works are key punched along with the titles of cited articles, separate sets of punched cards are prepared for each source item. The cards are verified and the date transferred to magnetic tape. Additional editing is done by computer program and the citation data are then framed and prepared for photo-offset printing.

A major advantage of the Science Citation Index is that it is produced by computer processing with very little expensive intellectual effort. Human indexes are not needed to select subjects, or create subject index entries. The references in the source articles have been found, revalued and verified by the author of the articles. Today the web-based version of that index covers 5,600 Journals across more than 150 scientific disciplines.

CHARACTERISTICS FEATURES OF CITATION INDEX

Citation index is based as on the simple concept that an authors references to previously recorded information identify much of the earliest work that as pertinent to the subject of his present document. The main features of citation index are of the following.

- **Simplicity:** Since citation indexing does not involve the services of indexers it is quicker and compilation is simple.
- **Search Effectiveness and Semantic Stability:** This quality has two components called search productivity and search efficiency. The search productivity is concerned with finding the largest possible number of relevant papers. Search efficiency is concerned with minimizing the retrieval of irrelevant papers.
- **Applicability in Different Languages:** The need for knowing the language in order to index the documents is altogether avoided in citation indexing. Documents in different languages get indexed in its most appropriate place automatically.
- **Extensive Time Reach:** A citation index search can start at any point in time when a pertinent paper on the subject was published.

REVIEW OF LITERATURE

The previous studies related to the present study are reviewed, some of the studies are, Lahiri (1996) reported in his book that citation analysis is one of the research methods, generally used for user studies. It is a technique to study the information needs of the scholars and the scientists. It is useful in providing a quantitative indication of some characteristics of literature used. Shafi (2002) studied 100 doctoral dissertations submitted to Kashmir University during 1980-2000 in the field of Natural Science. Around 8,000 citations were derived from these dissertations. It is found out that highest citations are from journal followed by Seminar proceedings. Anilkumar and Rajaram (2013) opined the most reliable way to know the contribution of research to the world knowledgebase is through publication and citation data. The research papers and doctoral thesis are the instruments through which results of the research are communicated to the outside world. Similarly, Devi and Sankar (2014) analysed the citations of doctoral dissertations in Commerce awarded by University of Kerala, India from 2001 to

2010. Data extracted include year of submission of thesis, title of the thesis, title of the cited work, year of the publication of the cited work type of the document and if journal title of the journals. The journal articles are cited more and it is 38.79% of total documents cited. Economic and Political Weekly (50) got highest number of citation. The latest document used is published in the year 2007 and the oldest one in the year 1923. The decade 1991-2000 got highest citation (35.73%). Similar findings were reported in the study of Borthakur (2015) conducted a study to find out the citation patterns used by students and scholars of Dibrugarh university in their thesis and dissertation. The study revealed that collaborative research is prevailing in the field of chemistry. It is found that journal contributes the highest no. of citation. Books and journals are found to be widely used format compared to web/internet resource and other form of literature.

RELEVANCE OF THE STUDY

Today Education is a well-organized discipline. Large amount of research is being done in this field. This analysis will be useful not only to the special Library Management. But also, specialist readers are selecting the relevant journals in their field of study. Analyses of citation have immense value in library management, especially in selection of periodical for subscription. Periodicals are primary sources of information have more important in Research libraries than other sources of information. The cost of the periodicals is increased day by day. Hence its judicious selection is essential. Citation analysis is a useful tool for modern Library in this regard. Now the role of the research oriented education is being fully realized and research was made an integral part of the curriculum. A Librarian must know the information needs and use pattern of the research scholars, so that we can make arrangements for the availability of the materials in the Library. It is therefore relevant to conduct a study based on Masters Dissertations submitted to the Manonmanium Sundaranar University, Tirunelveli. The investigator has selected 158 dissertations submitted during the period from 1996 to 2006 by the N.V.K.S.D. College of Education to the Manonmanim Sundaranar University.

TITLE OF THE STUDY

The title of the study is entitled as "Information Needs and Use Pattern of Educationists: A Citation Study based on Masters Dissertations in Education."

OBJECTIVES OF THE STUDY

1. To find out the most cited author in periodicals and books.
2. To identify authorship pattern in books and periodicals.
3. To compile Ranking order of Countries of cited books.
4. To compile a bibliography of the core books that has figured in Educational Research.
5. To identify the core books and periodicals that has figured in Educational Research.

METHODOLOGY

A total of 156 M.Ed dissertations are submitted from the N.V.K.S.D College of Education during 1996-2006 to Manonmaniam Sundaranar University were selected as a source for the study. Each thesis was examined and citations were extracted from the bibliography part. A total number of 6886 citations were identified. The present study reveals the nature and structure of the cited documents. In the analysis, bibliographic forms, rank list of periodicals cited, ranking of authors and authorship pattern (both periodicals and books) most cited Journals and also year-wise distributions of M.Ed. Dissertations in Education, Ranking of Books and Periodicals on the basis of published year were identified. To proceed with the work of the study, the necessary particulars of all citations were recorded. Data pertained to each citation was prepared on different coloured catalogue size cards. White card represents the book consists the name of the author, title, place of publications, publisher and year of publication. Pink card represents the journal article contains information like author's name, title of the article, name of the periodical and year of publication. Blue card represents the theses and yellow card represents the other forms (dictionary, encyclopedia, newspapers, reports and web studies). All the data were arranged. For comprehensive information, citations were recorded in the data sheet. A total number of 6886 citations were collected from the source documents. The citations were categorized broadly in the three bibliographic forms. The study becomes much easier by feeding the references in a computer and getting the desired details in a shortest period of time. As traditional methods proved to be much convincing in nature that has been adopted for the present study.

HYPOTHESES

1. Research in Education use a wide variety of document forms like journals, books, theses and newspapers.
2. The use of journals is very high among Researchers in Education.

SCOPE AND LIMITATIONS OF THE PRESENT STUDY

The scope of Citation Analysis focuses on citation about cited periodicals, books, other references such as authors, authorship pattern, gender and country. The study also aims to conduct a citation analysis in the field of Education research based on the M.Ed Dissertations submitted to Monomanium Sundaranar University. A total of 158 Masters Dissertations submitted to the Manonmanium Sundaranar University during the period 1996 to 2006 was used for the analysis.

ANALYSIS AND DISCUSSIONS

The information needs and seeking pattern of educationists are analysed and presented in various sub-headings.

1. Bibliographic Forms of Documents Cited

All the references cited in the dissertations are grouped broadly in to periodicals, books and miscellaneous forms (which includes Encyclopedia, dictionary, abstracts, reports, theses and data from websites). These broad divisions are represented in Table 1.

From the Table 1 it is found that "Periodical forms" references have the highest number of citations (2968) rank first with 43.10%. The second rank goes to Books with 2720 citations (39.50%) and third rank goes to miscellaneous forms with 1198 citations (17.40%). Table 1 can be diagrammatically represent in Figure 1

2. Distributions of Bibliographic Forms

Distributions of six bibliographical forms cited in the dissertation are shown in the Table 2. Detailed lists of miscellaneous forms are also shown here.

From the Table 2 it is found that after arranging all the six bibliographic forms, the highest number of citation is Periodicals (43.10%). The second rank goes to Books (39.50%) and third rank goes to theses /Dissertations (12.26%)

Table 1. Bibliographic Forms of Documents Cited

Sl. No	Bibliographic Forms	No. of Citation	% of Citation
1	Periodical	2968	43.10
2	Books	2720	39.50
3	Miscellaneous	1198	17.40
Total		**6886**	**100**

Figure 1. Bibliographic forms of documents cited

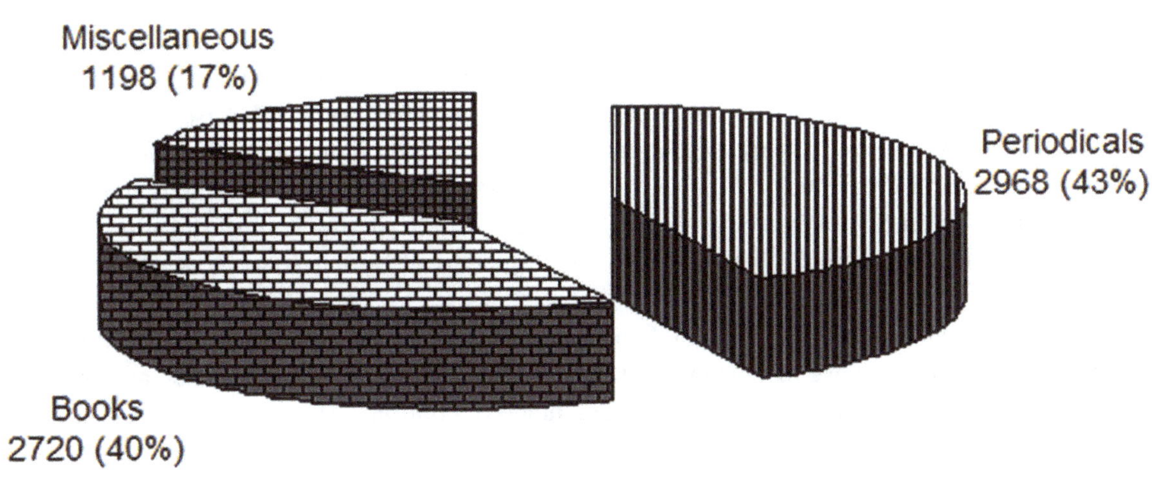

Table 2. Distribution of Bibliographical Forms

Sl. No	Rank	Distribution of Cited Documents by Their Types	No of Citation	% of Citation
1	1	Periodicals	2968	43.10
2	2	Books	2720	39.50
3	3	Theses/Dissertations	844	12.26
4	4	Encyclopedia/Dictionary	147	2.14
5	6	Web studies	68	0.98
6	5	Others	139	2.02
Total			**6886**	**100**

3. Analysis of Periodicals Cited

Journal literature is one of the most important and effective mode of publishing the results of research and improving the knowledge with in the scholarly community. Journals are essential to the continuous flow of intellectual information. It is found that journal articles are mostly cited in the theses. The journal articles are generally of high quality and it contains recent developments and information. The number of source articles citing in a journal can calculate the citation rate of a journal.

In this heading an attempt was made to analysis the following concepts of cited periodicals in education field. i.e., citation made by the researchers in the education field.

- Analysis of Ranked Periodicals.
- Analysis of ranked countries of the top ten core periodicals.
- Analysis of authorship pattern and most cited authors.
- Ranking of most cited authors in periodicals.
- Analysis of periodicals on the basis of the year of publishing.
- Gender-wise contribution of articles in periodicals.

3.1 Analysis of Periodicals Based on Tittles

Rank lists of cited journals are listed below, according to their ranking position. This list highlights the core periodicals in education, which got cited in the dissertations. The Table 3 shows the ranking order of periodical, title of periodical, no of citations and their cumulative count. The list included 175 periodicals, for the convenience the periodicals that have less than 3 citations are not included in the list.

It was observed from the table that the secondary periodical, "Dissertation abstracts International" with 716 citations (24.12%) ranked first. Second rank goes to "Experiments in Education" with 299 citations (10.07%) and third rank goes to "Journal of Educational Research and Extension" with 290 citations (9.77%).

The top ten best and most frequently cited core periodicals are listed separately in the Table 4. The core list of journals, which have received up to tenth position are given the details such as name of periodicals, country of origin, no of citations and percentage of citation.

Table 3. Rank List of Periodicals Cited

S.No.	Rank	Title of Periodicals	No. of Citations	Cumulative Citations
1	1	Dissertation Abstracts International	716	716
2	2	Experiments in Education	299	1015
3	3	Journal of Educational Research and Extension	290	1305
4	4	The Educational Review	274	1579
5	5	Indian Educational Review	258	1837
6	6	Journal of Indian Education	195	2032
7	7	Journal of the Institute of the Educational Research	162	2194
8	8	Asian Journal of Psychology and Education	86	2280
9	9	Indian Dissertation Abstracts	53	2333
10	10	Journal of Education and Psychology	51	2384
11	11	Australian Journal of Education and Psychology	46	2430
12	11	Journal of English Language Teaching	46	2476
13	12	School Science	44	2520
14	13	Science Education	43	2563
15	14	Journal of Research in Science Teaching	40	2603
16	15	P.T.A Magazine	39	2642
17	16	School Science and Mathematics	28	2670
18	17	British Journal of Education and Psychology	7	2677
19	17	Journal of the Research and Reflection on Education	7	2684
20	17	Journal of Perspectives in Education	7	2691
21	18	Edutracks	6	2697
22	19	University News	5	2702
23	19	Journal of Education and Psychology	5	2707
24	19	Journal of Applied Psychology	5	2712
25	19	The Science Teacher	5	2717
26	19	American Journal of Sociology	5	2722
27	19	Physical Education and Recreation including International sources	5	2727
28	20	Journal of School Health	4	2731
29	20	Herald of Health	4	2735
30	20	Indian Journal of Adult Education	4	2739
31	20	The Humanities and Social Science	4	2743
32	20	The Indian Journal of Social Work	4	2747
33	20	Research for Effective Teaching	4	2751
34	20	Journal of Higher Education	4	2755
35	20	Quest in Education	4	2759
36	20	Indian Journal of Psychology	4	2763
37	21	Exceptional Children	3	2766

continued on following page

Table 3. Continued

S.No.	Rank	Title of Periodicals	No. of Citations	Cumulative Citations
38	21	Journal of Social Psychology	3	2769
39	21	International Science Education	3	2772
40	21	Journal of Teacher Education	3	2775
41	21	Indian Journal of Clinical Psychology Journal of child Psychology	3	2778
42	21	Journal of Personal Psychology	3	2781
43	21	International Journal of Behavioral development	3	2784
44	21	British journal of educational studies	3	2787
45	21	Journal of Community Guidance and Research	3	2790
46	21	Research Abstracts of Dissertation in Education	3	2793
47	21	Indian Psychological Abstracts	3	2796
48	21	Teacher Plus	3	2799
49	21	Biology Education	3	2802
50	21	Psychological Reports	3	2805
51	21	Journal of Research and Computing in Education	3--	2808
Total			2968	

Table 4. Top ten best and most frequently cited core periodicals

S.No.	Rank	Title of Periodicals	Country	No. of Citations	% of Citation
1	1	Dissertation Abstracts International	U.S.A	716	30.03
2	2	Experiments in Education	India	299	12.54
3	3	Journal of Educational Research and Extension	India	290	12.16
4	4	The Educational Review	India	274	11.49
5	5	Indian Education Review	India	258	10.83
6	6	Journal of Indian Education	India	195	8.18
7	7	Journal of The Institute of The Educational Research	India	162	6.79
8	8	Asian Journal of Psychology and Education	India	86	3.62
9	9	Indian Dissertation Abstracts	India	53	2.22
10	10	Journal of Education And Psychology	India	51	2.14

Total Number of Citations in periodicals is 2968. Total Number of citations in top ten core periodicals is 2384 (80.32%). The core list of periodicals in the table 4.4 can be represented using pie diagram as shown in Figure 2.

3.2 Ranked Countries of Most Cited Periodicals Based on Top Ten Periodicals

Table 5 shows Analysis of ranked countries of the most cited periodicals, the core list of periodicals which have received up to 10th position and given the details such as name of the periodical, country of origin, no. of citation and percentage of citation among the top ten core periodicals.

Total citations of core periodicals are 2384. It is found that Indian periodicals ranked first with 1668 (70%) citations and U.S.A periodicals occupies second rank with 716 (30%) citations.

3.3 Authorship Pattern of Cited Periodicals

The authorship pattern, one of the prime aspects of citation analysis, needs to be worked out for better understanding of the study. The authorship pattern may be divided into four headings. These are the

Figure 2. Core list of periodicals

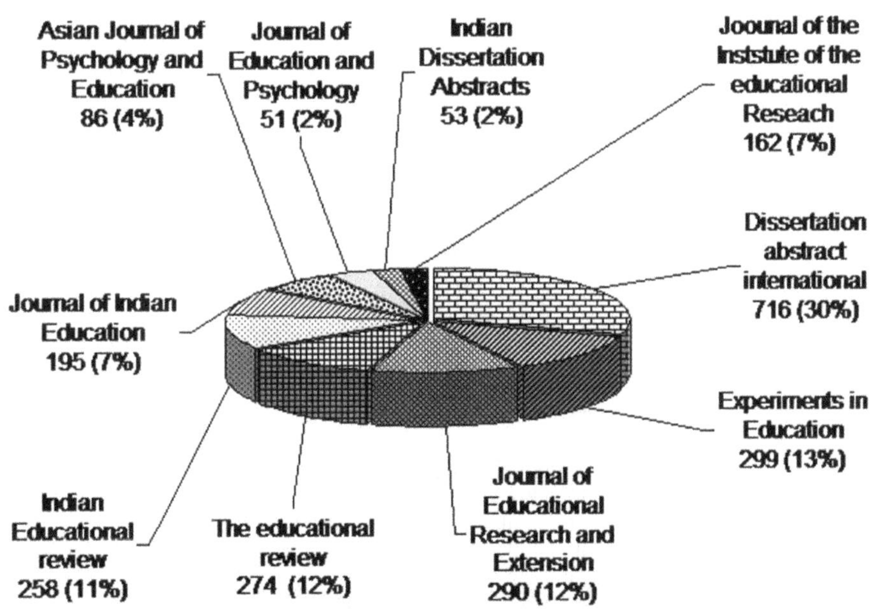

Table 5.Country-wise Distribution of top ten core periodicals

Sl. No	Rank	Name of the Country	No of Citations	% of Citations
1	1	India	1668	69.97
2	2	U.S.A	716	30.03

single author, two authors, three authors and more than three authors. Here the citations are grouped according to number of authors responsible for the work.

The Table 6 highlights the ranking of authorship pattern of periodical cited by the researchers. It seems from the analysis that single authorship pattern scores the highest position in the ranking order followed by two authors, three authors in the second and third position respectively.

The Table 6 shows the ranking of authorship pattern can be diagrammatically represented using pie diagram (Figure 3)

3.4 Ranking of Most Cited Author in Periodical Citation

The most prominent and most cited 10 authors of periodicals were considered. The top 10 authors have 216 citations Sundararajan,S who is a former Professor in Education, D.D.E. Annamalai University is identified as the most cited author with 82 citations. Second and third rank goes to Maslow Ediger and

Table 6. Ranking of authorship pattern in periodicals

Sl. No	Rank	Authorship Pattern	No. of Citations	Cumulative Citation	% of Citation
1	1	one	1690	1690	56.94
2	2	two	1166	2856	39.28
3	3	three	87	2943	2.94
4	4	more than three	25	2968	0.84

Figure 3. Ranking of authorship pattern in periodicals

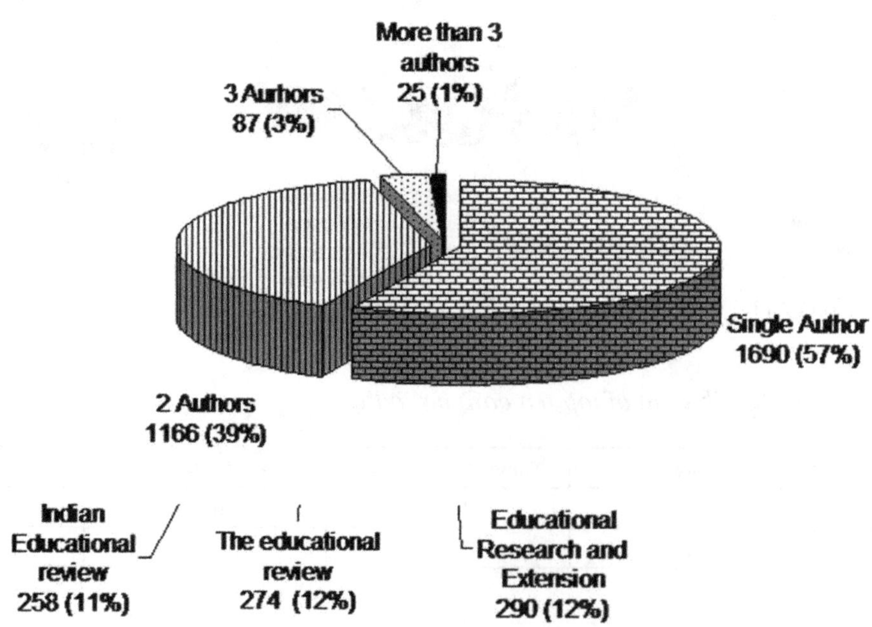

Kamala Pillai, S respectively. Out of the 10 authors, S. Sundararajan who ranks first was heavily concentrated his works on the core periodical "Experiments in Education".

Analysis of the Author citations revealed the fact that the 10 Authors are coming under the single authorship pattern. Table 7 which shows ranking of authors in periodicals can be diagrammatically represented using a pie diagram Figure 4.

Table 7. Ranking of authors in periodicals

SL. No	Rank	Name of the Author	No of Citation	Cumulative Citation
1	1	Sundararajan, S.	82	82
2	2	Maslow Ediger	48	130
3	3	Kamala Pillai, S.	34	164
4	4	Padhi, J.S.	29	193
5	5	Padmavathy, D.	23	216
6	6	Digumurthi Bhaskara Rao	21	237
7	7	Lalit Kishore	13	250
8	8	Anand S.P.	11	261
9	9	Chatterjee, S.	9	270
10	10	Jaganath, K.	4	274

Figure 4. Ranking of Author in Periodicals

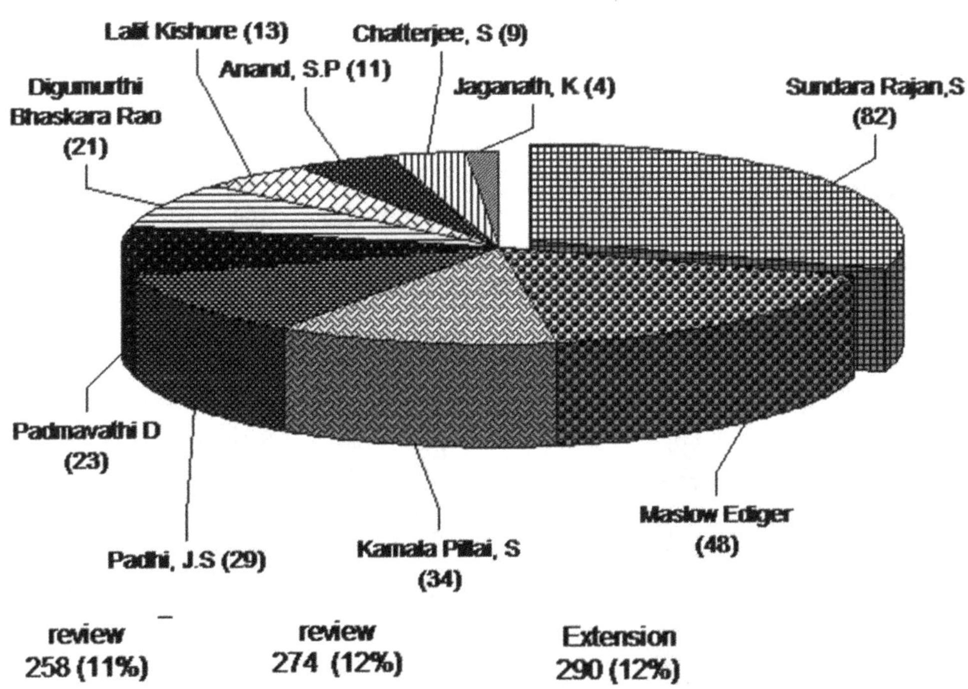

3.5 Analysis of Periodicals on the Basis of Year of Publication

Here the analysis reveals the year of origin of the most relevant periodical articles used by researchers in their dissertations. Only the ranking of 10 years is shown in Table 8.

From the above Table 8 we can conclude that the researcher's mostly cited periodicals are published in the year 1993 with 181 citations (14.23%). The second and third rank were occupied by the year 1991

Table 8. Ranking of periodicals on the basis of year of publication

S.No	Year	No. of Citations	% of Citations	Rank
1	1978	100	7.86	9
2	1979	120	9.43	5
3	1981	102	8.02	8
4	1983	138	10.85	4
5	1986	118	9.28	6
6	1987	115	9.05	7
7	1988	148	11.63	3
8	1991	158	12.42	2
9	1993	181	14.23	1
10	2001	92	7.23	10
	Total	**1272**	100	

Figure 5. Ranking of Periodicals on the Basis of Year

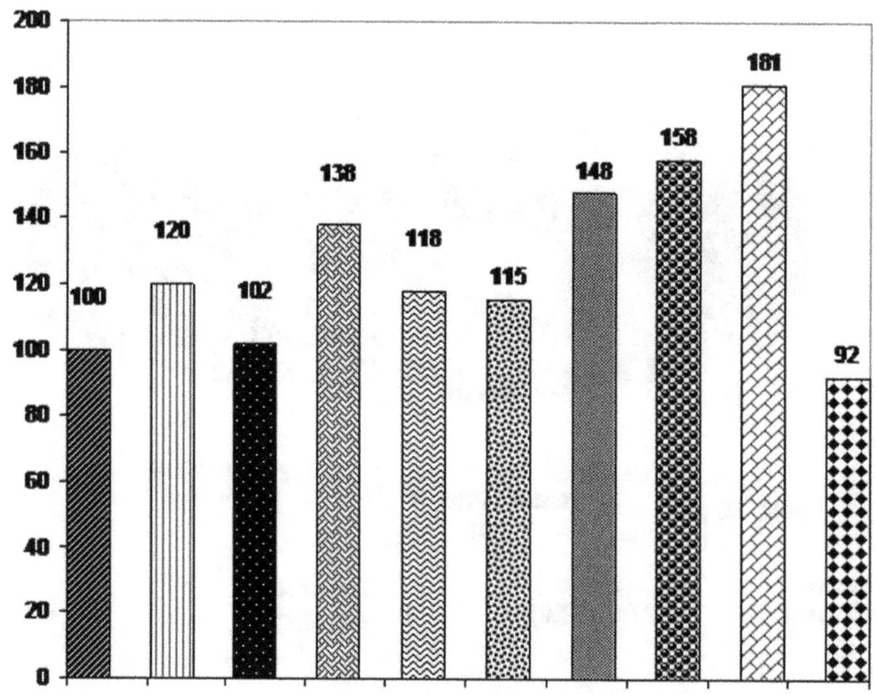

(12.42%) and 1988 (11.63%) respectively. Table 8 is diagrammatically represented using bar diagram (Figure 5).

3.6 Gender-Wise Contribution of Articles in Periodicals

There are a total of 2968 authors (including co-authors) out of them 2668(80.22%) are males and only 320(10.78%) are female contributions. This indicates that male dominancy is there even in respect of contribution of article in periodicals. The Table 9 shows that the Gender-wise contribution of articles in periodicals.

4. Analysis of Book Cited

Books are the second prominent bibliographic form cited in the area of Education by the researchers. The following aspects are taken for the analysis of book citation namely Author-wise analysis, Authorship pattern, most cited books, Country-wise analysis, Year-wise analysis and Gender-wise analysis

4.1 Author-Wise Analysis

Table 9 reflects the ranking of authors according to their frequency of citation. Out of a total of 2720 books a list of 10 most frequently cited authors (books) are listed in Table 10.

Table 9. Gender- wise Contribution of Articles in Periodicals

Gender	No of Citation	% of Citation
Male	2668	89.22
Female	320	10.78

Table 10. Ranked list of authors cited in books

S.No	Rank	Name of the Author	No. of Citations	Cumulative Citations
1	1	Buch,M.B	483	483
2	2	John W Best and James V Khan	128	611
3	3	Kulbir Singh Sidhu	89	700
4	4	Garrett,H.E	83	783
5	5	Agarwal,J.C	76	859
6	6	Kothari,C.R	40	899
7	7	LokeshKoul	34	933
8	8	Mankal,S.K	26	959
9	9	Guilford,J.P	18	977
10	10	Sharma,R.A	12	989

It is seen from the Table 10 that M.B.Buch has got the first rank with 483 citations. Second rank goes to John W. Best and James V. Khan with 128 citations and third rank goes to Kulbir Singh Sidhu with 89 citations.

4.2 Authorship Pattern of Cited Books

The author ship pattern of cited books are identified and analyzed. The authorship pattern is categorized in to one, two, three and more than three. Here the citations are grouped to number of authors responsible for the work. The Table 11 highlights the ranking of author ship pattern of books cited by the researchers.

It seems from the analysis that single author ship pattern scores the highest position in the ranking order with 1755 (64.52%) citations followed by two authors 810 (29.78%) citations and the third rank goes to corporate authors with 72 (2.65%) citations.

The Table 10, which shows the ranking of authorship pattern, can be diagrammatically represented using pie diagram in Figure 6.

The study reveals that single authorship pattern is most influenced and rank first with 1755 (64%) citations.

Table 11. Ranking of authorship pattern of cited books

S.No.	Rank	Authorship Pattern	No.of Citations	Cumulative Citations	% of Citations
1	1	one	1755	1155	64.52
2	2	Two	810	2565	29.78
3	4	Three	60	2637	2.21
4	5	More than three	23	2660	0.84
5	3	Corporate authors	72	2720	2.65

Figure 6. Ranking of Authorship Pattern

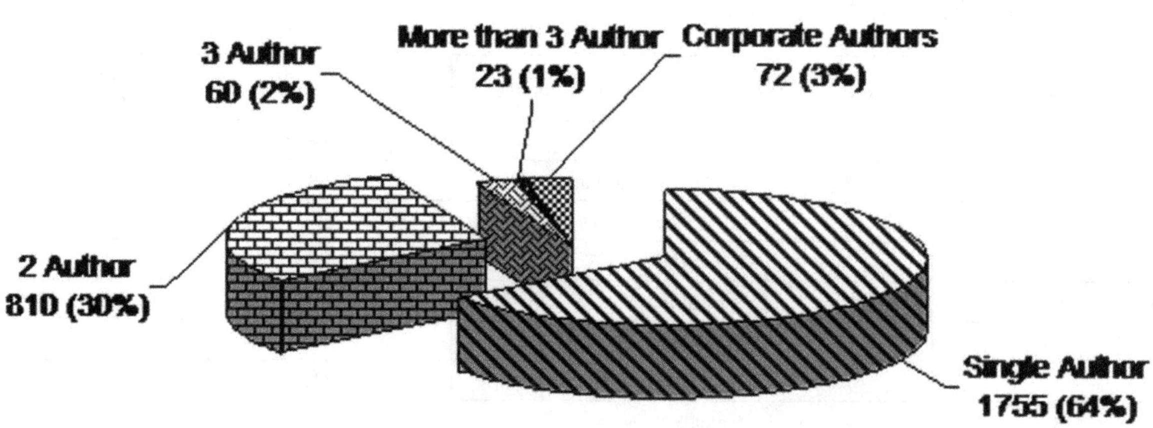

4.3 Most Cited Books

Out of the total 2720 book citation, most of the citations are concentrated on few books listed as follows. The Table 12 gives information about most cited books in the M.Ed dissertations.

Most cited book is "A survey of Research in Education" by Buch, M.B with 453 (46.94%) citations followed by "Research in Education" by John W Best and James Khan, V with 128 (13.27%) citations and Kulbir Singh Sidhu's book "Methodology of research in Education" with 89 (9.22%) citations and so on.

4.4 Country-Wise Analysis of Top Ten Books

Table 13 reflects the ranking order of country-wise top ten books with their distribution of citation.

It is found that books from Indian publishers ranked first with 736 (76.26%) citations, second and third ranks are occupied by U.S.A with 211 (21.87%) and U.K. with 18 (1.87%) citations respectively.

Table 12. Most cited books and its ranking

S.No	Title of the Books	Name of the Author	No.of Citations	% of Citations	Rank
1	A survey of research in Education	Buch, M.B	453	46.94	1
2	Research in Education	John W Best and James V Khan	128	13.27	2
3	Methodology of research in Education	Kulbir Singh Sidhu	89	9.22	3
4	Statistics in Psychology and Education	Garrett, H.E	83	8.6	4
5	Educational Research: An Introduction	Aggarwal,J.C	76	7.88	5
6	Research Methodology	Kothari,C.R	46	4.76	6
7	Methodology of Educational research	Lokeshkoul	34	3.52	7
8	Advanced Educational psychology	Mankal, S,K	26	2.70	8
9	Fundamentals of Statistics in Psychology and education	Guilford, J.P	18	1.87	9
10	Fundamentals of Educational research	Sharma, R.A	12	1.24	10
Total			**965**	**100**	

Table 13. Ranked order of countries of top ten books

SL. No	Name of the Country	No. of Citations	% of Citations
1	India	736	76.26%
2	U.S.A	211	21.87%
3	U.K	18	1.87%
Total		**965**	**100.00%**

4.5 Analysis of Books Based on Year of Publications

The analysis reveals the year of origin of the most relevant book used by researchers in their M.Ed dissertations. Only the top ten years of publication is given in the Table 14.

From the Table 14 we found that the researchers mostly cited the books published in the year 1996 with 71 (13.95%) citations. The second and third ranks goes to the year 1963 with 66 (12.97) citations and 1986 with 62 (12.18) citations respectively. Table 12 is diagrammatically represented using bar diagram as shown in below (Figure 7).

4.6 Gender-Wise Analysis of Books

Gender-wise analysis is carried out on contribution of books. The details are summarized in Table 15.

There are total of 2720 citations out of them 2660 (97.79%) are males and the remaining are by the female 60 (2.21%).

5. Distribution of Citations in the Masters Dissertations During 1996-2006

The number of citations in each dissertation is important while studying the change in citation behavior of the research scholar. Growth of literature results in the increased availability of literature of the subjects. Distribution of citation in Education in the M.Ed Dissertations is shown in the Table 16.

From the Table 16 it is clear that out of total 6886 citations 702 (10.19%) citations were cited during the year 1996 followed by 861 (12.50%) citations were cited during the year 1997and 729 (10.59%) citations were cited during 1998 respectively. Distribution of citation in the M.Ed dissertations during the period 1996-2006 are represented using the bar diagram Figure 8.

Study reveals that the number of citations is coming down in recent years.

Table 14. Analysis of books on the basis of year of publication

Sl. No	Year	No of Citations	% of Citation	Rank
1	1963	66	12.97	2
2	1964	40	7.86	9
3	1966	47	9.23	5
4	1973	39	7.66	10
5	1979	41	8.06	8
6	1982	44	8.64	6
7	1986	62	12.18	3
8	1990	57	11.20	4
9	1993	42	8.25	7
10	1996	71	13.95	1
Total		509	100	

Figure 7. Analysis of Books on the basis of Year of Publication

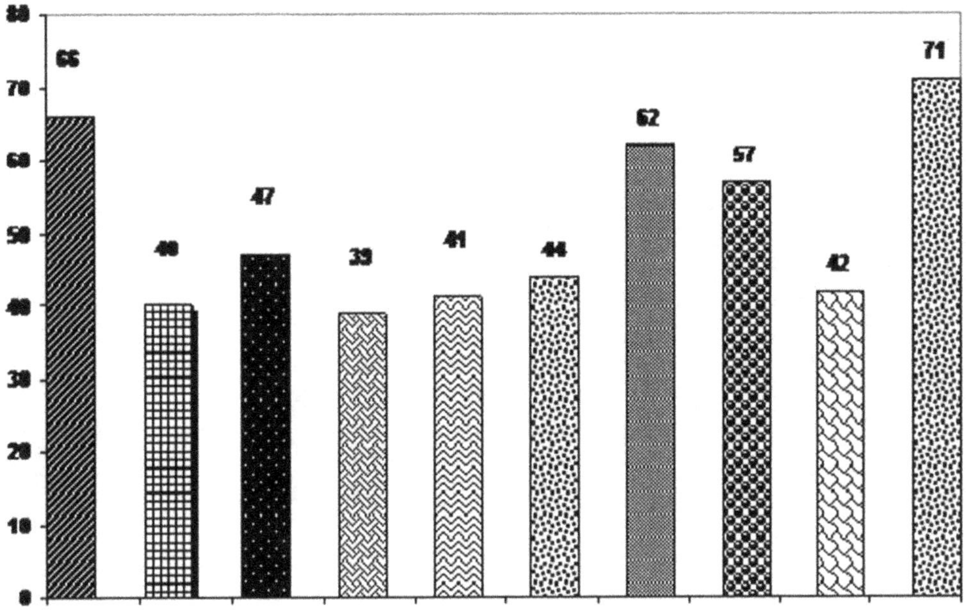

Table 15. Gender- wise contribution of books

Gender	No of Citation	% of Citation
Male	2660	97.79%
Female	60	2.21%
Total	**2720**	**100%**

Table 16. Distribution of citations in the masters dissertations during 1996-2006

S.No	Year	No. of Citations	Percentage	No. of Dissertations
1	1996	702	10.19	15
2	1997	861	12.50	15
3	1998	729	10.59	15
4	1999	692	10.05	15
5	2000	681	9.89	13
6	2001	337	4.89	10
7	2002	663	9.63	15
8	2003	669	9.72	15
9	2004	587	8.52	15
10	2005	551	8.00	15
11	2006	414	6.02	15
Total		**6886**	**100**	**158**

Figure 8. Distribution of Citations in the M.Ed Dissertations

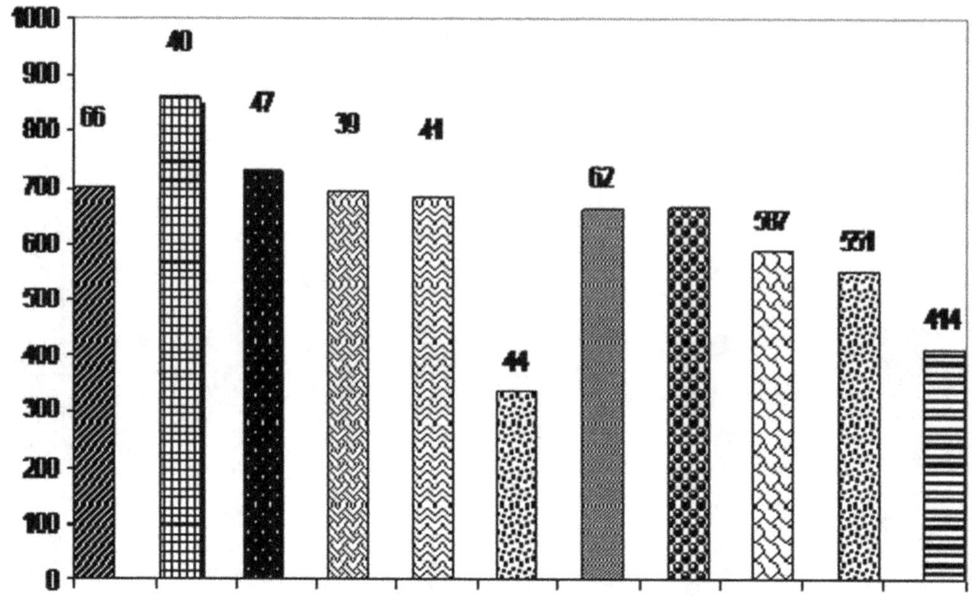

FINDINGS OF THE STUDY

The major findings of the study are summarized as

1. Out of the total 6886 citations spreading over 6 bibliographical forms which constitute the basis of this study, Periodicals had the highest number of citations (2968) ranked first with 43.10%. The second rank goes to Books with 2720 citations (39.50%) and third rank goes to miscellaneous forms which includes Encyclopedia, Dictionary, Theses, Dissertations, Web studies, abstract, report and others with 1198 citations (17.40%).
2. Periodical forms of publications occupy a prominent place rather than the books. The Analysis shows that web studies are used only 68 (0.98%).
3. Journal articles are mostly cited in the thesis. The periodical, "Dissertation abstracts International" with 716 citations (24.12%) ranked first. Second and third rank goes to "Experiments in Education" with 299 citations (10.07%) and "Journal of Educational Research and Extension" with 290 citations (9.77%)
4. Indian periodicals have first rank with 1668 (70%) citations and U.S.A periodicals comes second rank with 716 (30%) citations.
5. Single authorship pattern scores the highest position in the ranking order followed by two authors and three authors in the second and third position.
6. The top 10 authors have 216 citations. Sundararajan,S who is worked as a Professor in Education, D.D.E. Annamalai University is identified as the most cited author with 82 citations. Second and third position goes to Maslow Ediger and Kamala Pillai, S respectively. Out of the 10 authors, S.Sundararajan who ranks first position was heavily concentrated his works on the core periodical "Experiments in Education".

7. Mostly cited periodicals are published in the year 1993 with 181 citations (14.23%). The second and third ranks were occupied by year 1991 (12.42%) and 1988 (11.63%)
8. There are a total of 2968 authors (including co-authors). Out of them 2668(80.22%) are males and only 320(10.78%) are female contributions. This indicates that male dominancy is there even in respect of contribution of article in periodicals.
9. Books are the second prominent bibliographic form cited in the area of Education by the researchers
10. Author wise analysis shows that M.B.Buch has got the first place with 483 citations. Second rank goes to John W. Best and James V. Khan with 128 citations and KulbirsinghSidhu with 89 citations ranked third.
11. Single authorship pattern scores the first rank with 1755 (64.52%) citations followed by two authors 810 (29.78%) citations and the third position goes to corporate authors with 72 (2.65%) citations.
12. Out of the total 2720 book cited, most cited book is "A survey of Research in Education" by Buch,M.B with 453 (46.94%) citations followed by "Research in Education" by John, W. Best and James Khan,V with 128 (13.27%) citations and Kulbir Singh Sidhu's book "Methodology of research in Education" with 89 (9.22%) citations
13. Indian publishers ranked first with 736 (76.26%) citations, second and third positions are occupied by U.S.A with 211 (21.87%) and U.K. with 18 (1.87%) citations.
14. Mostly cited books were published in the year 1996 with 71 (13.95%) citations. The second and third position goes to the year 1963 with 66 (12.97%) citations and 1986 with 62 (12.18%) citations respectively.
15. Highest number of citations was in the year 1997 with 702(10.19%) citations and the lowest citations was during the year 2001 with 337(4.89) citations. The study reveals that the numbers of citations are coming down in recent years.

CONCLUSION

Citation analysis is an established research tool for bibliometric study for identifying variables such as most cited author, most cited periodicals and the aggregates of scientific activities involving the counting and analyses of citation. The present study analyses the citations appended to articles that appeared popular Indian education journals and attempts to identify the core periodicals in the field of education. In this respect, the results of present study would be interesting and useful to the users of literature in education in organizing and promoting research. Research in education use a wide variety of document forms like journals, books, theses and newspapers. The use of journal is very high among researchers in education. Thus, the hypothesis was accepted. This study will examine the nature and growth of research in education during 1996-2006. The study suggests that there is an urgent need for modernizing research through changes in the selection of topics and tools for research. We have to concentrate more in foreign journals and books so that we can raise our standards to compete with the International Society. We will get the knowledge of happenings around the globe only if we go through the foreign journals. It will be very much useful for the researchers who are going abroad for their career development.

REFERENCES

Anilkumar, N. & Rajaram, S. (2013). Theses submitted by doctoral students of physical research laboratory, India: a citation analysis. *Serials Review, 39*(2), 114–120.

Bhattacharya, G. (1978). *Information Science: A unified view through systems approach*. Calcutta: IASLIC.

Bothakur, P. (2015). Citation analysis of theses and dissertations in chemistry submitted to the LNB, library, Dibrugarh university, 2009-13. *International Journal of Research in Library Science, 1*(2), 33–41.

Deutsch, A. (1977). *The Librarians glossary*. New Delhi: Indian Book Company.

Devi, B. M., & Sankar, J. V. S. (2014). Information use pattern of researchers in commerce: A citation analysis of doctoral dissertations. *Library Philosophy and Practice*. Retrieved from http://digitalcommons.unl.edu/libphilprac/1105

Garfield, E. (1979). *Citation Indexing: Its theory and application in Science, Technology and Humanities*. USA: Wiley Inter Science Publications.

Lahiri, R. (1996). Citation Analysis: As a tool for collection Development in Libraries. Management of Libraries: Concepts and practices. New Delhi: Ess Ess publication.

Mookerjee, B. P. (1998). *Citation Indexing in Subject Indexing Systems: Concepts, Methods and Techniques*. Calcutta: IASLIC.

Narayana, G. J. (1991). *Library and Information Management*. New delhi: Prentice Hall of India Pvt. Ltd.

Rao, R. (1983). *Qualitative methods in library and Information science*: Newdelhi: Wiley eastern.

Reifz, J. M. (2004). *Dictionary for Library and Information Science*. USA: Libraries Unlimited.

Shafi, S. M. (2002). Citation Analysis of Ph.D. Thesis: A Study of Doctoral Theses submitted during, 1980-2000 in Natural Science. *Library Philosophy and Practice*.

Weinstock, N. (n.d.). [Citation indexes]. In Encyclopedia of Library and Information Science (Vol. 15). New York: Marcel Deccker.

Chapter 7
Visualizing Zika Virus Research Literature Through Bibliometric Mapping

Usharani Munuswamy
Tamil Nadu State Forensic Sciences Department, India

J. John Jeyasekar
Tamil Nadu State Forensic Sciences Department, India

ABSTRACT

This study outlines a significant relationship between a spurt of research activity in the field and the outbreak of the Zika disease, for the purpose of visualizing the phenomenon. By using bibliographic research, it is possible to correlate data with this disease. To identify the hotspots of research the data is evaluated using the following parameters: literature growth since 1975; most prolific authors and their contributions; top contributing institution, country, and journal; identify co-authorship clusters of the authors and institutions; and finally, to create and examine co-word maps of the keywords. It is also found that under-developed countries like Senegal and Zambia have contributed to Zika research along with the USA, France and Germany. Forty-five per cent of the contributions are concentrated in six journals.

INTRODUCTION

Viruses are infectious micro-organisms, which replicate inside the living cells of other organisms. Viruses can infect all types of life forms (Koonin et al., 2006). Viruses are found wherever there is life and have probably existed since living cells first evolved (Iyer et al., 2006). The Zika virus belongs to the family *Flaviviridae* and the genus *Flavivirus*. It is usually spread by the two *Aedes* mosquitoes, *A. aegypti* and *A. Albopictus* (Malone, et al., 2016). This virus was first isolated in 1947 from the monkeys found in the Zika Forest of Uganda. Zika virus is related to dengue, yellow fever, Japanese encephalitis, and West Nile viruses (Sikka et al., 2016).

DOI: 10.4018/978-1-5225-3457-0.ch007

The Zika disease caused by zika virus which affects the public health is emerging as a global threat. The disease though existed several years before, the term zika gained popularity only after its outbreaks in early 2015. The symptoms of zika fever are similar to that of dengue fever (Malone, et al., 2016) and often cause misdiagnosis. Zika fever was first found in the 1950s in a narrow equatorial belt from Africa to Asia. Later the virus was found spreading eastward across the Pacific Ocean. In the years 2013 and 2014, zika virus outbreaks occurred in Oceania to French Polynesia, New Caledonia, the Cook Islands, and Easter Island. In 2015 it has reached a pandemic level in Mexico, Central America, the Caribbean, and South America (Chastain, 2016). As of now, no medications or vaccines are available for its prevention. This virus is also found to spread from a pregnant woman to her baby. Microcephaly, Guillain-Barré Syndrome and other severe brain problems are found to be the result of zika (Rasmussen, 2016; CDC, 2016).

Scientometrics is the quantitative study of science and technology. It is the study of quantitative aspects of science as a discipline or economic activity (Tague-Sutcliffe, 1992). Scientometric techniques have wide applications in identifying the author productivity, authorship pattern, core periodicals, research trends in a subject, research collaboration and impact of research, etc. Hence it is of great use to identify the emerging research areas within a given subject. Science mapping or bibliometric mapping is a spatial representation of how disciplines, fields, specialties, and individual authors or documents are related to one another (Small, 1999). Science maps are useful tools to understand the state-of-the-art disciplinary structure within an academic field as well as to analyse the emergence of research networks among institutions and authorship collaborations.

REVIEW OF LITERATURE

Bibliometric mapping which is performed using available bibliographic information deposited at major journals and bibliographic databases helps to understand the existing scientific knowledge about the disease, the level of research and its impact on scientific community at different levels. Scoping reviews apart from bibliometric analysis, also plays a vital role in addressing broad, policy driven research questions by identifying all the relevant evidence concerning the issue and producing summaries of the findings (Levac et al., 2010; Colquhoun et al., 2014). These reviews follow a structured protocol for the identification and characterization of the literature in a manner that is both reproducible and updateable (Young et al., 2014; Colquhoun et al., 2014). The scoping review is also well-suited for the identification of evidence on a broad-topic, but does not include a quality assessment or in-depth data extraction stage that would be required for meta-analysis of studies. However, an important aspect of this review is the identification of where evidence is lacking or non-existent to help direct future research and use of resources (Waddell and Greig, 2016).

A bibliometric study conducted using indexing databases such as Science Citation Index (*SCI*), *Scopus* and *Medline* reviewed the current impact of zika virus on global scientific production. The study for its search strategy used the term 'zika' as a main operator to retrieve data on indexed articles from the databases (Martinez-Pulgarin et al., 2016). A scoping review was conducted in response to zika virus outbreaks and changes in its epidemiology, capturing all published literature addressing the aspects of zika virus infection in humans, zika virus pathogenesis, transmissions, molecular mechanisms, prevention strategies of zika virus infections, control of zika virus harbouring vectors etc. The same review has identified epidemiological studies examining risk factors for mosquito abundance, human exposure to mosquitoes and conditions for zika virus infected mosquitos (Waddell and Greig 2016). A retrospective and prospective study underway in Brazil and other zika virus affected countries helped in understanding the role of the virus in causing microcephaly, birth defects and negative pregnancy outcomes (Waddell and Greig 2016). The latter have also identified and characterized the global literature on zika virus and identified several knowledge gaps with respect to the epidemiology, the burden of disease in humans and complication related to zika virus infection. Charrel et al. (2016) have assessed the current state of knowledge on zika virus diagnostics and identified present knowledge gap for adequate zika virus diagnostics.

Likewise, to bibliometric studies on zika virus and disease, various authors have conducted scoping reviews, bibliometric and scientometric studies on similar topics of importance to scientific community. Using species distribution models, Pigott et al. (2014) predicted zoonotic transmission niche of Ebola virus in Africa. Vellaichamy (2016) performed a scientometric analysis of 20 years literature on Ebola virus and demonstrated the level of research, contributions by various journals and contributing countries etc in this field. The latter author along with Jeyshankar have also carried out bibliometric analysis on anaemia (Vellaichamy and Jeyshankar, 2014), cervival cancer (Jeyshankar and Vellaichamy, 2015), lung cancer (Jeyshankar and Vellaichamy, 2015), myeloma research (Vellaichamy and Jeyshankar, 2015), dengue research (Vellaichamy and Jeyshankar, 2015), autism (Jeyshankar and Vellaichamy, 2016), etc., and demonstrated the level of existing scientific knowledge and the need for different levels of research in the respective fields.

Gupta et al. (2011) performed a scientometrics analysis of publications output during the period 1999-2008 on mapping of *AIDS/HIV* research in India. A Bibliometric analysis of scientific output on *H1N1* influenza in Mexico for the period 2000-2012 was conducted by Castillo-Perez et al. (2015). Bibliometric analysis of severe acute respiratory syndrome (*SARS*) has also been carried out by Chiu et al. (2004). A scientometric study on the growth of poliomyelitis research literature during 1999 to 2014 using the techniques of relative growth rate, doubling time, etc., was performed by Keshava and Thimmaiah (2015).

OBJECTIVES OF THE STUDY

The objectives of this study are

1. To analyse the growth of zika research literature since 1975
2. To examine the most prolific authors and their contributions
3. To observe the top contributing institution, country, and journal
4. To visualise the co-authorship clusters of the authors and also that of the institutions
5. To create and examine co-word maps of the keywords

MATERIALS AND METHODS

The bibliographic data pertaining to zika virus for the period 1976 to 2015 were downloaded from *PubMed*. *PubMed* is a search engine accessing primarily the Medline database of references and abstracts on life sciences and biomedical topics. A total of 111 records were obtained. These data were cleaned and then subjected to further analysis using MS-Excel spreadsheets and *VOSviewer* software. *VOSviewer* is a program developed for constructing and viewing bibliometric maps. The program is freely available to the bibliometric research community (www.vosviewer.com). *VOSviewer* can be used to construct maps of authors or journals based on co-citation data or to construct maps of keywords based on co-occurrence data.

RESULTS AND DISCUSSIONS

1. Literature Growth

The year-wise growth of literature, cumulative growth and also their percentage are calculated and presented in Table 1. It is found growth of literature was very slow until the year 2013. A total of only 54 documents had been published from 1976 to 2013. This works out to an average of 1.38 papers per year. However, in the year 2014 a sudden spurt has been observed and the growth trend continues on the exponential growth mode. This rapid growth rate of literature coincides with the pandemic outbreak of the virus. This pattern may possibly mean that the researchers begin to show interest after the occurrence of the event. Significant damage could have been averted had sufficient work been done much earlier.

The growth of the zika virus literature is visualised with the aid of the pictorial representation in Figure 1. The year-wise number documents as-well-as the cumulative numbers of documents are plotted as graphs in this figure. The year-wise growth rate is a 'Sleeping Beauty' (*SB*) until 2013. The concept of *SB* was first applied in citation analysis by van Raan (2004) to denote publications that respond delayed with regard to attracting citations. However, this principle is used in studying the publication count and zika virus research literature is found to be a *SB* until awaken by the outbreak of the virus to pandemic level.

Table 1. Literature Growth

Year	Documents	Percentage	Cumulative Growth	Percentage
1976-1980	9	8.11	9	8.11
1981-1985	10	9.01	19	17.12
1986-1990	5	4.50	24	21.62
1991-1995	3	2.70	27	24.32
1996-2000	3	2.70	30	27.03
2001-2005	5	4.50	35	31.53
2006-2010	10	9.01	45	40.54
2011	1	0.90	46	41.44
2012	5	4.50	51	45.95
2013	3	2.70	54	48.65
2014	23	20.72	77	69.37
2015	34	30.63	111	100
Total	111	100		

The cumulative growth rate is also analysed with respect to the Price law of exponential growth of research literature. It is found from the same figure that the growth of zika research literature fits the exponential growth model.

2. Most Prolific Authors

The authors are ranked according to their number of contributions and presented in Table 2. The percentage of their contribution with respect to the total publication and the total number of authorship are also calculated.

It is found from the table that at least 10 authors have contributed more than 5 papers. The total contribution of these ten authors together account for about fifty percent of the total literature produced. Faye O is the top-ranking author with 12 papers to his credit, followed by Sall A A who has 7 papers to his credit. Five authors namely, Musso D, Schmidt-Chanasit J, Tappe D, and Weaver S C share the third position and each of them have contributed five papers. It is also found that 467 authorships are involved in the production of the 111 papers. Hence, the average author per paper is found to be as high as 4.21. It is also found that more than 57% of the authors have contributed just one paper. It will be highly beneficial for the society if these researchers are more active in publishing too.

3. Top Contributing Institution

The top ranking contributing institutions are ranked and presented in Table 3. This table also provides the number of papers contributed by these institutions and their percentage.

Figure 1. Literature Growth

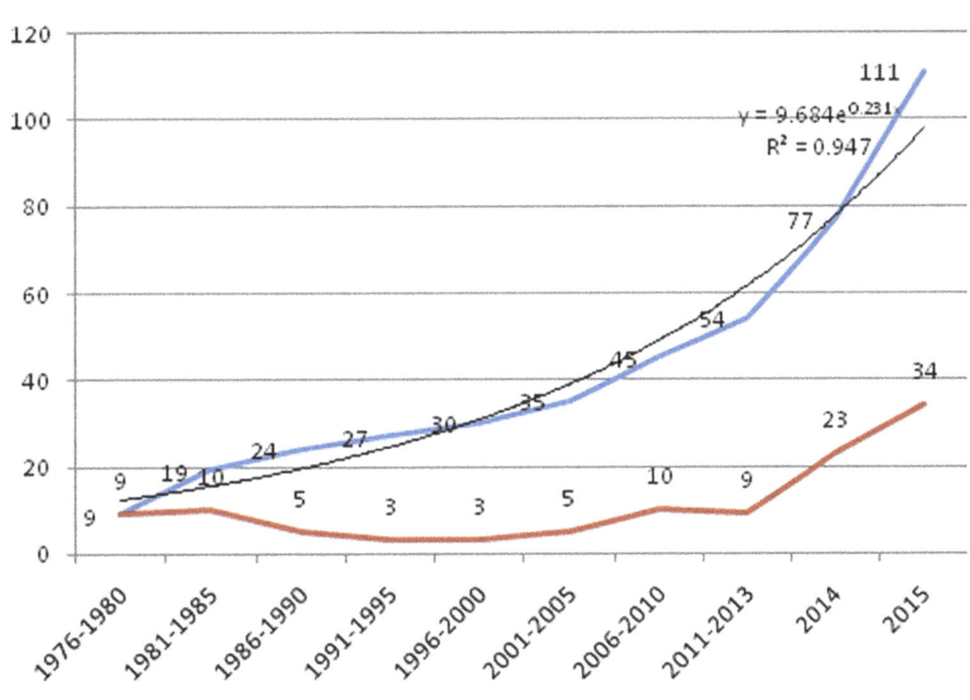

Table 2. Most prolific authors

Author	Documents	Percentage	Percentage of 467	Rank
Faye O	12	10.81	2.57	1
Sall A A	7	6.30	1.50	2
Musso D	6	5.40	1.28	3
Schmidt-Chanasit J	6	5.40	1.28	3
Tappe D	6	5.40	1.28	3
Weaver S C	6	5.40	1.28	3
Cao-Lormeau V M	5	4.50	1.07	7
Diallo D	5	4.50	1.07	7
Roche C	5	4.50	1.07	7
Teissier A	5	4.50	1.07	7
9 Authors X 4 papers	36		7.71	
12 Authors X 3 papers	36		7.71	
32 Authors X 2 papers	64		13.70	
268 Authors X 1 paper	268		57.39	
Total	467	49.55	100	

Table 3. Leading Institutions

Institution	Documents	Percentage	Rank
Institut Pasteur, Senegal	7	6.31	1
Institut Louis Malarde, French Polynesia	6	5.41	2
Centers For Disease Control And Prevention, USA	5	4.50	3
Institut Pasteur, France	4	3.60	4
University Of Texas, USA	4	3.60	4
Bernhard Nocht Institute For Tropical Medicine, Germany	3	2.70	6
Universita Degli Studi Di Firenze, Italy	3	2.70	6
Total	32	28.83	

The table reveals that Insitut Pasteur, Senegal is the highest contributor with 7 papers followed by Insitut Louis Malarde, French Polynesia with 6 papers. The top seven institutions together account for about 29 percent of the total papers published.

4. Top Contributing Country

The leading countries, their contributions and percentage of the contribution, and their rank are enumerated in Table 4. The eight countries listed in this table together account for 82% of the total.

The table also shows that USA with 21 papers, which is about 19%, is ranked first. USA is closely followed by France with about 17%. French Polynesia and Germany with 10 papers each share the

Table 4. Leading Countries

Country	Documents	Percentage	Rank
USA	21	18.92	1
France	19	17.12	2
French Polynesia	10	9.01	3
Germany	10	9.01	3
Senegal	9	8.12	5
Zambia	8	7.21	6
Brazil	7	6.31	7
Canada	7	6.31	7
Total	91	82	

third position. USA, the world leader in almost all the spheres of life clearly dominates every field of research. This pattern is found to be true in this case too. However, lesser developed nations such as French Polynesia, Senegal and Zambia too find a place in this ranked list. This is obviously due to fact that high incidence of zika virus infection reported in these parts of the world.

5. Top Contributing Journal

The ranked listed of source journals contributing zika virus research papers are enumerated in Table 5. This table also details the number of papers contributed by these journals, their percentage and also the cumulative percentage.

The table reveals that the top-ranked journal 'Emerging Infectious Diseases' accounts for about 14% of the total papers. The second rank is shared by two journals which together have contributed about 14% of the total. That means the second ranked journals have produced just half the first ranked one. It is also observed from the table that the top ranked 6 journals together have contributed 50 papers. Hence

Table 5. Top Contributing Source Journals

Journal	Frequency	Percentage	Cumulative Percentage	Rank
Emerging Infectious Diseases	16	14.41	14.41	1
American Journal of Tropical Medicine & Hygiene	8	7.21	21.62	2
Transactions of the Royal Society of Tropical Medicine & Hygiene	8	7.21	28.83	2
Euro Surveillance	7	6.31	35.14	4
PLoS Neglected Tropical Diseases	6	5.41	40.55	5
Bulletin de la Société de Pathologie Exotique et de ses Filiales	5	4.50	45.05	6
2 Journals X 3 papers	6	5.41	50.46	
6 Journals X 2 papers	12	10.81	61.27	
43 Journals X 1 paper	43	38.74	100	
Total	111	100		

45% of the papers are concentrated among these 6 journals and the rest 55% papers are found scattered among 51 journals. A journal published in French language also finds a place among the top six journals.

6. Co-Authorship Maps

A co-authorship map is constructed with the aid of the software VOSviewer. The unit of analysis selected is 'authors' and the counting method selected is 'full counting'. The threshold value of minimum number of documents of an author selected is 2. Out of the total of 467 authors, 92 authors fulfill this criterion. However, out of these 92 authors only 50 are connected, i.e., only 50 authors have co-author relationship. The co-authorship map of these 50 authors is presented in Figure 2.

The fifty authors found in the map are found in 6 clusters. Cluster number 1 has 12 items or authors; cluster number 2 has 11 items; cluster number 3 has 9 items; cluster number 4 has 7 items; cluster number 5 has 6 items; and cluster number 6 has 5 items. The authors who have co-authorship value of 10 and above 10 are listed in Table 6. Thirty-two authors have co-authorships of above 10.

Similarly, a co-authorship map is created with the 'organization' as unit of analysis. A total of 105 organizations are found in the analysis and out of these 105, only 13 are connected. The map of the connected items is created and is illustrated in Figure 3.

It is found from the figure, the 13 items belong to three clusters. Cluster number 1 consists of 6 items / organizations; cluster number 2 consists of 4 items; and cluster number 3 consists of 3 items.

Figure 2. Co-authorship map

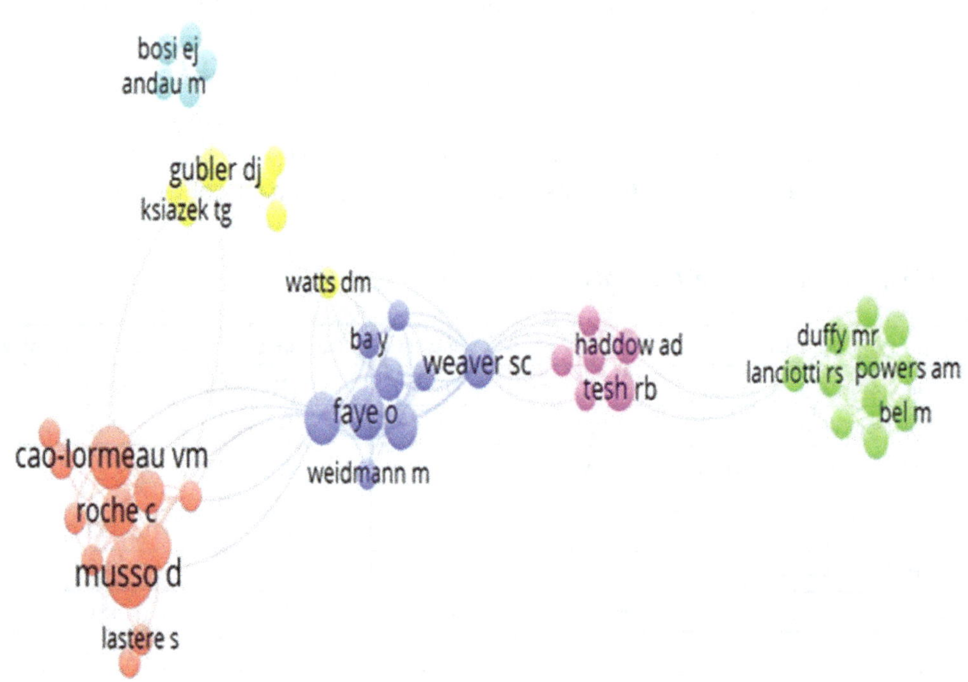

Table 6. Authors and their co-authorship values

Author	Co-Authorships	Author	Co-Authorships
Musso D	32	Guillaumot L	17
Sall A A	32	Powers A M	17
Cao-Lormeau V M	30	Pretrick M	17
Weaver S C	28	Hanley K A	14
Faye O	27	Lanciotti R S	14
Diallo M	26	Tesh R B	14
Roche C	25	Diagne C T	13
Diallo D	24	Haddow A D	12
Teissier A	22	Lambert A J	12
Ba Y	20	Gubler D J	11
Robin E	20	Broult J	10
Bel M	19	Hayes E B	10
Duffy M R	19	Heang V	10
Hancock W T	19	Kasper M R	10
Marfel M	19	Nhan T	10
Griggs A	17	Yasuda C Y	10

Figure 3. Organizational co-authorships

7. Keyword Co-Occurrence Map

Cluster Analysis is a multivariate procedure of detecting natural groupings in data (Wulder, 2014). Cluster Analysis is used to group objects, people, countries or other entities on the basis of shared characteristics (McCain, 1990). Grouping the entities together on the basis of similarities and differences is possible by Cluster Analysis (Tryon and Bailey, 1970). The similarity strength between the entities, in this case the key terms used in the abstracts of the bibliographic records, are analysed and then they are represented graphically.

Four clusters are observed in the cluster map of keywords presented in Figure 4. Each cluster is represented by a different colour. Cluster number 1 has 24 keywords, which include alphavirus, antibodies, arbovirus, infections, chikungunya, Nigeria, and Uganda. The second cluster has 11 keywords including dengue virus, flavivirus, haplorhini, hemagglutination inhibition, yellow fever, aedes, togaviridae infections, culicidae, dengue, and Micronesia. Cluster number 3 includes keywords like fever, viral RNA, reverese transcriptase polymerase chain reaction, zika virus, travel, and Polynesia. Altogether this cluster has 11 terms. Cluster number 4 has 6 terms including phylogeny, molecular sequence data, and viral proteins.

CONCLUSION

Though the literature growth pattern fits the exponential model, a spurt is seen after the virus reached the pandemic level. It is concluded that the researchers should be prepared before a disease becomes endemic. The analysis of authorship trend implies most of the researchers are not active in publishing their results. The study also indicates that lesser known countries like French Polynesia, Senegal, and Zambia are as active as the USA, France and Germany. The study of source journals reveals that forty-five percent of the papers are concentrated among six journals.

Figure 4. Cluster Map of MESH Keywords

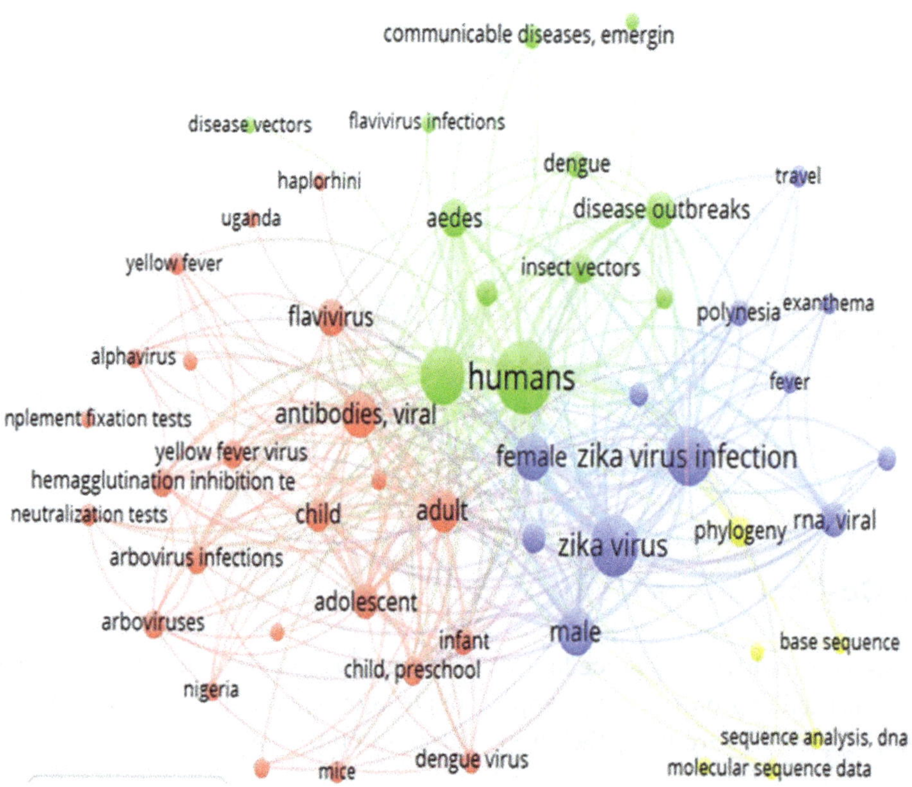

REFERENCES

Castillo-Pérez, J. J., Muñoz-Valera, L., García-Gómez, F., & Mejía-Aranguré, J. M. (2015). Bibliometric analysis of scientific output on influenza in Mexico 2000-2012. *Revista Medica del Instituto Mexicano del Seguro Social*, *53*, 294–301. PMID:25984614

Centers for Disease Control and Prevention. (2016, April 13). *CDC concludes* zika *causes microcephaly and other birth defects.* News Room Release.

Charrel, R. N., Leparc-Goffart, I., Pas, S., de Lamballerie, X., Koopmans, M., & Reusken, C. (2016). State of knowledge on zika virus for an adequate laboratory response. *Bulletin of the World Health Organization*. doi:10.2471/BLT.16.171207

Chastain, M. (2016). National Institutes of Health: zika virus is a pandemic. Breitbart; Retrieved 13 February 2016.

Chiu, W.-T., Huang, J.-S., & Ho, Y.-S. (2004). Bibliometric analysis of Severe Acute Respiratory Syndrome-related research in the beginning stage. *Scientometrics*, *61*(1), 69–77. doi:10.1023/B:SCIE.0000037363.49623.28

Colquhoun, H. L., Levac, D., O'Brien, K. K., Straus, S., Tricco, A. C., Perrier, L., & Moher, D. et al. (2014). Scoping reviews: Time for clarity in definition, methods, and reporting. *Journal of Clinical Epidemiology*, *67*(12), 1291–1294. doi:10.1016/j.jclinepi.2014.03.013 PMID:25034198

Gupta, A. B., & Har, K. (2011). Mapping of AIDS/HIV research in India: A scientometrics analysis of publications output during 1999-2008. *Collnet Journal of Scientometrics and Information Management*, *5*(2), 185–203. doi:10.1080/09737766.2011.10700912

Iyer, L. M., Balaji, S., Koonin, E. V., & Aravind, L. (2006). Evolutionary genomics of nucleo-cytoplasmic large DNA viruses. *Virus Research*, *117*(1), 156–184. doi:10.1016/j.virusres.2006.01.009 PMID:16494962

Jeyshankar, R., & Vellaichamy, A. (2015). A bibliometric analysis of global literature on cervical cancer. *International Journal of Information Dissemination and Technology*, *4*, 298–302.

Jeyshankar, R., & Vellaichamy, A. (2015). Research productivity of lung cancer by Indian scientist during 1984 – 2013. *Productivity*, *56*, 24–35.

Jeyshankar, R., & Vellaichamy, A. (2016). Scientometric analysis of autism research output during 2007-2011. *SRELS Journal of Information Management*, *53*(1), 55–64. doi:10.17821/srels/2016/v53i1/86768

Keshava & Thimmaiah, B. N. (2015). Growth of poliomyelitis research literature during 1999-2014: A scientometric study. *Indian Journal of Research*, *4*, 199–201.

Koonin, E. V., Senkevich, T. G., & Dolja, V. V. (2006). The ancient virus world and evolution of cells. *Biology Direct*, *1*(1), 29. doi:10.1186/1745-6150-1-29 PMID:16984643

Levac, D., Colquhoun, H., & O'Brien, K. K. (2010). Scoping studies: Advancing the methodology. *Implementation Science; IS*, *5*(1), 69. doi:10.1186/1748-5908-5-69 PMID:20854677

Malone, R. W., Homan, J., Callahan, M. V., Glasspool-Malone, J., Damodaran, L., Schneider, A. D. B., & Wilson, J. et al. (2016). Zika virus: Medical countermeasure development challenges. *PLoS Neglected Tropical Diseases*, *10*(3), e0004530. doi:10.1371/journal.pntd.0004530 PMID:26934531

Martinez-Pulgarin, D. F., Acevedo-Mendoza, W. F., Cardona-Ospina, J., Rodriguez-Morales, A. J., & Paniz-Mondolfi, A. E. (2016). A bibliometric analysis of global zika research. *Travel Medicine and Infectious Disease*, *14*(1), 55–57. doi:10.1016/j.tmaid.2015.07.005 PMID:26257029

McCain, K. W. (1990). Mapping authors in intellectual space: A technical overview. *Journal of the American Society for Information Science*, *41*(6), 433–443. doi:10.1002/(SICI)1097-4571(199009)41:6<433::AID-ASI11>3.0.CO;2-Q

Pigott, D. M., Golding, N., Mylne, A., Huang, Z., & Henry, A. J. (2014). Mapping the zoonotic niche of Ebola virus disease in Africa. *eLife*, *04395*. doi:10.7554/eLife.04395 PMID:25201877

Rasmussen, S. A., Jamieson, D. J., Honein, M. A., & Petersen, L. R. (2016). Zika virus and birth defects - reviewing the evidence for causality. *The New England Journal of Medicine*, *374*(20), 1981–1987. Retrieved15April2016. doi:10.1056/NEJMsr1604338 PMID:27074377

Sikka, V., Chattu, V., Popli, R. K., Galwankar, S. C., Kelkar, D., Sawicki, S. G., & Papadimos, T. J. et al. (2016). The emergence of zika virus as a global health security threat: A review and a consensus statement of the INDUSEM Joint working Group (JWG). *Journal of Global Infectious Diseases*, *8*(1), 3–15. doi:10.4103/0974-777X.176140 PMID:27013839

Small, H. (1999). Visualizing science by citation mapping. *Journal of the American Society for Information Science*, *50*(9), 799–813. doi:10.1002/(SICI)1097-4571(1999)50:9<799::AID-ASI9>3.0.CO;2-G

Tague-Sutcliffe, J. (1992). An introduction to informetrics. *Information Processing & Management*, *28*(1), 13. doi:10.1016/0306-4573(92)90087-G

Tryon, R. C., & Bailey, D. E. (1970). *Cluster analysis*. New York: McGraw-Hill Book Co.

van Raan, A. F. J. (2004). Sleeping beauties in science. *Scientometrics*, *59*(3), 461–466. doi:10.1023/B:SCIE.0000018543.82441.f1

Vellaichamy, A. (2016). A scientometric analysis of 20 years literature on Ebola virus. *International Journal of Innovative Knowledge Concepts.*, *4*, 255–262.

Vellaichamy, A., & Jeyshankar, R. (2014). *Anemia research in India: a bibliometric analysis of publications output during 1993–2013*. Library Philosophy and Practice.

Vellaichamy, A., & Jeyshankar, R. (2015). Dengue research in India and China: A comparative study using bibliometrics. *International Journal of Library Science and Information Management*, *1*, 1–9.

Vellaichamy, A., & Jeyshankar, R. (2015). Myeloma research: A bibliometric assessment of world publications output during 2008-2012. *International Journal of Digital Library Services*, *5*, 13–25.

Waddell, L. A., & Greig, J. D. (2016). Scoping review of the Zika virus literature. *PLoS One*, *11*(5), e0156376. doi:10.1371/journal.pone.0156376 PMID:27244249

Wulder, M. (2014). A practical guide to the use of selected multivariate statistics. Retrieved 15 July 2014 from http://www.psych.yorku.ca/lab/psy6140/DataScreeningChecklist.pdf

Young, I., Waddell, L., Sanchez, J., Wilhelm, B., McEwen, S. A., & Rajic, A. (2014). The application of knowledge synthesis methods in agri-food public health: Recent advancements, challenges and opportunities. *Preventive Veterinary Medicine*, *113*(4), 339–355. doi:10.1016/j.prevetmed.2013.11.009 PMID:24485274

Chapter 8
Research Trends in Forensic Sciences:
A Scientometric Approach

Rajshree Raghunath
Tamil Nadu State Forensic Sciences Department, India

ABSTRACT

This chapter outlines how scientometric tools and techniques are increasingly used to find the growth and direction of research. Forensic sciences are the application of science and technology in the administration of law and justice. Forensic Sciences encompass a wide range of subspecialties which are applied in adjudication of civil and criminal legal issues. Five subspecialties, viz., Forensic Anthropology (FA), Forensic Genetics (FG), Forensic Medicine (FM), Forensic Psychology (FP), and Forensic Toxicology (FT) were comparatively analyzed to find the growth of literature, source documents contributing to the literature growth and also the keywords used to find the research trends in these subspecialties. Data pertaining to these subspecialties documented from 2006 to 2015 were retrieved from the Science Direct platform and studied using MS-Excel and VOSviewer. The cumulative growth of literature showed a linear trend. The study found Forensic Science International as the top contributing journal. Significant association was observed between FA, FG, FM and FT literature while FP was not.

INTRODUCTION

Research is careful investigation or inquiry especially through search for new facts in any branch of knowledge (Advanced Learner's Dictionary of Current English, 1952). Scientific research is an extraordinary endeavor and so the substance of it is highly complex. It requires specialized skills and tools because personal knowledge and experience alone are not enough to measure the trends and quantify the results.

Research funding agencies, whether governmental or non-governmental need to evaluate a research project quantitatively as well as qualitatively. Standard bibliometric or scientometric tools are being increasingly used to evaluate research. Recently, scientometrics is used as a quantitative analysis method for scientific research. The statistics derived from the scientometric analyses that measure the contribu-

DOI: 10.4018/978-1-5225-3457-0.ch008

tion of scientific publications within a given topic could represent the current research trends and can also be used to identify the course of future research.

Bibliometric and scientometric studies of scientific disciplines and the rendering of results as knowledge domain visualizations aim to find the answers for some pertinent questions related to the major research areas, experts, institutions, regions, nations, grants, publications, journals in a specific research field, the relationships between domains of knowledge, their dynamics and impact (Börner, Chen, & Boyack, 2003; Cronin & Atkins, 2000; White & McCain, 1989; Murray, Ke & Börner, 2006).

Domain visualisation is an emerging field of study. Visualisation techniques are used to map the ever-growing domain structure of scientific disciplines. They also support information classification and retrieval. In domain visualisation interrelationships between research fronts are represented through spatial representations. Such spatial representations allow users to navigate the scientific literature based on the spatial patterns depicted. Domain visualisation aims to reveal realms of scientific communication as reflected through scientific literature and citation paths interwoven by individual scientists in their publications.

The field of domain visualisation is also called scientography (Garfield, 1994), though not widely used. Garfield (1994) introduced the concept of longitudinal mapping. In longitudinal mapping, a series of chronologically sequential maps can be used to detect the advances of scientific knowledge. Domain experts use longitudinal maps to forecast emerging trends for a subject domain. Scientometrics is a distinct discipline that has emerged from citation-based domain visualisation. Scientometrics is the quantitative study of scientific communications, which applies the principles and techniques of bibliometrics to scientific literature. This bibliometric mapping concept has been further developed by researchers in the Netherlands, particularly, Noyons and Van Raan.

Visualisation provides an ability to comprehend huge amounts of data on large scale as well as small scale. It gives a better understanding of a complex data set. Data-set can be seen in from several different perspectives simultaneously. Visualisation exploits visual perceptions and therefore can enable visual comprehension. It is an effective source of communication. The process of analyzing and transforming non-spatial data into an effective visual form can improve the interaction with large amount of data.

These bibliometric, scientometric, mapping, and visualization studies are generally conducted based on bibliographic data downloaded from digital libraries or bibliographic and abstract databases. In order to map a specific discipline, keyword based searches for relevant phrases are run or cited reference search is used to retrieve all papers that are cited by or are citing a set of seminal papers (Murray, Ke & Börner, 2006).

FORENSIC SCIENCE AND ITS SPECIALTIES

Forensic science can be defined as the application of scientific principles in the adjudication of law and justice. Forensic science is a wondrous intersection where science, medicine and the law meet, with the final disposition being the adjudication of criminal cases (Pyrek, 2007). Forensic science uniquely shares its applications to legal issues for resolution in a public forum (James & Nordby, 2003). Forensic science operates in the service of the public, represented through the rule of law by the courts. Different functions, but all necessary for the common good (Inman & Rudin, 2001).

Forensic science is multidisciplinary, encompassing a wide spectrum of subspecialties that are steeped in the traditional sciences (Pyrek, 2007). A plethora of sciences have application to law and justice and hence there is an endless list of subspecialties in Forensic Sciences perhaps starting from Forensic Accounting and ending with Forensic Zoology. In between these two are a number of specialties such as, Forensic Art, Forensic Anthropology, Forensic Chemistry, Forensic Ballistics, Forensic Biology, Forensic Entomology, Forensic Linguistics, Forensic Medicine, Forensic Psychology, Forensic Odontology, Forensic Serology, Forensic Taphonomy, Forensic Toxicology and so on and so forth.

KEY CONCEPTS OF FORENSIC SCIENCE

According to Inman and Rudin (2001) the fundamental concepts of Forensic Sciences are transfer, individualisation, identification, association and reconstruction. Edmund Locard is universally credited with articulating the concept of 'exchange' or transfer as "every contact leaves a trace". He undertook an extensive study of 'dusts', 'mud' and 'grime' and proposed that the study of dust on clothing or body parts can be an indicator of occupation and whereabouts.

Individualisation is unique to Forensic Science. It refers to the demonstration that a particular sample is unique, even among members of the same class. The nature of these individual characteristics varies from one type of evidence to another, but Forensic scientists try to take advantage of them in efforts to individualise a piece of physical evidence by some type of comparison process (De Forest, Gaensslen & Lee, 1983). Fingerprint is an example of individualisation.

According to Saferstein (1998), identification has as its purpose the determination of the physical or chemical identity of a substance with as near absolute certainty as existing analytical techniques will permit. The determination of identity is of importance to the criminalist in two ways: in establishing between two objects an identity of origin; and in determining the nature of a specimen of evidence. The first category is ordinarily the most significant because on it rests the final determination of the value of the evidence. It is, for example, more valuable to say that two bullets, one of known origin, were fired from the same gun, than to say that the bullets are of the same type of ammunition. It is more valuable to be able to say that two hairs have come from the same head, than say that both hairs are human in origin (Kirk, 1953). Personal identification and physical identification of objects are very important in Forensic Science. Identification of handwriting may lead to the writer, the forger of a questioned document and identification of a murder weapon may lead to its possessor.

Association is inferring contact between two objects. Osterburg (1968) describes associative evidence as "linking a person to a crime scene." Stoney (1984) describes it as "linking a suspect with an offense." Further, Stoney (1991) states that "association is an inference of contact between two objects, and is the only way in which legally relevant evidence can be generated by a scientist. Mere source determination is not legally relevant, *per se*." A fingerprint found at the scene of a burglary after elimination of the current residents can be associated to that of the burglar.

Reconstruction refers to the process of putting the 'pieces' of a case or situation together with the objective of reaching an understanding of a sequence of past events (De Forest, Gaensslen& Lee, 1983). Reconstruction is the ordering of events in relative space and time based on the physical evidence. Reconstruction attempts to answer the questions of 'where?, how? and when?'. Reconstruction is useful in cases in which eyewitness evidence is absent or unreliable and also in automobile accidents.

BACKGROUND OF THE STUDY

Scientific research culminates in the form of publication – monographs, journal articles, case studies, etc., patents, and thesis. These literatures reflect scientific activity of a country or of a particular field of study. Hence examining the scientific literature provides the structure of the field and the quantity and quality of the scientific activity of the geographical area studied.

Sensabaugh (1986) after studying the Forensic Science literature and discussing with the colleagues presented the research trend in Forensic Science in four different classified areas such as, Forensic Biology, Trace, Toxicology and Pathology. He predicted that in Forensic Biology, DNA technology would open a new dimension in genetic-marker analysis. In the area of trace evidence, he visualised chemical instrumentation such as high-pressure liquid chromatography, scanning electron microscopy and X-ray fluorescence to dominate. In toxicological analysis according to Sensabaugh new instrumental approaches to analysis would continue to be developed. Research in Forensic Pathology is traditionally descriptive at both gross and microscopic level. Small amount of work has been done to address post-mortem chemical changes and chemical changes associated with wound. This would probably continue leading to new methods for assessing the cause and time of death.

According to Gualdi-Russo and Fonti (2013) the progressive increase in Forensic Anthropology (FA) publications during the last 40 years is a clear indication of the increasing interest of biological anthropologists in FA research. They may enhance the potential of the Forensic Sciences by developing new approaches to a wide range of topics. While skeletal biology will remain the main area of investigation by Forensic Anthropologists, major new contributions will come from research on living individuals, with particular emphasis on identification conducted directly on the subject (i.e., biological profile and, in particular, age determination) and indirectly on images from video-surveillance systems. Moreover, the identification of a specific biological profile may be crucial to build the virtual models needed to analyse the case in a virtual reconstructed crime scene.

Jeyasekar and Saravanan (2015a) conducted a bibliometric study of Indian Forensic Science literature indexed in *PubMed*. The result of the cluster analysis revealed that Forensic Medicine, Pathology, Forensic Genetics, Forensic Anthropology, Forensic Odontology, Age Determination, Sex Determination and Mass Disasters as the recent trends or major areas of Forensic Science research in India.

The impact factors of forensic journals, prolific forensic authors and highly cited articles were discussed by Jones (1999, 2002, 2003, 2004, 2005a, 2005b & 2007). Jeyasekar and Saravanan (2013, 2014a, 2014b & 2014c) have conducted single journal studies in the specialties of Forensic Sciences using different data sources. Savageau, Desnoyers and Godin (2009) have comparatively mapped two Forensic journals. Four Forensic sub-specialties, viz., Wildlife Forensics (Jeyasekar & Saravanan, 2016), Forensic Odontology (Jeyasekar & Saravanan, 2015a), Forensic Anthropology (Gauldi-Russo & Fonti, 2013) and Forensic Psychology (Black, 2012) have also been scientometrically mapped. Scientometric Studies on global and Indian Forensic Science using Indian Citation Index (ICI) (Jeyasekar & Saravanan, 2012a), SCOPUS (Jeyasekar & Saravanan, 2012b), PubMed (Jeyasekar & Saravanan, 2015b), and Web of Science (WoS) have also been conducted (Kumbar & Biradar, 2015; Amsaveni & Ramesh, 2016; Rao, Raghavan & Bhargav, 2017).

METHODOLOGY

Ferrara, Bajanowski, Cecchi, Boscolo-Berto and Viel (2011) examined the Bio-medicolegal sciences literature in Europe and found the most frequent topics were the description of new analytical methods in Forensic Toxicology (5.7% of all papers), the analysis of short tandem repeat systems (STR, 5.6%) as well as the analysis of injury mechanisms in Forensic Pathology and Clinical Forensic Medicine (4.9%).

The study by Jeyasekar and Saravanan (2015b) revealed that Forensic Medicine (FM), Forensic Anthropology (FA), Forensic Toxicology (FT) as the happening area of research among the Forensic Scientists in the recent four decades. A new area of study found from the key terms of the last decade is Brain Mapping, which is part of Forensic Psychology (FP). DNA fingerprinting made an impact in global Forensic Science arena after the Colin Pitchfork case in 1986 and henceforth Forensic Genetics (FG) is a dominant specialty of Forensic Science research. Hence these five specialties, viz., FA, FG, FM, FP, and FT were selected for this comparative study to find the trends and future direction.

Bibliometric / scientometric studies require a source of bibliographic data. These data can be extracted from the materials available in the library. However, a comprehensive bibliographic database can provide a much better platform to conduct an exhaustive study (Jeyasekar & Saravanan, 2015a). The creativity of the scientometrician is as important as the database. Numerous analytical studies have been conducted using Web of Science, SCOPUS, PubMed, etc. In this study an attempt has been made to study the research trends using the data obtained from Science Direct. Software such as *MS-Word, MS-Excel, and VOSviewer* have been used to analyse and visualize the data.

RESULTS AND DISCUSSION

Growth of Forensic Literature

The data pertaining to number of documents published in the five forensic subspecialties viz., FA, FG, FM, FP and FT obtained from Science Direct is tabulated in Table 1. A percentage analysis was also carried out and the results are presented in the same.

The FA literature data available in Table 1 revealed that 2342 records of publication were available during the period of study. It revealed a steady growth from 129 records in 2006 to 351 records in 2015. It also showed negative growth trend in 2010. In the year 2006 the FG literature records available in the platform were 866 and in 2015 it became 1444. However the growth trend kept fluctuating throughout the period of study. The FM bibliographic records grew from 553 in 2006 to 851 in 2015. But downward trend was observed in 2008, 2010 and 2011. The FP documents showed a growth from 226 records to 500 records with negative trend in 2007 and 2014. The FT bibliographic records in 2006 were 387 in number. It became almost double in 2015. However, negative trend had also been observed several times. These trends have been illustrated in Figure 1. The year-wise percentage of contribution of each subspecialty were plotted as a graph and presented in Figure 1. Although the five subspecialties showed linear growth trend, FA, FP and FT literature showed a slight increase whereas FG and FM literature showed a decrease. However, the highest percentage of contribution was seen in FG literature throughout the period of study whereas the lowest one was observed in FA. The FM literature revealed to be the all-time second.

Table 1. Year-wise Growth of Literature

Year	FA (% of the specialty) [% of the Year's total]	FG (% of the specialty) [% of the Year's total]	FM (% of the specialty) [% of the Year's total]	FP (% of the specialty) [% of the Year's total]	FT (% of the specialty) [% of the Year's total]	Year-wise Total [Percentage]
2006	129(5.5)[6.1]	866(8.3)[41.0]	553(8.1)[26.2]	226(7.3)[10.7]	337(6.3)[16.0]	2111[7.5]
2007	167(7.1)[7.7]	764(7.3)[35.0]	626(9.1)[28.7]	187(6.0)[8.6]	438(8.2)[20.1]	2182[7.8]
2008	185(7.9)[8.3]	907(8.7)[40.5]	503(7.3)[22.5]	218(7.0)[9.7]	427(8.0)[19.1]	2240[8.0]
2009	187(8.0)[7.3]	976(9.3)[38.2]	699(10.2)[27.3]	220(7.1)[8.6]	474(8.9)[18.5]	2556[9.1]
2010	159(6.8)[7.0]	786(7.5)[34.7]	593(8.6)[26.2]	273(8.8)[12.1]	453(8.5)[20.0]	2264[8.1]
2011	221(9.4)[8.1]	1170(11.2)[43.1]	619(9.0)[22.8]	259(8.4)[9.6]	444(8.3)[16.4]	2713[9.7]
2012	272(11.6)[9.3]	984(9.4)[33.6]	722(10.5)[24.7]	391(12.6)[13.3]	560(10.5)[19.1]	2929[10.4]
2013	343(14.6)[9.3]	1351(12.9)[36.6]	841(12.2)[22.8]	419(13.5)[11.3]	740(13.9)[20.0]	3694[13.1]
2014	328(14.0)[9.1]	1191(11.4)[33.2]	860(12.5)[23.9]	408(13.2)[11.4]	805(15.1)[22.4]	3592[12.8]
2015	351(15.0)[9.2]	1444(13.8)[37.9]	851(12.4)[22.3]	500(16.1)[13.1]	666(12.5)[17.5]	3812[13.6]
Total	2342(100)[8.3]	10439(100)[37.2]	6867(100)[24.4]	3101(100)[11.0]	5344(100)[19.0]	28093[100]

Figure 1. Year-wise Percentage of Literature Growth

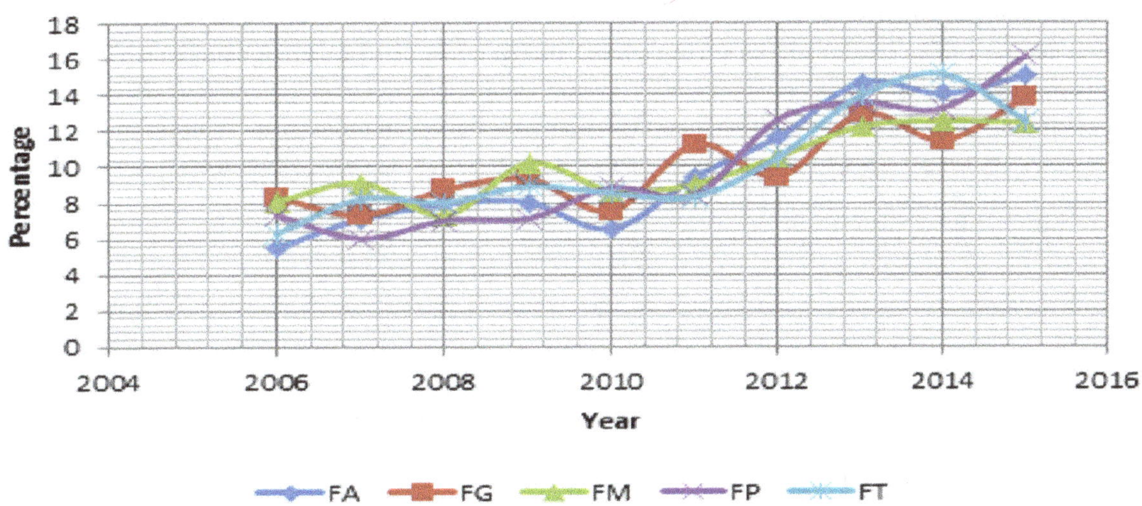

The year-wise percentage of the literature of FA, FG, FM, FP, and FT in relation to the percentage of the total literature produced was plotted as a graph in Figure 2. It was found that FA, and FP literature was below the average of the all put-together, while FT literature was at average level and FM, and FG were above average.

Cumulative Growth

In all a total of 28093 bibliographic records of these five subspecialties were found during the period of study. The year-wise data revealed that a total of 2111 documents have been indexed in the year 2006.

Figure 2. Percentage Analysis of Literature Growth

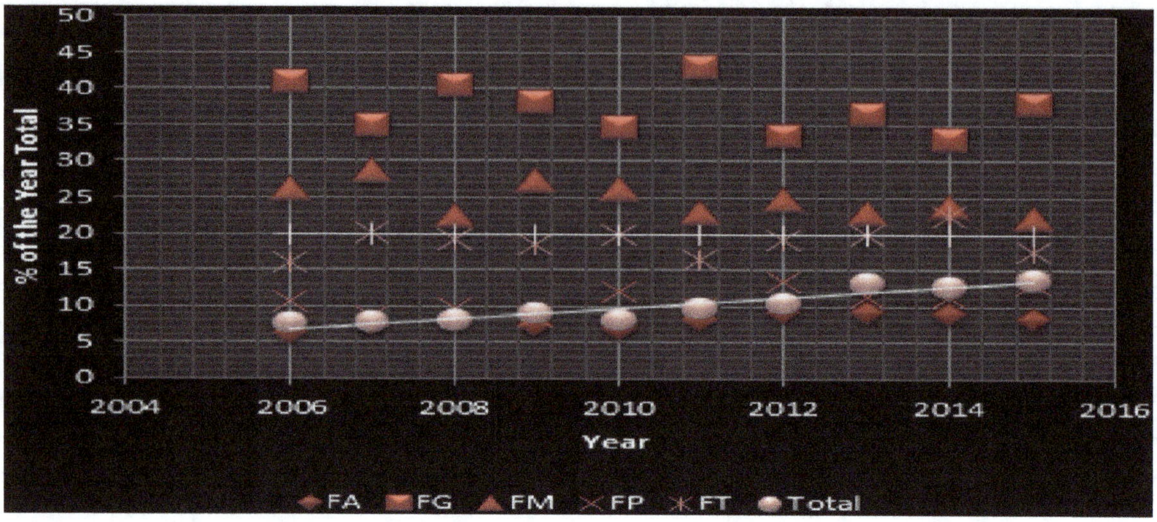

During the year 2007 the number of total documents stood at 2182, then grew steadily till it reached 3812 in 2015. However there was a dip in the growth trend during the years 2010 and 2014.

The cumulative growth of the numbers of FA, FG, FM, FP and FT research literature were calculated and tabulated in Table 2 and also plotted as a graph in Figure 3. The percentage analysis done is given within parentheses in the table.

The cumulative growth showed linear trend rather than the exponential growth propounded by Price (1963). The cumulative growth of all the five subspecialties FA, FG, FM, FP and FT put together also showed linear trend line as seen in Figure 3.

Table 2. Cumulative Growth

Year	FA (%)	FG (%)	FM (%)	FP (%)	FT (%)	Total (%)
2006	129(5.5)	866 (8.3)	553(8.1)	226(7.3)	337(6.3)	2111(7.5)
2007	296(12.6)	1630(15.6)	1179(17.2)	413(13.3)	775(14.5)	4293(15.3)
2008	481(20.5)	2537(24.3)	1682(24.5)	631(20.3)	1202(22.5)	6533(23.3)
2009	668(28.5)	3513(33.7)	2381(34.7)	851(27.4)	1676(31.4)	9089(32.4)
2010	827(35.3)	4299(41.2)	2974(43.3)	1124(36.2)	2129(39.8)	11353(40.4)
2011	1048(44.7)	5469(52.4)	3593(52.3)	1383(44.6)	2573(48.1)	14066(50.1)
2012	1320(56.4)	6453(61.8)	4315(62.8)	1774(57.2)	3133(58.6)	16995(60.5)
2013	1663(71)	7804(74.8)	5156(75.1)	2193(70.7)	3873(72.5)	20689(73.7)
2014	1991(85)	8995(86.2)	6016(87.6)	2601(83.9)	4678(87.5)	24281(86.4)
2015	2342(100)	10439(100)	6867(100)	3101(100)	5344(100)	28093(100)

Figure 3. Cumulative Growth

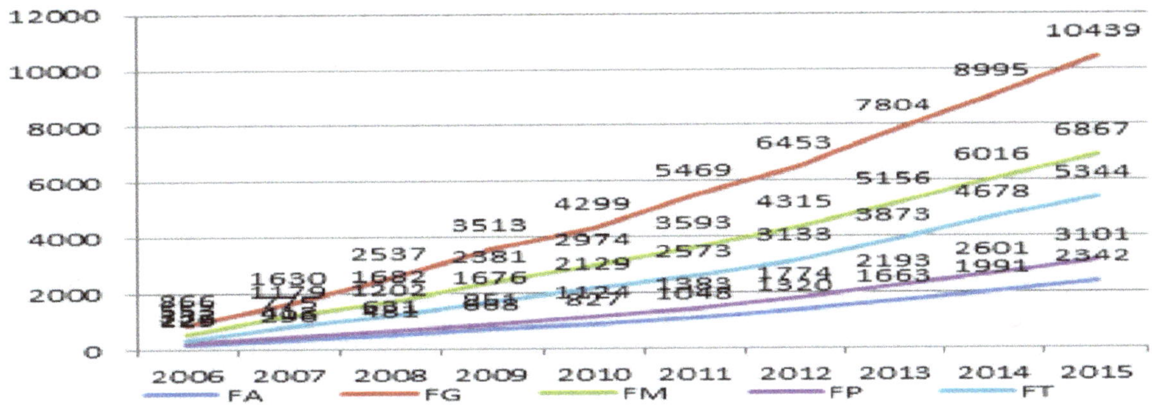

Source Documents

The source documents that had contributed to the five subspecialties were studied for their contributions and listed in Table 3.

Forensic Science International (FSI) was the top most contributing source journal in the case of FA, FM, and FT literature, while it ranked third in case of FG literature. It had contributed 27% of FA, 12.5% of FM, 16.1% of FT and 7.8% of FG literature. The second highest contributor of FA literature was *Journal of Forensic & Legal Medicine*, and the third was *HOMO – Journal of Comparative Human Biology*. The highest contributor for FG literature was *Forensic Science International – Genetics* and second highest was *Forensic Science International – Genetics Supplement* followed by *Journal of Forensic & Legal Medicine*. The second highest for FM literature was *Journal of Forensic & Legal Medicine* and the third was *Legal Medicine*. Among FT literature the second highest contributor was *Journal of Forensic & Legal Medicine* and third was *Toxicology Letters*. The highest contributor for FP was *International Journal of Law & Psychiatry* followed by *Aggression & Violent Behavior* and *Personality & Individual Differences* in that order.

Keyword Analysis

The author and title keywords provide reasonable details of article subjects (He, 1999). The cluster maps of keywords as visualized from the VOSviewer output is presented in Figures 4 to 8 Scientometrics studies are not only based on citations but also on thetext of the publications or itsmetadatalike titles, abstract, keywords. These can be analysed and measured instead of citations to detect the association between scientific concepts for identifying the growing subfields or disciplinary patterns.

The cluster map of the keyword occurrences as visualized from the VOSviewer has been presented in Figure 4. A total of 3026 keywords were found in the analysis of keywords. The cluster map presented below was created with a threshold value of 50, i.e., terms used for more than 50 times alone were considered for the analysis and 195 terms qualified the threshold value.

Table 3. Source Documents

Journal	FA (% of specialty)	FG (% of specialty)	FM(% of specialty)	FP (% of specialty)	FT(% of specialty)	Total
Forensic Science International	631(27)	815(7.8)	855(12.5)	60(1.9)	861(16.1)	3222(11.5)
Journal of Forensic & Legal Medicine	181(7.7)	174(1.7)	434(6.3)		388(7.3)	1177(4.2)
HOMO - Journal of Comparative Human Biology	134(5.7)		71(1.0)			205(0.7)
Legal Medicine	116(5.0)	379(3.6)	362(5.3)		160(3.0)	1017(3.6)
Journal of Archaeological Science	95(4.0)					95(0.3)
Forensic Science International: Genetics	77(3.3)	1193(11.4)	109(1.6)			1379(4.9)
Science & Justice	39(1.7)	233(2.2)				272(1.0)
Forensic Science International: Genetics Supplement	37(1.6)	1171(11.2)	105(1.5)			1313(4.7)
Pathology	34(1.5)		354(5.2)			388(1.4)
International Congress Series		254(2.4)				254(0.9)
Biosensors & Bioelectronics		136(1.3)				136(0.5)
Journal of Pharmaceutical & Biomedical Analysis		115(1.1)			226(4.2)	341(1.2)
Journal of Forensic Radiology & Imaging			105(1.5)			105(0.4)
Journal of Comparative Pathology			72(1.0)			72(0.3)
International Journal of Law & Psychiatry				178(5.7)		178(0.6)
Aggression & Violent Behavior				139(4.5)		139(0.5)
Personality & Individual Differences				124(4.0)		124(0.4)
Procedia - Social & Behavioral Sciences				89(2.9)		89(0.3)
Schizophrenia Research				74(2.4)		74(0.3)
European Psychiatry				62(2.0)		62(0.2)
Psychiatry Research				52(1.7)		52(0.2)
European Neuropsychopharmacology				50(1.6)		50(0.2)
Journal of Criminal Justice				47(1.5)		47(0.2)
Toxicology Letters					296(5.5)	296(1.1)
Journal of Chromatography B					181(3.4)	181(0.6)
ToxicologieAnalytique et Clinique					137(2.6)	137(0.5)
Journal of Chromatography A					127(2.4)	127(0.5)
Food and Chemical Toxicology					76(1.4)	76(0.3)
Total	1344 (59.4)	4470 (42.8)	2467 (35.9)	875 (28.2)	2452 (45.8)	11608 (41.3)

Among the 195 keywords 180 terms were found to be connected and hence these terms alone were taken into account for creating the map. A total of 13 clusters were found in the map. Cluster no. 1 consisted of 33 items or keywords, cluster no. 2 and 3 consisted of 19 items each, cluster no. 4 had 15 words. The keywords decreased when the number of the cluster increased and finally ended with 7 words in cluster no. 13.

Figure 4. Keyword Map of FA Literature

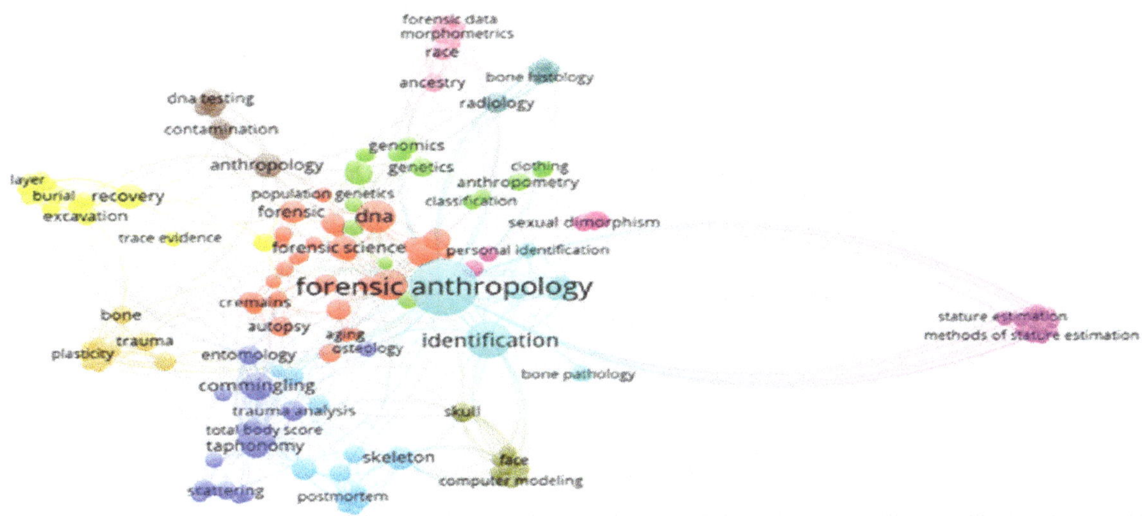

The data extracted from the map in Figure 4 showed that FA literature had keywords such as, archaeology, pathology, taphonomy, DNA, odontology, population genetics, radiology, fingerprints. Hence it is inferred FA research literature during the study period had more association with these subspecialties. It was also found that words like bone, bone measurement, craniofacial measurement, anthropometry, facial reconstruction, age estimation, etc. are found indicating the major thrust areas. Computer modeling is another important term found from the analysis implying the application of computers in FA.

The cluster map of the keywords used in FG documents was visualized in Figure 5. In this keyword analysis 3009 words were found. The maximum time a keyword used was 33 and only three words qualified for this threshold value. They are 'Short Tandem Repeat', 'DNA' and 'Forensic Science'. However, a keyword analysis with the threshold value of 5 yielded 89 terms. Among these terms only 84 terms were connected and the figure was created using these 84 terms.

In the cluster map of FG keywords in the figure above, 8 clusters were observed. Cluster no. 1 had 18 items, while cluster no. 2 had 14 items and last one had 2 items. Apart from the commonly used keywords like STR, Genetics and DNA / DNA typing, other widely used words are mtDNA, Y-STR, haplotype, SNP, autosomal STR, DNA extraction, DNA methylation, human identification, sexual assault, forensic entomology, forensic pathology, forensic medicine, forensic anthropology, forensic odontology, kinship analysis, paternity testing, and phylogeny.

The cluster map of keywords obtained from FM documents is seen in Figure 6. This map is obtained with the threshold value of 5. The highest threshold value was 32, and with this only 5 items were obtained, which included 'incision', 'laceration', 'adipocere', 'fetus' and 'forensic anthropology'. Hence it can be presumed much work had been done during the period in these areas and significant association existed between FM and FA.

A total of 3218 keywords were found in FM literature. With the threshold value 5, four hundred and ninety-three words were found and of these 381 are connected. A total of 216 keywords are found with the threshold value of 10 and among these 157 were connected. The Figure 6 was created with the threshold value of 10 and the number of clusters found was 11. Cluster no. 1 had 27 words, cluster no.

Figure 5. Keyword Map of FG Literature

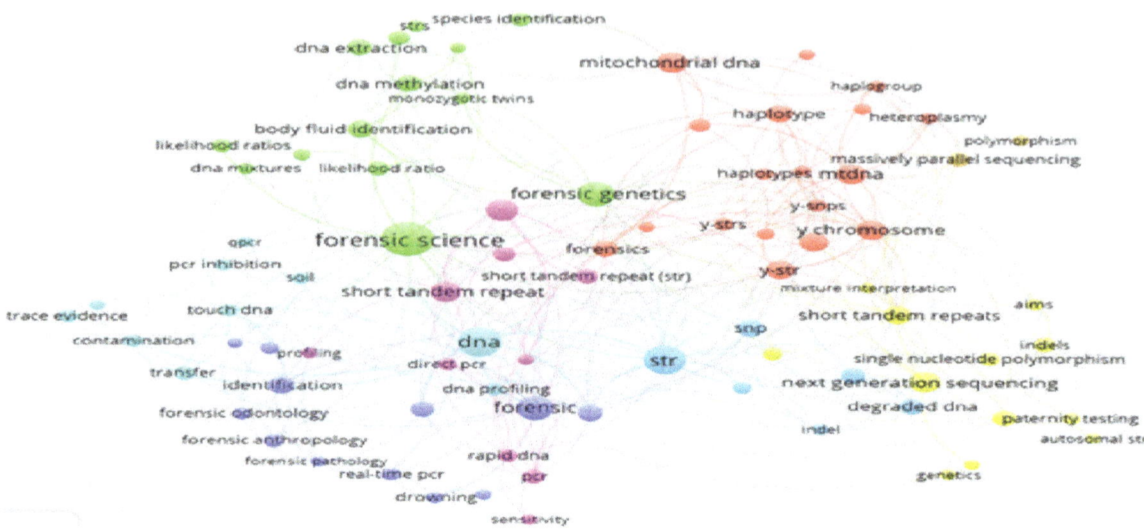

Figure 6. Cluster Map of FM Keywords

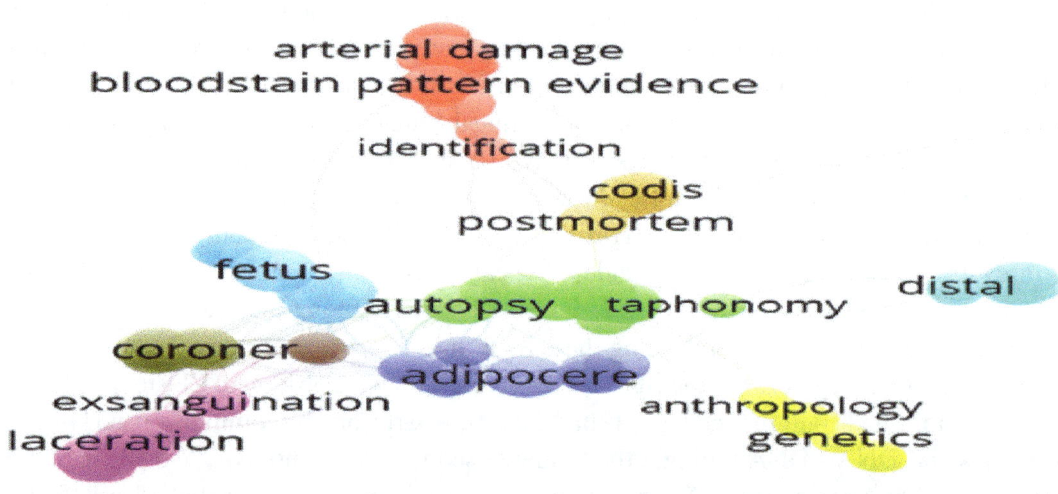

2 had 21 words and while cluster no. 11 had 5 words.The commonly used keywords in FM literature were "cause of death, forensic pathology, autopsy, manner of death, chain of custody, coroner, medical examiner, bloodstain pattern evidence, postmortem, Alphonse Bertillon, Locard's exchange principle, model postmortem, mummification, quality assurance, contusion, saponification, exsanguination, femoral blood, PCR, subclavian blood, trace evidence, virchow technique, forensic anthropology population data genetics, arterial damage, asphyxia, blunt-force injury, cyanide, dicing wounds, distal, bloodstain, genomics, embalming, eugenics, exhumation, hesitation wound, stab wound, toxicology, alveolus, atrophy, biotechnology, embolus, forensic odontology, identification, jaundice, myocardium, bloodstain pattern interpretation, cyanotic, hematoidin, hypoxia, meconium, positional asphyxia, skeletal survey, anoxia, c4 binding protein (c4bp), carbon monoxide, child abuse, compression asphyxia, mitochondrial

DNA, population genetics, taphonomy, lethal injection, virtopsy, chromosome mapping, impact spatter, schizophrenia, y chromosomes, forensic DNA, forensic radiology, forensic engineering, forensic imaging, questioned documents, ricin, rigor mortis, warfare agents, and wildlife" in that order. The frequent areas of study in FM can be deduced from these keywords. Association between FA, FG, FT, Forensic radiology, Forensic odontology, Forensic imaging, Forensic taphonomy Wildlife forensics could also be found. However, the surprising factor is the association between FM and Forensic engineering and also Questioned documents.

A cluster map of FP literature was created and illustrated in Figure 7. The number of keywords found in FP was 3089 and with a threshold value of 50 seventeen terms were found. Among these 'neurobiology', 'schizophernia', 'psychosis', 'diagnosis', 'dementia praecox', 'epistemology', 'psychopathology' and 'depression' indicating most popular areas of research in FP.

In Figure 7, the cluster map was constructed with the threshold value of 5 and 660 words were found in this map. Among these 500 words used for maximum number of time and 391 words were connected. These words were found as 24 clusters. Cluster no. 1 had 31 words while cluster no. 24 had 8 words. Social psychiatry, trauma, criminology, law, deception detection, sociology, political science, post-traumatic stress disorder (ptsd), economics, eye-tracking, heat-maps, Adam Smith, Charles Booth, Dubois, Durkheim, Freud, Galton, Marx, Munsterberg, Quetelet, Weber, Wundt, behavior analysis, behavior therapy, Daubert, Nepal, Sierra Leone, Uganda, genetics, anthropology, DNA, genomics, neuroimaging, antidepressant, anxiety disorder, computerized polygraph analysis, directed-lie test, friendly polygraph examiner, polygraph decision methods, polygraph examinations, polygraph screening, probable-lie questions, psychiatric comorbidity, substance use disorder, event-related potentials (ERPS), Adam Walsh act, criminal profiling, deviant sexual arousal, internet sex offenders, juvenile sex offenders, complex post-traumatic stress disorder (CPTSD), narrative exposure therapy (NET), aggressive driving, posttraumatic beliefs, posttraumatic emotions, psychological trauma, groth typology, loneliness, robots, service dogs, socializing, stalking typology, and therapy dogs. The comparison of these words with that of FA, FG, FM, and FT literature showed that FP was connected FA and FG literature. Personality-oriented names such as Adam Smith, Charles Booth, Freud, Galton and Marx indicate that there were many studies

Figure 7. Cluster Map of FP Keywords

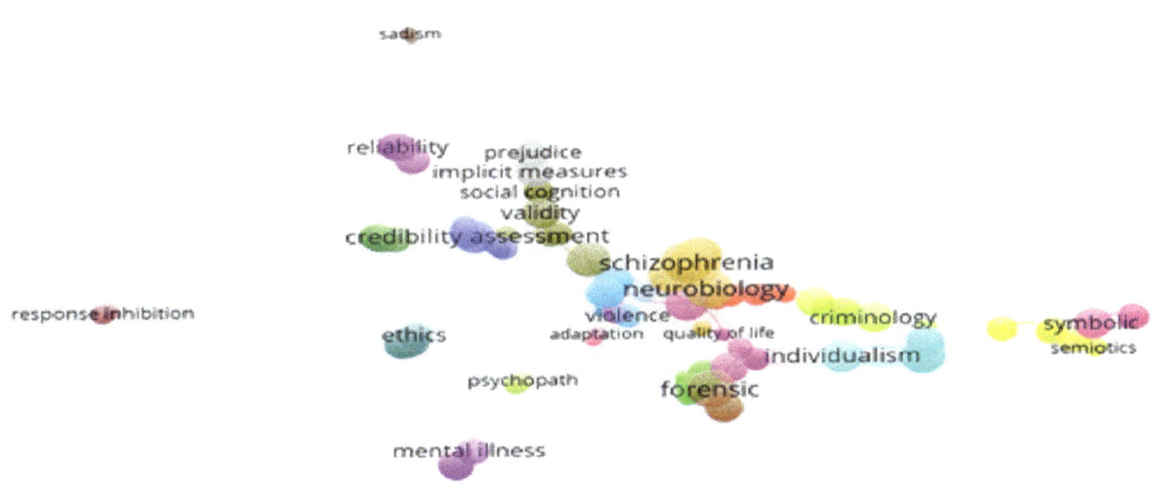

pertaining to the theories of these personalities. The presence of geographic terms like Nepal, Sierra Leone, and Uganda could lead to conclude that much FP research were carried out in these countries or relate to these countries.

The total number of FT keywords found was 2922. The highest threshold value in this analysis was 50 and 6 items met the threshold value. They were ‘chemical warfare agents’, ‘nerve agents’, ‘autopsy’, ‘cause of death’, ‘Toxicology’, and ‘Forensic Toxicology’. However, a cluster map (Figure 8) was created with the threshold value of 5.

A total of 449 words met the threshold value of 5 and among them 376 words were found to be connected. The popular among these keywords arranged according to the highest number of occurrences were “alcohol, coroner, forensic pathology, medical examiner, chain of custody, manner of death, mass spectrometry, cyanide, synthetic cannabinoids, fetus, contusion, hair analysis, asphyxia, tardieu spots, drug abuse, embalming, exhumation, toxicity, forensic anthropology, histology, ricin, drugs, immunoassay, organophosphates, adipocere, hair, autopsy cases, biomarkers, blood, fluid cases, intoxications, pharmacokinetic studies, pharmacokinetics, sulfur mustard, forensic odontology, medico-legal autopsy, behavioral sciences, blunt force trauma, cocaine, forensic engineering, hypothermia, livor mortis, questioned documents, algor mortis, autolysis, carboxyhemoglobin, contact gunshot wound, distant gunshot wound, exsanguination, hematoma, hydrogen sulfide, hyperthermia, incised wounds, ligatures, liquid chromatography, mechanical trauma, microtome, petechiae, postmortem clock, putrefaction, rigor mortis, sharp force trauma, exit wound, apnea, barbiturate, coma, conjunctivitis, controlled substance, drug, Harrison act, hydrosulfide, irritant gas, neurotoxicology, olfactory fatigue, olfactory paralysis, pulmonary edema, stimulant, toxidrome, pharmacology, Alphonse Bertillon, biological warfare agents, chemical weapons of mass destruction, Locard’s exchange principle, mustard in WW1, organophosphate, anticholinesterases, chemical warfare, cyclosarin (gf), sarin (gb), soman (gd), tabun (ga), chemical weapon, gas chromatography, mummification, trace evidence, amyl nitrite, asphyxiant, bioanalysis, laceration, livor, rigor, saponification, sodium nitrite, biological weapons, bite-mark ruler, chromatography, femoral

Figure 8. Cluster Map of FT Literature

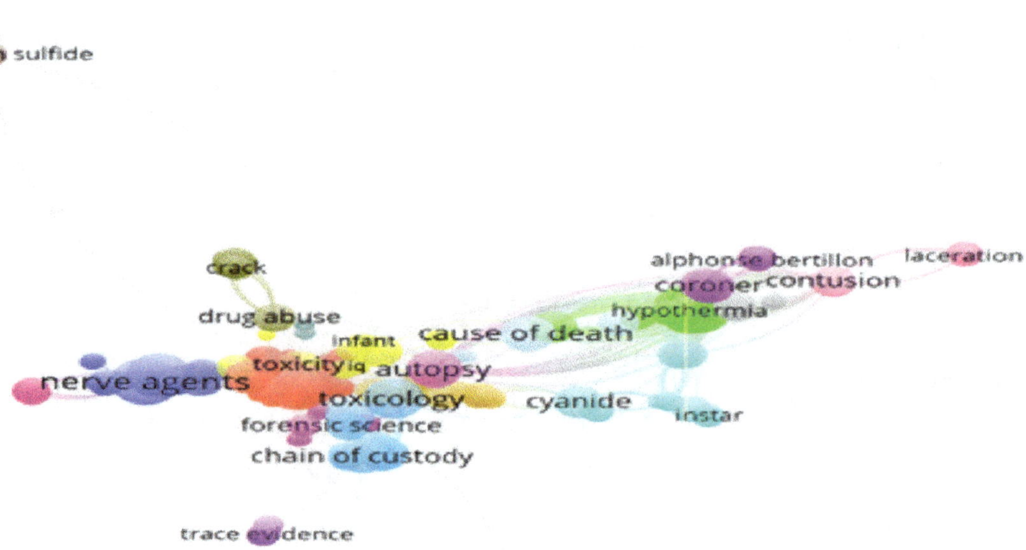

blood, fracture, lectin, n-95 mask, polymerase chain reaction, ribosome-inactivating proteins, rokitansky technique, subclavian blood, suicide, toxin, virchow technique, wildlife, blunt-force injury, dicing wounds, incision, sodium fluoroacetamide, sodium monofluoroacetate, tetramethylenedisulfotetramine, aging, benzodiazepines, cholinesterases, enzyme inhibition, ethanol, hair physiology, hydrolysis, poisoning, babygram, forensic autopsy, lead, meconium, anoxia, carbon monoxide, chlorpyrifos, cyanotic, dichlorvos, drugs of abuse, forensic medicine, hematoidin, cholinesterase activity, compression asphyxia, cortisol, heroin, hypoxia, positional asphyxia, cannabis, chemical terrorism, HPLC, lethal injection, pediatrics, sudden cardiac death, fibrosis, pesticides, postmortem toxicology, psychosis, capillary electrophoresis, depleted uranium, amphetamines, ketamine, mephedrone, drowning, forensic radiology, hair testing, heavy metals, India, medical genetics, methadone, methamphetamine, narcotics, SIDS, silica, sudden unexplained death, and synthetic cannabinoid".

CONCLUSION

The analysis of growth of literature in FA, FG, FM, FP, and FT documents showed linear trend. Forensic Science International was the top contributing journal during the period of study. The study revealed that significant association existed between FA, FG, FM, and FT literature. Top areas of FA research were archaeology, pathology, taphonomy, DNA, odontology, population genetics, radiology, and fingerprints. STR, DNA, and Forensic Science were the frequently used keywords in FG literature. In the FM documents incision, laceration, adipocere, fetus, and forensic pathology were most commonly used keywords. Neurobiology, schizophrenia, psychosis, dementia praecox, epistemology and depression were the frequently used keywords in FP literature. Among FT documents, chemical warfare, nerve agents, autopsy, cause of death, and Toxicology were frequently used.

REFERENCES

Advanced Learner's Dictionary of Current English. (1952). Oxford: Oxford University Press.

Börner, K., Chen, C., & Boyack, K. (2003). Visualizing Knowledge Domains. In B. Cronin (Ed.), *Annual Review of Information Science & Technology*. Medford, NJ: American Society for Information Science and Technology. doi:10.1002/aris.1440370106

Cronin, B., & Atkins, H. B. E. (2000). *The Web of Knowledge: A Festschrift in Honor of Eugene Garfield*. ASIST.

De Forest, P. R., Gaensslen, R. E., & Lee, H. C. (1983). *Forensic Science: an introduction to criminalistics*. New York: McGraw-Hill Book Company.

Garfield, E. (1994). Scientography: Mapping the tracks of science. *Current Contents: Social and Behavioural Sciences*, *7*(45), 5–10.

Gualdi-Russo, E., & Fonti, G. (2013). Recent trend and perspectives in forensic anthropology: A bibliometric analysis. *Collegium Antropologicum*, *37*(2), 595–599. PMID:23941009

He, Q. (1999). Knowledge discovery through co-word analysis. *Library Trends*, *48*(1), 133–159.

Inman, K., & Rudin, N. (2001). *Principles and Practice of Criminalistics: The profession of Forensic Science*. Boca Raton: CRC Press.

James, S. H., & Nordby, J. J. (2003). *Forensic Science: An introduction to Scientific and investigative techniques*. Boca Raton: CRC Press.

Jeyasekar, J. J., & Saravanan, P. (2012a). Scientometric study of forensic science: a study based on *SCOPUS* database. In *Proceedings of the UGC Sponsored National Seminar on Scientometrics and Informetrics, Annamalai Nagar* (pp. 75-78).

Jeyasekar, J. J., & Saravanan, P. (2012b). Scientometric analysis of Indian forensic science literature based on ICI database. *Journal of Library Advancements*, *2*(1), 1–5.

Jeyasekar, J. J., & Saravanan, P. (2013). Journal of forensic sciences: a bibliometric study for the period 2006 to 2010. In *Proceedings of the Second National Conference of Scientometrics and Knowledge Management, Dharwad* (CD-ROM version).

Jeyasekar, J. J., & Saravanan, P. (2014a). Scientometric analysis of the journal medicine, science and the law from 2001 to 2012. In *Proceedings of UGC Sponsored National Seminar on Application of Quantitative and Qualitative Indicators for the Excellence of Scholarly Communication, Annamalai Nagar,* Annamalai Nagar (pp. 308-313).

Jeyasekar, J. J., & Saravanan, P. (2014b). A scientometric analysis of the Journal of Forensic Identification. *KELPRO Bulletin*, *18*(2), 34–47.

Jeyasekar, J. J., & Saravanan, P. (2014c). A scientometric portrait of the journal Digital Investigation. *Journal of Advances in Library and Information Science*, *3*(2), 155–162.

Jeyasekar, J.J., & Saravanan, P. (2014d). A scientometric analysis of global forensic science research publications. *Library Philosophy & Practice*.

Jeyasekar, J. J., & Saravanan, P. (2015a). Indian forensic science research literature: A bibliometric study of its growth, authorship and publication patterns. *SRELS Journal of Information Management*, *52*(1), 67–75.

Jeyasekar, J. J., & Saravanan, P. (2015b). Mapping forensic odontology literature using open source bibliographies and software: a case study. In S. Thanuskodi (Ed.), *Handbook of Research in Inventive Digital Tools for Collection Management and Development in Modern Libraries*. Philadelphia: IGI Global. doi:10.4018/978-1-4666-8178-1.ch011

Jeyasekar, J.J., & Saravanan, P. (2016). Science maps of global and Indian wildlife forensics: a comparative map. *Library Philosophy and Practice*.

Jones, A. W. (1993). The impact of forensic science journals. *Forensic Science International*, *62*(3), 172–178. doi:10.1016/0379-0738(93)90206-P PMID:12742682

Jones, A. W. (1999). The Impact of alcohol and alcoholism among substance abuse journals. *Alcohol and Alcoholism*, *34*(1), 25–34. doi:10.1093/alcalc/34.1.25 PMID:10075398

Jones, A. W. (2002). JAT's impact factor--room for improvement? journal of analytical toxicology. *Journal of Analytical Toxicology*, *26*(1), 2–5. doi:10.1093/jat/26.1.2 PMID:11888013

Jones, A. W. (2003). Impact factors of forensic science and toxicology journals: What do the numbers really mean? *Forensic Science International*, *133*(1-2), 1–8. doi:10.1016/S0379-0738(03)00042-2 PMID:12742682

Jones, A. W. (2004). Impact of JAT publications 1981-2003: The most prolific authors and the most highly cited articles. *Journal of Analytical Toxicology*, *28*(7), 541–545. doi:10.1093/jat/28.7.541 PMID:16078379

Jones, A. W. (2005a). Creme de la creme in forensic science and legal medicine the most highly cited articles, authors and journals 1981-2003. *International Journal of Legal Medicine*, *119*(2), 59–65. doi:10.1007/s00414-004-0512-x PMID:15645298

Jones, A. W. (2005b). Which articles and which topics in the forensic sciences are most highly cited? *Science & Justice*, *45*(4), 175–182. doi:10.1016/S1355-0306(05)71661-0 PMID:16686268

Jones, A. W. (2007). The distribution of forensic journals, reflections on authorship practices, peer-review and role of the impact factor. *Forensic Science International*, *165*(2-3), 115–128. doi:10.1016/j.forsciint.2006.05.013 PMID:16784827

Kessler, M. M. (1963). Bibliographic coupling between scientific papers. *American Documentation*, *14*(1), 10–25. doi:10.1002/asi.5090140103

Kirk, P. L. (1953). *Crime Investigation*. New York: John Wiley.

Kumbar, P., & Biradar, N. (2015). Research trends in forensic science: A study of scientometric analysis. *International Journal of Research in Library Science*, *1*(2), 42–48.

Murray, C., Ke, W., & Börner, K. (2006). Mapping Scientific Disciplines and Author Expertise Based on Personal Bibliography Files. In *Proceedings of the KDVis Symposium at Information Visualisation*, London, UK.

Noyons, E. C. M., & van Raan, A. F. J. (1998). Monitoring Scientific Developments from a Dynamic Perspective: Self-Organized Structuring to Map Neural Network Research. *Journal of the American Society for Information Science*, *49*(1), 68–81.

Osterburg, J. W. (1968). *The Crime Laboratory: Case studies of scientific criminal investigation*. Bloomington: Indiana University Press.

Price, D. (1963). *Little Science, Big Science*. New York: Columbia University Press.

Pyrek, K. M. (2007). *Forensic Science Under Siege: The challenges of forensic laboratories and the medico-legal investigation system*. Amsterdam: Elsevier.

Rao, I. K. R., Raghavan, K. S., & Bhargav, K. N. (2017). Quantitative analysis of scientific literature in forensic science. *COLLNET Journal of Scientometrics and Information Management*, *11*(1), 59–72. doi:10.1080/09737766.2016.1251561

Saferstein, R. (1995). *Criminalistics: an introduction to forensic science* (7th ed.). New Jersey: Prentice-Hall.

Sauvageau, A., Desnoyers, S., & Godin, A. (2009). Mapping the literature in forensic sciences: A bibliometric study of North-American journals from 1980 to 2005. *The Open Forensic Science Journal*, *2*(1), 41–46. doi:10.2174/1874402800902010041

Sensabaugh, G. F. (1986). Forensic Science Research: Who Does it and Where is it Going? In G. Davies (Ed.), *Forensic Science* (2nd ed., pp. 129–140). Washington, DC: American Chemical Society.

Stoney, D. A. (1984). Evaluation of associative evidence: Choosing the relevant question. *Journal - Forensic Science Society*, *24*(5), 473–482. doi:10.1016/S0015-7368(84)72326-7

Stoney, D. A. (1991). What made us ever think we could individualise using statistics? *Journal - Forensic Science Society*, *3*(2), 197–199. doi:10.1016/S0015-7368(91)73138-1 PMID:1940832

vanRaan, A. F. J. (1991). Fractal geometry of information space as represented by co-citation clustering. *Scientometrics*, *20*(3), 439–449. doi:10.1007/BF02019764

White, H. D., & McCain, K. W. (1989). Bibliometrics. In M. E. Williams (Ed.), *Annual Review on Information Science and Technology* (Vol. 24). Amsterdam, Netherlands: Elsevier Science Publishers.

Chapter 9
Dynamics of Indian Forensic Science Research

J. John Jeyasekar
Tamil Nadu State Forensic Sciences Department, India

ABSTRACT

Science and technology are continuously changing as new discoveries and inventions are made. Research funding agencies, project directors, and individual researches need to keep a tab on these dynamics. This chapter tracks the research directions of forensic science for a period of thirty-eight years starting from 1975. Data for the analysis was obtained from SCOPUS bibliographic and citation database. Over the study period there was an exponential growth of forensic science literature and documentation. The United States of America contributed about one fourth of the research papers published while the most prolific author was Bruce Budowle (University of Texas). The majority of the contributors were from the non-governmental sector. The Journal of Forensic Sciences was the most productive journal during the study period in terms of number of published papers. It was also found that internationally collaborated papers attracted more citations.

INTRODUCTION

The dynamics of the academic world of science and technology keeps on changing the landscape characterized by the continuous emergence and development of new research directions, funding initiatives, scientific publications, and communication and collaboration networks (Etzkowitza & Lydesdorffb, 2000). Science maps aid to represent and analyze the dynamic and changing scientific landscape. Science maps are multi-dimensional domain visualizations that represent specific characteristics like authorship, collaborations, citations, etc. related to a particular scientific domain. Maps representing collaborative networks such as, authorship collaboration, institutional collaboration, and international collaboration, scientometric analysis of scientific publications, research funding, and the evolution of scientific domains and its likely future course are some examples of science maps. Scientific and research publications provide a relevant point of entry to study the nature, trajectory, and structure of scientific fields (Noyons & van Raan, 1998).

DOI: 10.4018/978-1-5225-3457-0.ch009

Science maps are generally constructed using bibliometric data consisting of a wide array of bibliographic information viz., authors, title, year of publication, source journal, page numbers, keywords, citations, publication type etc. These entities are used to generate unique maps representing different aspects of the relationships between and within scientific disciplines and research communities. Science map based on publication types (differentiating between types of journals for example) helps showcase the dominant disciplinary and research fields while a map based on co-occurrence of authors (two or more authors co-authoring a publication) represents the formulation and evolution of collaborative networks between disciplines and research communities (Bettencourta & Kaur, 2011). Science maps, thus, are useful tools to understand the state-of-the-art and disciplinary structure within an academic field as well as to analyze the emergence of research networks and collaborations. The focus of these maps is to offer insights into the specific subspecialties and the ways in which collaborative research networks can be sustained and developed further.

Scientometrics is one of the measuring and evaluating techniques used by the Library and Information Science professionals. Scientometricians use various mathematical and statistical techniques for the evaluation of scientific research. Scientometrics is a quantitative and qualitative measuring technique for evaluation and interpretation of science and its different activities such as productivity, progress, organization and management. Scientometric study is a statistical method of counting to evaluate and quantify the growth of a subject. The research trend during the said time span would be clearly understood from this study and a predictive projection may be made for anticipatable future.

EVOLUTION OF FORENSIC SCIENCE

The evolution of the practice of Forensic Science dates back to prehistoric period. Fingerprint patterns were used as the first means of establishing personal identity. The complex patterns inherent in fingerprints were noticed even by primitive man, as evidenced by their incorporation in prehistoric paintings and rock carvings (Ashbaugh, 2000). The use of fingerprints for identification by the Babylonians and later by the Chinese seems clear from the archaeological relics of clay tablets and other legal documents bearing the prints of interested parties (Morland, 1950). Indians studied various patterns of the papillary lines, thousands of years ago. It is presumed that they knew about the persistency and individuality of fingerprints, which they used as signatures (Nanda & Tewari, 2001). In the ancient *Sangam Age* Tamil literature '*Chilapathigaram*', Kannagi, wife of the wrongly convicted Kovalan confronts the Pandya King with an anklet as physical evidence and proved his innocence. In *The Holy Bible*, Solomon, the king of Israel, used the principles of psychology to adjudicate when two women claimed to be the mother of the same child.

However, Forensic Sciences in the present form have evolved during the end of 18th century, which marked the beginnings of modern chemistry. This opened the way for the birth of toxicology. The most illustrious figure in the history of toxicology was probably M.J.B.Orfila (1787-1853). One of the first systems of personal identification was based upon a series of body measurements. Its major proponent, Alphonse Bertillon (1853-1914) developed this system of anthropometry in order to establish an identification file suitable for use in criminal investigations. This system is also called Bertillonage. Employment of fingerprints as a device for personal identification grew out of the early efforts of William Herschel, a British civil servant in India in the 1870s, and of Henry Faulds, a Scottish physician who was working in Japan at around the same time (De Forest, Gaensslen & Lee, 1983).

Complete documentation of the development and pioneers in the field of Forensic Science is beyond the scope of this study. Nevertheless, some of the important names are:

1. In 1901 Sir. Edward Richard Henry was appointed as head of Scotland Yard. He forced the adoption of fingerprint identification to replace anthropometry.
2. Henry P. De Forrest, in 1901, pioneered the first systematic use of fingerprints in the United States by the New York Civil Service Commission.
3. In 1902, Prof. R.A. Reiss, a pupil of Bertillon, set up the first academic curricula in Forensic Science. His Forensic photography department grew into Lausanne Institute of Police Science.
4. In 1910, Edmund Locard, Professor of Forensic medicine at the University of Lyons, France, established the first police crime laboratory.
5. Paul Kirk assumed leadership at the University of California at Berkley in 1937. In 1945, he formalized a major in technical criminology.
6. In 1975 the Federal Rules of Evidence, originally promulgated by the U.S. Supreme Court, were enacted as a congressional statute.
7. Sir. Alec Jeffreys developed the first DNA profiling test in 1984 and in 1986 he first used DNA to solve a crime in which two young girls were murdered by Colin Pitchfork in the English Midlands.

LITERARY BEGINNINGS OF FORENSIC SCIENCE

Literature related to Forensic Science existed earlier than its practice. Some references to the use of footprints can be seen in *The Holy Bible*. *Thirukurral*, an ancient Tamil poetry says in the chapter xxxvi and couplet-355 translated by G.U. Pope states "whatever thing, of whatsoever kind it be, it is wisdom's part in each the very thing to see" (Pope, 1886). Kautilya's *Arthasastra* has detailed reference about various patterns of papillary lines. The old Chinese book, *His DUan Yu* (The Washing Away of Wrongs) written in 1248 contains a description of how to distinguish drowning from strangulation. This was the first recorded application of medical knowledge to the solution of crime.

In 1813, Mathieu Orfila, a Spaniard, who became professor of medical jurisprudence at the University of Paris, published "*Traite des Poisons Tires des Regnes Mineral, Vegetal et Animal, ou Toxicologie General l.*" Orfila is considered as the father of modern toxicology. In 1880, Henry Faulds, a Scottish physician working in Tokyo, published a paper in the journal '*Nature*' suggesting that fingerprints discovered at the scene of a crime could be used to identify the offender. Hans Gross, Professor of criminal law at the University of Graz, Austria in the year 1891, published a book titled "*Criminal Investigation,*" which represents the first comprehensive book on the uses of physical evidence in solving crimes. The same book was translated into English by John Adam and John Collyer Adam in the year 1906 and published from Madras. Sir Francis Galton in the year 1892 published "*Fingerprints*", the first comprehensive book on the nature of fingerprints and their use in solving crimes (Galton, 1892).

In 1904, Locard published "*L'enquete criminelle et les methodes scientifique*", in which he advocates the Forensic precept that "Every contact leaves a trace." Albert S. Osborne published the book "*Questioned Documents*" in the year 1910. This book is still considered as the bible for document examiners. In 1930, *American Journal of Police Science* was founded by the staff of Goddard's Scientific Crime Detection Laboratory in Chicago. In 1932, this periodical was absorbed by *Journal of Criminal Law and Criminology*, becoming the *Journal of Criminal Law, Criminology and Police Science*. The

American Academy of Forensic Sciences (AAFS) was founded in St Louis in the year 1948 and AAFS began publishing the *Journal of Forensic Science (JFS) in 1956*. Paul Kirk published the book "*Crime Investigation*" in the year 1953. This book is considered as one of the authentic textbook in crime investigation. Sir Alec Jeffreys developed the first DNA profiling test in 1984 and published his findings in *Nature* in the year 1985. DNA profiling commonly known as DNA fingerprinting revolutionized the field of Forensic Sciences (Gill, Jeffreys & Werrett, D. J. (1985).

FORENSIC SCIENCE IN INDIA

The application of science and technology to the detection of crime and administration of justice is not new to India. The practice of affixing the thumb print as signature has been proven to be scientifically sound and by some considered infallible. However, modern Forensic Science owes its development to the formation of the Chemical Examiner's Laboratory. During the 19th century, the investigation of poisoning deaths posed a problem for the law enforcement agencies, became proof of poisoning required isolating, detecting and estimating the amount of various poisons absorbed in the human system. The first Chemical Examiner's Laboratory was, therefore, set up for this purpose at the then Madras Presidency, under the Department of Health, during 1849 (Nanda & Tewari, 2001). This laboratory later on became the State Forensic Science Laboratory and is now renamed as Forensic Sciences Department. The first Anthropometric Bureau for maintaining the Bertillon's anthropometric records of criminals was established at Calcutta in 1892. The first fingerprint bureau in the world was established in Calcutta in 1887, thanks to the efforts of William Herschel, the then District Collector of Hooghly, Bengal. The Department of Explosives started functioning in the year 1898 with Nagpur as its headquarters.

Figure 1. The earliest traceable report of the Chemical examiner's laboratory

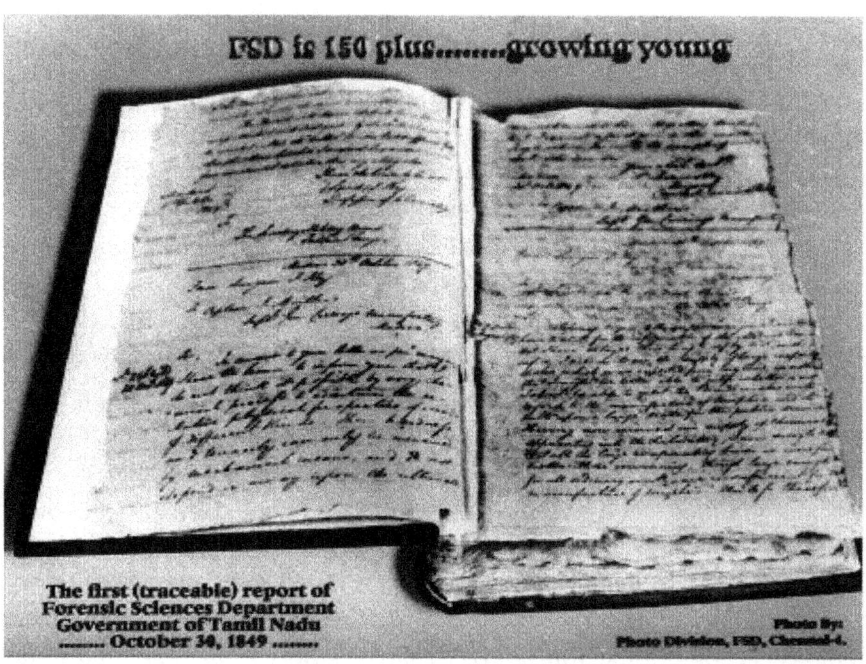

Box 1. Content Of The Report (Figure 1)

Madras 30th October 1849 From Surgeon T.Rey To Captain I.Maitland Superintendent, Government Carriage Manufactory, Madras. Sir, In answer to your letter as per No.286 dated 20th October 1849 may I have the honour to inform you that I do not think it possible by any chemical process to ascertain the relative physical properties of wood of different kinds. Their hardness and tenacity can only be measured by mechanical means and it must depend in a way upon the ultimate … and return herewith the samples of wood sent and remain. Yours obedient servant, P.Thomas Rey Surgeon (Chemical Examiner) Madras 30th October 1849

Source: http://tn.gov.in/tamilForensic

The British Government of Bengal felt the necessity to identify the handwritings on the secret documents connected with the Indian independence movement and, therefore, created a post of Government Handwriting Expert of Bengal and C.R. Hardless was appointed to the post (Nanda & Tewari, 2001). This post was later renamed as Government Examiner of Questioned Documents (GEQD) and re-located to Shimla, the summer capital of British India. In 1910, the Serology Department was established in Calcutta under the Imperial Serologist to examine the human blood samples as evidence of crime. The Footprint section, Forgery Section and the Ballistics Laboratory of the Criminal Investigation Department of Government of Bengal were established during 1915, 1917, and 1930, respectively.

The State Forensic Science Laboratory was established in the year 1952 at Calcutta. The first Central Forensic Science Laboratory was established in Calcutta in 1957. At present, there are four Central Forensic Science Laboratories, and three Central Document Examination Laboratories. Almost all the states of Indian Union have their own State Forensic Science Laboratories and some states have developed Regional Forensic Science Laboratories and Mobile Forensic Science Laboratories.

Apart from these establishments, various other units exist, such as the Forensic Science Laboratory of Central Bureau of Investigation (CBI), Lok Nayak Jayaprakash Narayan National Institute of Criminology and Forensic Science (LNJN NICFS), Bureau of Police Research and Development (BPRD), Indian Academy of Forensic Science (IAFS), Centre for DNA Fingerprinting (CDFD), Research institutions like Bhaba Atomic Research Centre (BARC), Universities, Departments of Forensic Medicine of Medical Colleges, and Dental Colleges contribute to the growth of Forensic Science literature in India.

REVIEW OF LITERATURE

Several scientometric studies have been carried out in the specialties of science, social science and arts. Studying a single journal specializing in a particular subject area or discipline gives the researcher in that area a wide understanding of the scope, coverage, and visibility of the journal. Such studies in forensic sciences had been carried out by Jeyasekar and Saravanan (2013, 2014a, 2014b, 2014c). Sauvageau, Desnoyers, and Godin, A. (2009) mapped the literature of two North-American forensic science journals. Jones (1993, 1999, 2003, 2004, 2005a, 2005b & 2007) had comprehensively studied the impact factor and highly cited papers in forensic sciences for long period and this can be considered as the pioneering work in this subject categoty. Scientometric studies on global and Indian forensic science using SCOPUS, Indian Citation Index (ICI), PubMed, and Web of Science (WoS) have also been conducted (Jeyasekar & Saravanan, 2012a, 2012b, 2014d, & 2015a; Kumbar & Biradar, 2015; Amsaveni & Ramesh, 2016; Rao, Raghavan & Bhargav, 2017). Forensic sub-specialties, viz., Forensic Odontology (Jeyasekar & Saravanan, 2015b), Forensic Anthropology (Gauldi-Russo & Fonti, 2013), Forensic Psychology (Black, 2012), and Wildlife Forensics (Jeyasekar & Saravanan, 2016) have also been scientometrically mapped. In this chapter, the Indian Forensic Science research literature output during the period of thirty-eight years since 1975 was quantified in a global perspective and also evaluated for its progress as an economic activity.

MATERIALS AND METHOD

Science mapping aims to find representations of intellectual connections within the dynamically changing system of scientific knowledge. This evolving scientific data is put into public domain in the form research papers. The data related to global as well as Indian forensic science research articles and their citations were downloaded from SCOPUS. *SCOPUS* is a bibliographic database containing abstracts and citations for academic journal articles. According the data retrieved, in the specialty of forensic sciences a total of 116660 articles have been published globally and a total of 2096 papers have been published by Indian authors during the period of study. These data were further analyzed, tabulated and visualized with tools such as *MSWord, MSExcel, VOSviewer, Pajek* software packages.

RESULTS AND DISCUSSION

The objective of scientometrics is to evaluate quantitatively recent growth of any basic scientific discipline, and the factors responsible for the steady growth in research activity in that area of knowledge in a particular period, a particular discipline or a particular geographic area. The quantitative analysis of science along with the inter and intra comparison of scientific activities including its productivity and progress may be of great use to the science administrators in planning proper utilization of available financial and manpower resources more economically.

Literature Growth

Growth in the number of publications can be an indicator of growth of knowledge. The fact that knowledge grows exponentially was propounded by Doyle (1905) in the short story "The Great Keinplatz Experiment," which states, "Knowledge begets knowledge as money bears interest." Knowledge growth is compared to compound interest - the increase at any time is a fixed percentage of the current amount. This type of growth is described mathematically by an exponential function.

The growth pattern of global Forensic Science literature was calculated using the standard methods. Relative Growth Rate (RGR) is the measure to study the increase in number of articles of a particular period of time (Mahapatra, 1985). Doubling Time (Dt) is the period of time required for a quantity to double in size or value.

RGR is expressed mathematically as,

$$1-2\bar{R}=\frac{\log_{e2}W-\log_{e1}W}{2^{T}-1^{T}}$$

where

$1-2\bar{R}$ = mean relative growth over the specific period of interval
$\log_{e1}W$ = log of initial number of articles/ pages
$\log_{e2}W$ = log of final number of articles/ pages after a specific period of interval
$2^{T}-1^{T}$ = the unit difference between the initial time and the final time

Doubling Time (Dt) is the time required for articles / citations to become double of the existing amount (Mahapatra, 2009).

$$Dt(p)=\frac{\log_{e}2}{\bar{R}(p)}=\frac{0.693}{\bar{\bar{R}}(p)}$$

where,

p is the number of publications, and $\bar{R}\, is\, the\, relative\, growth\, rate$

The year-wise growth of global Forensic Science literature, its cumulative growth, RGR and Dt were calculated using the data obtained from SCOPUS. In the year 1975 the global Forensic Science literature output was 2034. The annual literature output showed a decreasing and increasing trend alternatively till 1996. In 1996 the literature output reached 2267 and thereafter kept on increasing till it reached 6795 in the year 2012. The cumulative total output of global Forensic Science literature during the period of study stands at 116660. The growth of global Forensic Science literature showed negative trends in the years 1976, 1981, 1983, 1986, 1987, 1988, 1990, 1991 and 2010. Rest of the years it was positive.

The cumulative growth of global Forensic Science literature was also calculated. It was observed that the Forensic Science literature conforms to the exponential pattern of literature growth. Similarly, the cumulative growth of Indian Forensic Science literature during the period of study was also analyzed. The trend line shows the exponential growth pattern. However, it was observed the growth trend is rather steep from the year 2005 onwards.

It was observed that the Indian Forensic Science literature's Dt was between a minimum of 5.19 and a maximum of 19.42. The mean Dt during the first decade was 5.19; during the second decade was 19.42; during the third decade was 11.80; and during the fourth decade was 5.36. Price (1963) states that "once in ten years the number of articles in a field (particularly in science) doubles". Nevertheless, from the observation made in the study, Price statement on the growth of scientific literature is not true in the case of Forensic Science literature.

Country-Wise Productivity

The countries were ranked according to their productivity to find the countries which contributed the highest number of Forensic Science research papers. It was found that 27 countries have contributed 0.05% (i.e., 580 articles) or more of the total 116660 articles. The contributions of these 27 countries together account for 84047 articles, which was 72.05 percent of the total. Among these countries, the United States of America was the numero uno nation with 28242 contributions equivalent to 24.21 percent of the total. The United Kingdom was ranked second with 9714 contributions, which is 8.33 percent of the total. Germany was ranked third for its 8386 contributions, which is 7.19 percent. India, with 2096 contributions i.e., 1.80 percent of the total to its credit was placed at 10th rank globally.

Top Ranking Authors

Twenty authors have contributed at least 0.10 percent or more of the total research output of 116660 articles. Bud Budowle tops the list with 306 contributions to his name. His contribution is 0.26 percent of the total. Burkhard Madea had 281 papers and Klaus Püschel accumulated 263 papers, these two forensic pathologists from Germany being the second and third top contributors, respectively. While checking the research profile of Budowle, the top most contributor, it is found that he is engaged in research activities related to DNA Fingerprinting. DNA Fingerprinting is one of the latest and hot subject areas of Forensic Sciences. Hence there is no doubt that the hot subject areas always find prominence in any research activity.

Among the authors who contributed the 2096 papers of Indian origin, forty authors had contributed at least 0.50 per cent or more of the total Indian contribution. Among them, Tanuj Kanchan was the top most contributor with 89 contributions (4.25%). Ritesh G. Menezes, and O. P. Murty were respectively the second and third highest contributors. However, none of the Indian top contributors have found a place in the ranked global top contributors. This means the top ranking Indian author has not contributed even 0.01% of the global Forensic Science research literature output.

Jeyasekar and Saravanan (2015a) studied the authorship pattern of forensic science authors as covered in the *PubMed* database for the period 2001 to 2013. A rank list of authors who have contributed a minimum of 10 papers was made in the study. In this study too, Tanuj Kanchan, and Ritesh G. Menezes were the most prolific Indian Forensic Science authors ranked first and second respectively.

Author Network Maps

An author network map (Fig. 2) of the authors of the highly cited 2000 papers was created using *Pajek*. A total of 104 vertices, each representing an author, were observed in this network map. The closest vertices were Cook and Rogers. The smallest angle was Good and Harris. The shortest line was between Jackson and Day and the longest line was between Rogers and Chang.

A map was created using VOSviewer to visualize the most prolific Indian Forensic Science authors. The map was created with a threshold value of 5 i.e., authors who have contributed a minimum of 5 papers alone were taken into account to create the map. Figure 3 is the label view of cluster map of the Indian authors.

Figure 2. Network Map of Top Cited Global Authors

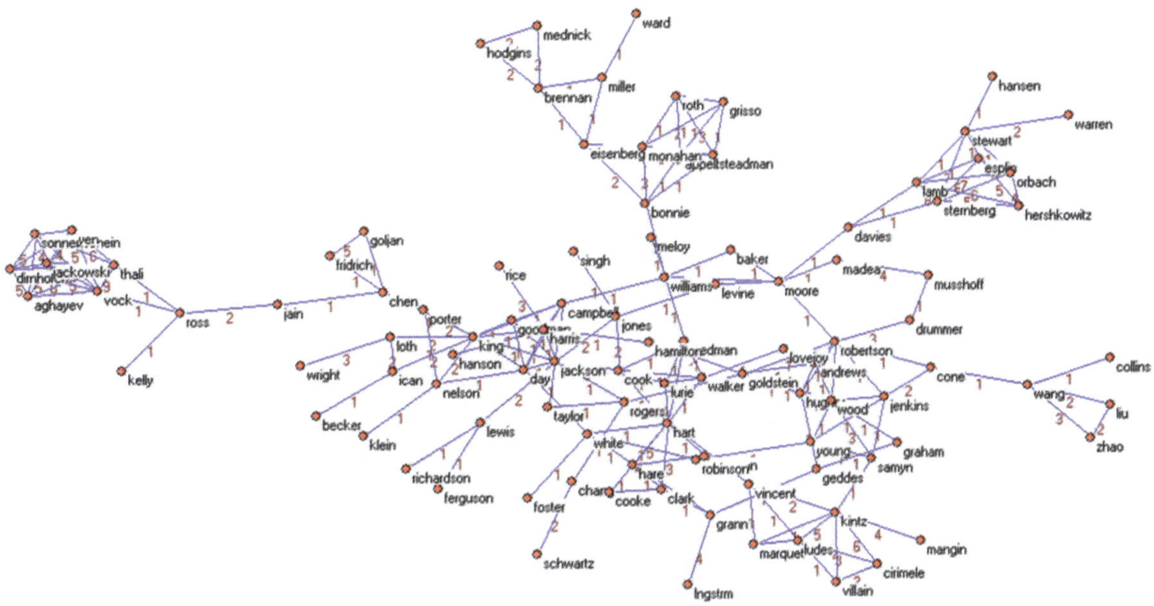

Figure 3. Label View of Cluster Map of Indian Forensic Science Authors

VOSviewer is an algorithm to visualize co-occurrences on the basis of their similarities. Here authors who co-author papers are linked and visualized together. In the cluster map of Indian Forensic Science authors, 84 items or authors in 23 clusters were found. Since the map was created with a threshold value of 5, this can mean 84 authors have contributed a minimum of 5 papers and have some relationship as co-authors.

A network map of all the authors who have contributed to the Indian Forensic Science during the study period is created with *Pajek*. This network map is illustrated in Figure 4.

Prolific Authors

Prolific means abundant. Prolific authors refer to authors who have produce a huge quantity of published papers. Authors who have contributed more number of papers are enumerated in this analysis. This examination was done with regard to the quantity of literature produced and does not reflect the quality.

One hundred and fifty-seven institutions had contributed a minimum of 0.1% of the total forensic science literature. The majority of these institutions were either academic or research institution and none of these institutions was a government controlled Forensic Science Laboratory. Jones (2005) studied the working of Forensic Science laboratories extensively and has commented that government Forensic scientists are under less pressure to publish. This statement was found to be true with the available data.

Indian institutions contributing 1 percent or more of the total number of papers were studied for their contributions to forensic science research. A total of eighteen institutions qualified to find a place in this ranked list. It was found that Kasturba Medical College, Mangalore was the top productive institution with 131 papers, which is 6.25 percent of the total papers. The All India Institute of Medical Sciences came second with 5.15 percent of the papers. The number of papers contributed by the AIIMS was 108. Among these eighteen institutions, the third ranked Central Forensic Science Laboratory and the eighth ranked Forensic Sciences Department, Tamil Nadu were the government controlled Forensic laboratories.

Figure 4. Indian Forensic Science Authors Network Map

Rest of the institutions were either academic or research institutions both publicly funded and private. Thus Jones statement that government Forensic scientists are under less pressure to publish was found to be partially true in the case of Indian institutions.

Productive Journals

The journals which had contributed at least 0.5 percent or more of the total 116,660 articles were ranked according to their productivity and a total of 16 journals found to qualify for this ranked list. It was observed that the *Journal of Forensic Sciences* (*JFS*) which had contributed 6538 (5.6%) articles tops the list. The second highest contributor was *Forensic Science International (FSI)* with 4744 (4.1%) articles, followed by the third highest contributor, the German journal *Sudebno Meditsinskaya Ekspertiza*, with 2124 (1.82%) articles. All these 16 journals together had contributed 25,193 articles, which was 21.59 percent of the total. Four journals in this list were non-English journals. The top two journals (*JFS* and *FSI*) and the German journal *Archiv für Kriminologie* alone were the ones which cover all aspects of Forensic Sciences. The *Journal of American Academy of Psychiatry and Law* and *Lancet* were journals which were from other fields such as Psychiatry and Medicine. *Journal of Analytical Toxicology (JAT)* represents a specialty journal covering analytical aspects of forensic toxicology. The remaining ten journals were the ones which cover forensic and legal medicine.

Similarly, a ranked list of journals which published articles by authors affiliated to Indian institutions was investigated. Journals contributing at least a minimum of one percent of the total of 2096 articles were ranked according to the number of articles contributed. It was observed that a total of 1191 articles which was more than half of the total contribution were found to be concentrated in these 16 journals. *FSI* tops the list with 144 (6.87%) articles followed by *Medico Legal Update* with 137 (6.54%) articles and *Journal of Forensic and Legal Medicine* with 123 (5.87%) articles. 8 journals in this list were Indian journals.

A cluster map of the source journals was created using VOSViewer for better visualization. The cluster map is presented as Figure 5.

Bibliographic Format

The bibliographic format of global and Indian Forensic Science literature output was examined to find the most preferred form of communication by Forensic Scientists. The results are presented in Table 1.

It was found that the most preferred form of research communication was the journals. Globally 94.05 percent of the literature and in India 92.46 percent of the literature were published in the journals. Conference proceedings were the other popular medium but not as popular as the journals. Conference proceedings were used by 4.30 percent of the researchers globally and 5.96 percent of the Indian researchers. Much difference could not be found in the preferred source type by the Indians and others as illustrated in Figure 6.

Relative Quality Index of Leading Institution

Papers that received more than twice the average citations are considered as high quality papers (NHQ). Relative Quality Index (RQI) is the ratio of the proportion of high quality papers to the total number of the publications (TNP%). The measure relates to the incidence of high quality papers for a country or an

Figure 5. Cluster Map of Source Journals

Table 1. Bibliographic Format of Forensic Science Literature

Bibliographic Format	No of Records – Global (Percent)	No of Records - India (Percent)
Journals	109,715 (94.05)	1938 (92.46)
Conference Proceedings	5017 (4.30)	125 (5.96)
Book Series	1321 (1.13)	31 (1.48)
Trade Publications	566 (0.49)	1 (0.05)
Books	41 (0.04)	1 (0.05)
Total	116660 (100)	2096 (100)

institution. Value of RQI greater than 1 indicates higher than average value, whereas the value of RQI less than 1 indicates lower than average quality. The NHQ, NHQ% and RQI were calculated for the 18 top ranking institutions and listed in Table 2.

The thirteenth ranked institution, SDM College of Dental Sciences & Hospital has the highest RQI of 3.28 followed by Government Medical College & Hospital, Chandigarh with the RQI value 2.90 and the Panjab University with 2.87. Maulana Azad Medical College has the least RQI value of 0.25. In all, 9 institutions have RQI greater than 1. They are Kasturba Medical College, Mangalore; Central Foren-

Figure 6. Bibliographic Type

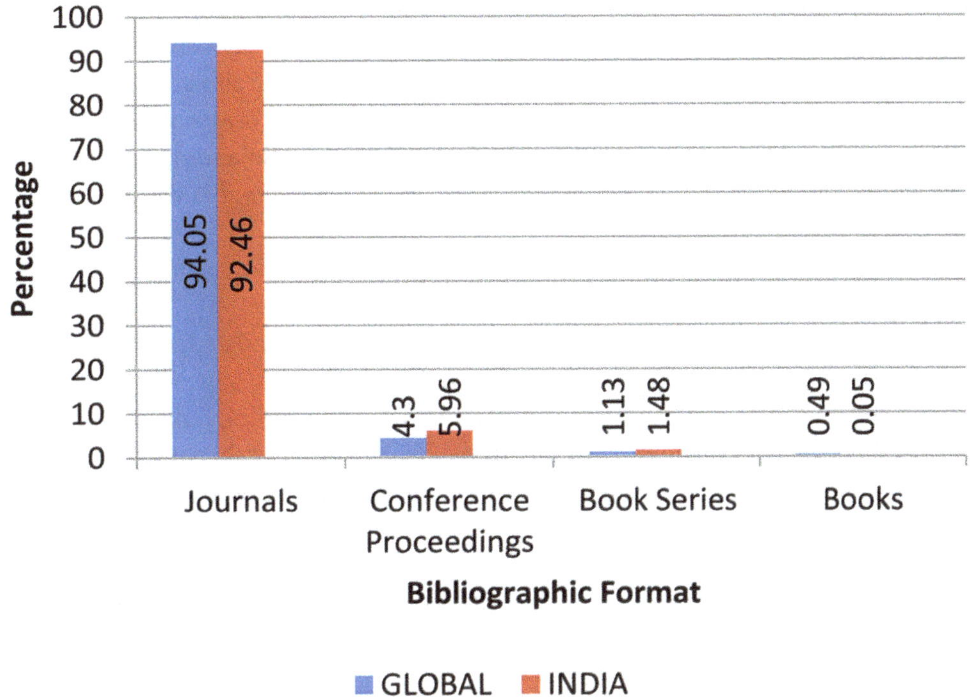

Table 2. RQI of Leading Institutions

Institution	NHQ	NHQ%	TNP%	RQI
Kasturba Medical College Mangalore	33	11.30	6.25	1.81
All Indian Institute of Medical Sciences	8	2.74	5.15	0.53
Central Forensic Sciences Laboratory	27	9.25	3.91	2.37
Punjabi University	14	4.79	3.82	1.25
Kasturba Medical College Manipal	19	6.51	6.58	1.82
Panjab University	16	5.48	1.91	2.87
Government Medical College & Hospital Chandigarh	15	5.14	1.77	2.90
Forensic Sciences Department Chennai	4	1.37	1.77	0.77
Bhabha Atomic Research Center	2	0.68	1.67	0.41
Maulana Azad Medical College	1	0.34	1.34	0.25
Government Medical College Nagpur	2	0.68	1.24	0.55
Pandit Bhagwat Dayal Sharma Postgraduate Institute of Medical Sciences	1	0.34	1.24	0.27
S.D.M. College of Dental Sciences & Hospital	11	3.77	1.15	3.28
Indian Statistical Institute, Kolkata	8	2.74	1.05	2.61
Armed Forces Medical College	2	0.68	1.00	0.68
Regional Institute of Medical Science, Imphal	2	0.68	1.00	0.68
J.S.S. University, Mysore	2	0.68	1.00	0.68
University of Delhi	3	1.03	1.00	1.03

sic Sciences Laboratory; Punjabi University; Kasturba Medical College, Manipal; Panjab University; Government Medical College & Hospital, Chandigarh; SDM College of Dental Sciences & Hospital; Indian Statistical Institute, Kolkata; and University of Delhi. The following 9 institutions have RQI value less than 1. They are All India Institute of Medical Sciences; Forensic Sciences Department, Chennai; Bhabha Atomic Research Center; Maulana Azad Medical College; Government Medical College, Nagpur; Pandit Bhagwat Dayal Sharma Postgraduate Institute of Medical Sciences; Armed Forces Medical College; Regional Institute of Medical Science, Imphal; and JSS University, Mysore.

RCR of Leading Institutions

The Relative Citation Rate (RCR) of the leading institutions was calculated, ranked and tabulated in Table 3.

Panjab University had RCR of 3.70 and ranked number one. The second ranked institution was Government Medical College and Hospital, Chandigarh, which had RCR of 2.38. SDM College of Dental Sciences and Hospital which had RCR value 2.20 was ranked third. Out of the 18 institutions examined, 8 institutions have RCR less than 1. Pandit Bhagwat Dayal Sharma Postgraduate Institute of Medical Sciences, which has RCR of 0.32 is ranked 18.

The RQI and RCR values of the top 18 institutions are plotted as a graph in Figure 7.

Table 3. RCR of Leading Institutions

Institution	CPP	RCR	Rank
Kasturba Medical College Mangalore	4.41	1.62	6
All Indian Institute of Medical Sciences	2.09	0.77	12
Central Forensic Sciences Laboratory	4.94	1.81	4
Punjabi University	3.96	1.45	8
Kasturba Medical College Manipal	4.15	1.52	7
Panjab University	10.10	3.70	1
Government Medical College & Hospital Chandigarh	6.51	2.38	2
Forensic Sciences Department Chennai	1.07	0.62	13
Bhabha Atomic Research Center	1.63	0.60	14
Maulana Azad Medical College	1.39	0.51	15
Government Medical College Nagpur	2.85	1.04	10
Pandit Bhagwat Dayal Sharma Postgraduate Institute of Medical Sciences	0.88	0.32	18
S.D.M. College of Dental Sciences & Hospital	6.00	2.20	3
Indian Statistical Institute, Kolkata	3.77	1.38	9
Armed Forces Medical College	2.33	0.85	11
Regional Institute of Medical Science, Imphal	1.14	0.42	17
J.S.S. University, Mysore	1.33	0.49	16
University of Delhi	4.81	1.76	5

Figure 7. RQI & RCR of Leading Institutions

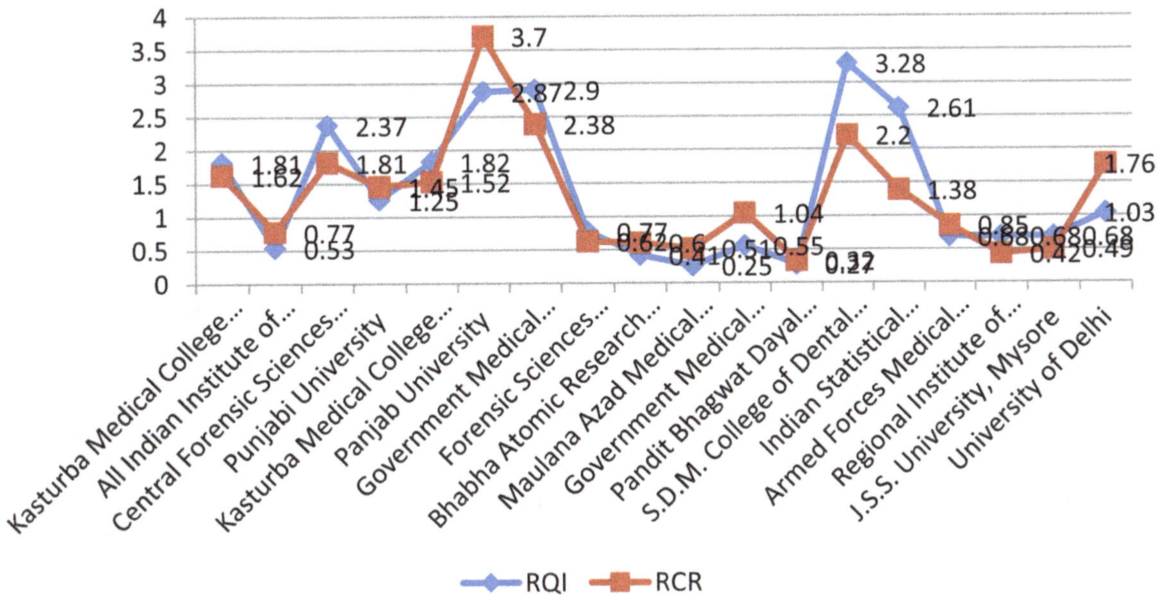

Research Collaboration

'Research collaboration' could be defined as the working together of researchers to achieve the common goal of producing new scientific knowledge (Katz & Martin, 1997). Collaboration is an essential component of research. Over recent years, there has been increasing interest among researchers and within science policy circles in the notion of research collaboration. It is widely assumed that collaboration in research is 'a good thing' and that it should be encouraged (Beaver & Rosen, 1978). Rev. Dr. Ignacimuthu states in an interview that "... - (collaboration) is what makes you move towards path breaking research. Cooperation is also very important. No one individual today can become successful, unless he is able to get the cooperation of so many others... In India people are a little slow in understanding the implication of the industry-educational institution collaboration" (Dharmapalan, 2014).

Impact of Research Collaboration

Research collaboration in multidisciplinary subject like forensic science is a necessity. Numerous attempts have been made to bring together individual researchers. The collaboration among researchers may be at the institution level or between different institutions or with industries. Similarly, collaboration may be at national level or international level. Here an attempt is made to study the volume of research collaboration and also the impact of research collaboration.

The single author papers are considered as zero collaborative ones. When two or more authors of the same institution or different institutions collaborated, it is considered as national collaboration. International bilateral collaborative papers are the ones where an author or authors of another country collaborate with an Indian author or authors. International multilateral collaborative papers are the ones with authors of two or more countries collaborating with the national author or authors.

The various levels of collaboration, their volume, percentage of volume, the number of publication cited, the number of citations, and the percentage of citation, citedness, citation per paper and the RCI are calculated and given in Table 4.

The number of papers or volume of the zero level collaborative papers is 342, which 16.42% of the total papers. Out of these, 126 papers have received a total of 606 citations, which is 10.59% of the total citations. The citedness, CPP, and RCI of zero collaborative papers stand at 36.84, 1.77, and 0.65 respectively. The volume of national level collaborative papers is the highest with a frequency of 1593, which is 76% of the total. 711 papers out the 1593 papers have received citations. The number of citations received is 4466, which is 78% of the total citations. This is the highest among the various levels of collaborative papers. However, citedness, CPP and RCI values are relatively low when compared with those of international bilateral as well as multilateral collaborative papers.

The volume of international bilateral papers is 28. Among these, 18 papers have received 134 citations, which is 2.34% of the total. The citedness, CPP and RCI values of this category are 64.29, 4.79, and 1.75 respectively. 133 papers, which are 6.35% of the total papers, have international multilateral collaborative authorship. Among them, 80 papers have received 519 citations, which is 9.07% of the total citations. The citedness is 44.61; the CPP is 6.49 and RCI is 2.38 for international multi-lateral papers.

The cluster map created with the aid of *VOSviewer* was converted into a network map with *Pajek* and given in Figure 9. The figure shows 43 vertices, each indicating a collaborating country. The collaboration or link strength is also indicated in the connecting lines. The vertices are arranged alphabetically.

Bibliographic Coupling

Citation analysis is a quantitative study of citations and the relationships between the cited and citing documents. Research on co-citation network started as early as 1973. Marshakova and Small suggested studying the relationships between bibliographies using co-citations (Egghe & Rousseau, 1990). Co-citation means that when document A cites documents B and C, B and C are co-cited documents (van Raan, 1991). Egghe and Rousseau found that the frequency of co-citation between documents B and C grows higher when these two documents are more related. Kessler (1963) grouped technical and scientific papers together using bibliographic coupling. When the same reference is cited by two papers, it was defined as a unit of coupling between two papers. The coupling strength between them is measured by the numbers of coupling units. In other words, when two documents cite the same reference, these two articles are related. The strength of this association is determined by the frequency of coupling (Egghe & Rousseau, 1990). Figure 10 is the map created of bibliographically coupled Indian forensic science authors.

Table 4. Volume & RCI of Research Collaboration

Collaboration	Volume	%	Tpc	Citation	%	Citedness	CPP	RCI
Zero	342	16.32	126	606	10.59	36.84	1.77	0.65
National	1593	76.00	711	4466	78.00	44.63	2.80	1.03
International Bilateral	28	1.34	18	134	2.34	64.29	4.79	1.75
International Multilateral	133	6.35	80	519	9.07	60.15	6.49	2.38
Total	2096	100	935	5725	100	44.61	2.73	1.00

Figure 8. CPP & RCI of Research Collaboration

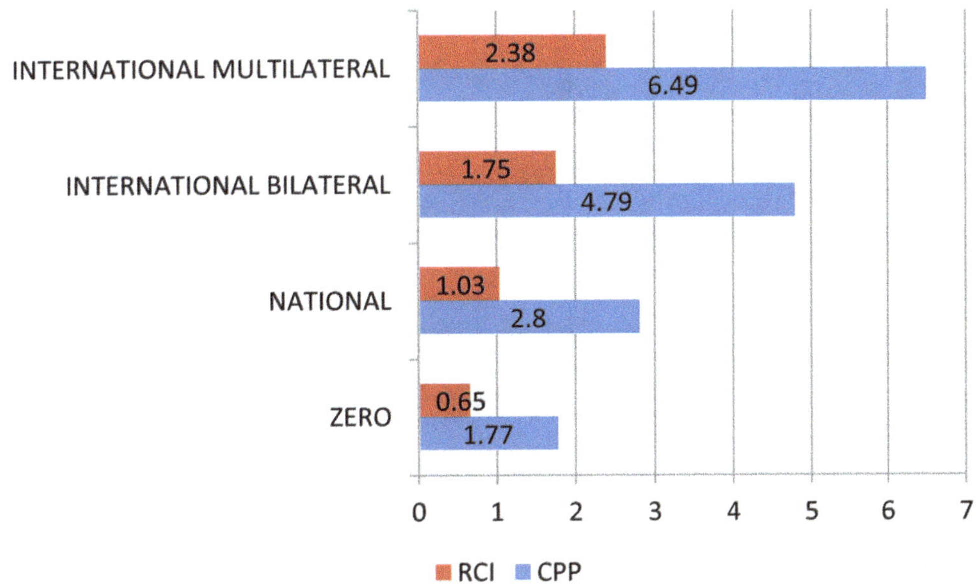

Figure 9. Network Map of Collaborating Countries

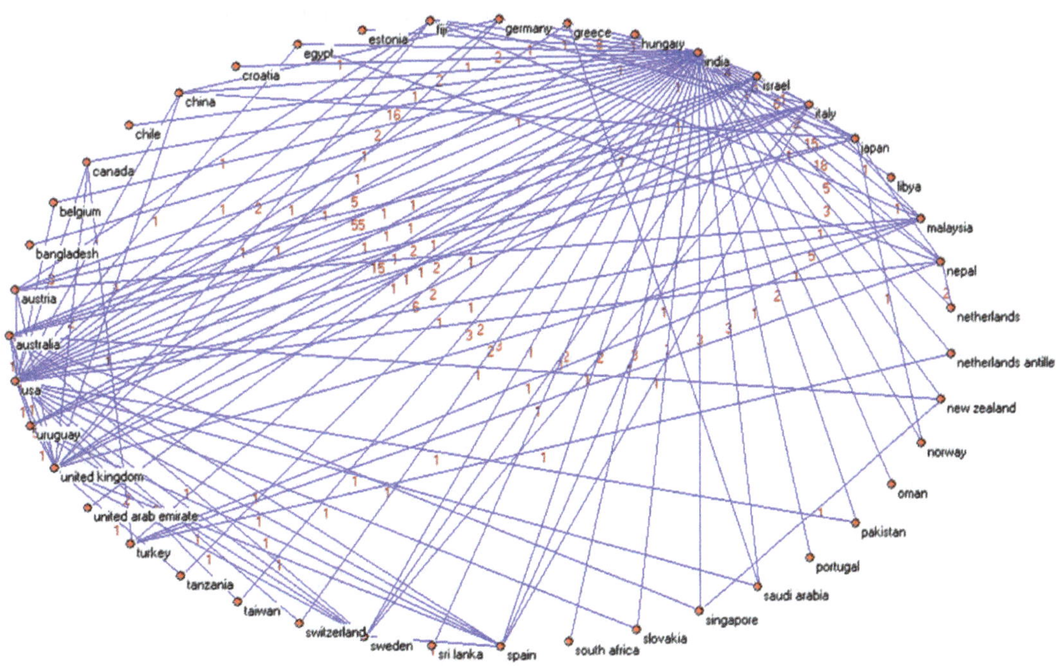

Figure 10. Bibliographically coupling among Indian authors

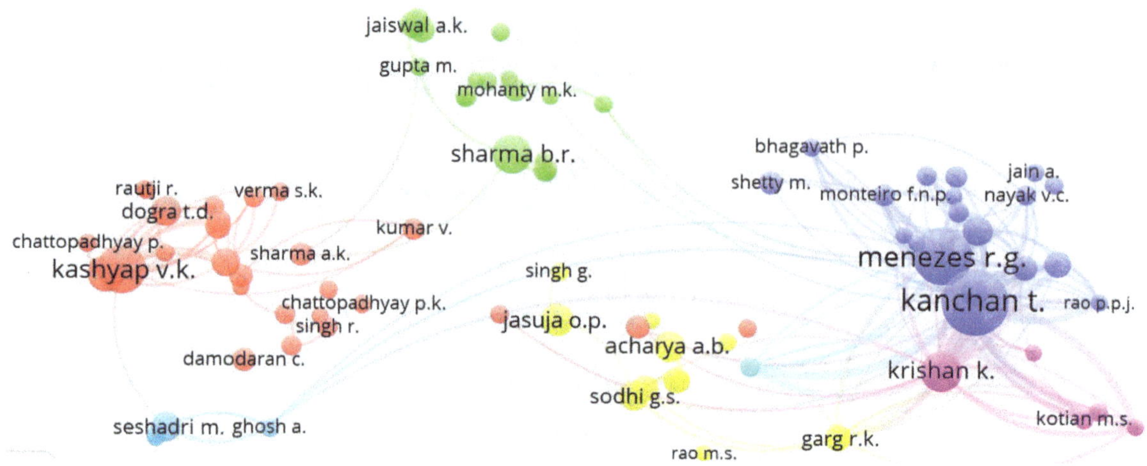

CONCLUSION

Though exponential growth pattern was found in the global and Indian forensic science literature, there is much scope for improvement in the Indian forensic science domain. Further, it was found that the scientific community from the governmental agencies is not inclined to publishing their works. Incentives for publication can be awarded to boost further research. Since internationally collaborative papers were useful and informative and such joint research project should also be encouraged in the future.

REFERENCES

Amsaveni, N., & Ramesh, K. (2016). Mapping of research productivity in forensic science: A scientometric analysis. *Journal of Advances in Library and Information Science*, *5*(4), 372–378.

Ashbaugh, D. R. (2000). *Quantitative-Qualitative Friction Ridge Analysis: An introduction to basic and advanced ridgeology*. Boca Raton: CRC Press.

Beaver, D., & Rosen, R. (1978). Studies in scientific collaboration: Part I--The professional origins of scientific co-authorship. *Scientometrics*, *1*(1), 65–84. doi:10.1007/BF02016840

Bettencourta, L. M. A., & Kaur, J. (2011). Evolution and structure of sustainability science. *Proceedings of the National Academy of Sciences of the United States of America*, *108*(49), 19540–19545. doi:10.1073/pnas.1102712108 PMID:22114186

Black, S. (2012). Frequently cited journals in forensic psychology. *Psychological Reports*, *110*(1), 276–282. doi:10.2466/09.13.17.PR0.110.1.276-282 PMID:22489393

De Forest, P. R., Gaensslen, R. E., & Lee, H. C. (1983). *Forensic Science: an introduction to criminalistics*. New York: McGraw-Hill Book Company.

Dharmapalan, B. (2014). To be successful in science one needs to follow a disciplined way of life – Dr. Ignacimuthu. *Scientific Reports*, *51*(9), 42–44.

Doyle, A. C. (1905). *The Great Keinplatz Experiment*. New York: George H. Doran Co.

Egghe, L., & Rousseau, R. (1990). *Introduction to Informetrics: Quantitative Methods in Library, Documentation and Information Science*. New York: Elsevier.

Etzkowitza, H., & Leydesdorffb, L. (2000). The dynamics of innovation: From national systems and 'mode2' to a triple helix of university-industry-government relations. *Research Policy*, *29*(2), 109–123. doi:10.1016/S0048-7333(99)00055-4

Galton, F. (1892). *Fingerprints*. London: Macmillan.

Gill, P., Jeffreys, A. J., & Werrett, D. J. (1985). Forensic application of DNA 'fingerprints'. *Nature*, *318*(6046), 577–579. doi:10.1038/318577a0 PMID:3840867

Gualdi-Russo, E., & Fonti, G. (2013). Recent trend and perspectives in forensic anthropology: A bibliometric analysis. *Collegium Antropologicum*, *37*(2), 595–599. PMID:23941009

Jeyasekar, J. J., & Saravanan, P. (2012a). Scientometric study of forensic science: a study based on *SCOPUS* database. In *Proceedings of the UGC Sponsored National Seminar on Scientometrics and Informetrics, Annamalai Nagar* (pp. 75-78).

Jeyasekar, J. J., & Saravanan, P. (2012b). Scientometric analysis of Indian forensic science literature based on ICI database. *Journal of Library Advancements*, *2*(1), 1–5.

Jeyasekar, J. J., & Saravanan, P. (2013). Journal of forensic sciences: a bibliometric study for the period 2006 to 2010. In *Proceedings of the Second National Conference of Scientometrics and Knowledge Management, Dharwad* (CD-ROM version).

Jeyasekar, J. J., & Saravanan, P. (2014a). Scientometric analysis of the journal medicine, science and the law from 2001 to 2012. In *Proceedings of UGC Sponsored National Seminar on Application of Quantitative and Qualitative Indicators for the Excellence of Scholarly Communication, Annamalai Nagar*, Annamalai Nagar (pp. 308-313).

Jeyasekar, J. J., & Saravanan, P. (2014b). A scientometric analysis of the Journal of Forensic Identification. *KELPRO Bulletin*, *18*(2), 34–47.

Jeyasekar, J. J., & Saravanan, P. (2014c). A scientometric portrait of the journal Digital Investigation. *Journal of Advances in Library and Information Science*, *3*(2), 155–162.

Jeyasekar, J.J., & Saravanan, P. (2014d). A scientometric analysis of global forensic science research publications. *Library Philosophy & Practice.*

Jeyasekar, J. J., & Saravanan, P. (2015a). Indian forensic science research literature: A bibliometric study of its growth, authorship and publication patterns. *SRELS Journal of Information Management*, *52*(1), 67–75.

Jeyasekar, J. J., & Saravanan, P. (2015b). Mapping forensic odontology literature using open source bibliographies and software: a case study. In S. Thanuskodi (Ed.), *Handbook of Research in Inventive Digital Tools for Collection Management and Development in Modern Libraries*. Hershey, PA: IGI Global. doi:10.4018/978-1-4666-8178-1.ch011

Jeyasekar, J.J., & Saravanan, P. (2016). Science maps of global and Indian wildlife forensics: a comparative map. *Library Philosophy and Practice.*

Jones, A. W. (1993). The impact of forensic science journals. *Forensic Science International, 62*(3), 172–178. doi:10.1016/0379-0738(93)90206-P PMID:12742682

Jones, A. W. (1999). The Impact of alcohol and alcoholism among substance abuse journals. *Alcohol and Alcoholism, 34*(1), 25–34. doi:10.1093/alcalc/34.1.25 PMID:10075398

Jones, A. W. (2002). JAT's impact factor--room for improvement? journal of analytical toxicology. *Journal of Analytical Toxicology, 26*(1), 2–5. doi:10.1093/jat/26.1.2 PMID:11888013

Jones, A. W. (2003). Impact factors of forensic science and toxicology journals: What do the numbers really mean? *Forensic Science International, 133*(1-2), 1–8. doi:10.1016/S0379-0738(03)00042-2 PMID:12742682

Jones, A. W. (2004). Impact of JAT publications 1981-2003: The most prolific authors and the most highly cited articles. *Journal of Analytical Toxicology, 28*(7), 541–545. doi:10.1093/jat/28.7.541 PMID:16078379

Jones, A. W. (2005a). Creme de la creme in forensic science and legal medicine. the most highly cited articles, authors and journals 1981-2003. *International Journal of Legal Medicine, 119*(2), 59–65. doi:10.1007/s00414-004-0512-x PMID:15645298

Jones, A. W. (2005b). Which articles and which topics in the forensic sciences are most highly cited? *Science & Justice, 45*(4), 175–182. doi:10.1016/S1355-0306(05)71661-0 PMID:16686268

Jones, A. W. (2007). The distribution of forensic journals, reflections on authorship practices, peer-review and role of the impact factor. *Forensic Science International, 165*(2-3), 115–128. doi:10.1016/j.forsciint.2006.05.013 PMID:16784827

Katz, J. S., & Martin, B. R. (1997). What is research collaboration? *Research Policy, 26*(1), 1–18. doi:10.1016/S0048-7333(96)00917-1

Kessler, M. M. (1963). Bibliographic coupling between scientific papers. *American Documentation, 14*(1), 10–25. doi:10.1002/asi.5090140103

Kumbar, P., & Biradar, N. (2015). Research trends in forensic science; A study of scientometric analysis. *International Journal of Research in Library Science, 1*(2), 42–48.

Mahapatra, G. (2009). *Bibliometric Studies in the Internet Era.* New Delhi: Indiana Publishing House.

Mahapatra, M. (1985). On the validity of the theory of exponential growth of scientific literature. In *Proceedings of the 15th IASLIC conference, Bangalore* (pp. 61-70).

Morland, N. (1950). *An Outline of Scientific Criminology*. New York: Philosophical Library.

Nanda, B. B., & Tewari, R. K. (2001). *Forensic Science in India: A vision for the twenty-first century*. New Delhi: Select Publishers.

Noyons, E. C. M., & van Raan, A. F. J. (1998). Monitoring Scientific Developments from a Dynamic Perspective: Self-Organized Structuring to Map Neural Network Research. *Journal of the American Society for Information Science*, *49*(1), 68–81.

Pope, G. U. (1886). *The Sacred Kurral of Tiruvalluva Nayanar*. New Delhi: Asian Education Services.

Price, D. (1963). *Little Science, Big Science*. New York: Columbia University Press.

Rao, I. K. R., Raghavan, K. S., & Bhargav, K. N. (2017). Quantitative analysis of scientific literature in forensic science. *COLLNET Journal of Scientometrics and Information Management*, *11*(1), 59–72. do i:10.1080/09737766.2016.1251561

Sauvageau, A., Desnoyers, S., & Godin, A. (2009). Mapping the literature in forensic sciences: A bibliometric study of North-American journals from 1980 to 2005. *The Open Forensic Science Journal*, *2*(1), 41–46. doi:10.2174/1874402800902010041

van Raan, A. F. J. (1991). Fractal geometry of information space as represented by co-citation clustering. *Scientometrics*, *20*(3), 439–449. doi:10.1007/BF02019764

ADDITIONAL READING

Bhattacharya, S., & Basu, P. K. (1998). Mapping a research area at the macro level using co-word analysis. *Scientometrics*, *43*(3), 359–372. doi:10.1007/BF02457404

Boyack, K. W., Wylie, B. N., & Davidson, G. S. (2002). Domain visualization using VxInsight for science and technology management. *Journal of the American Society for Information Science and Technology*, *53*(9), 764–774. doi:10.1002/asi.10066

Callon, M., Courtial, J., & Laville, F. (1991). Co-word analysis as a tool for describing the network of interactions between basic and technological research — the case of polymer chemistry. *Scientometrics*, *22*(1), 155–205. doi:10.1007/BF02019280

Callon, M., Courtial, J. P., Turner, W. A., & Bauin, S. (1983). From translations to problematic networks: An introduction to co-word analysis. *Social Sciences Information. Information Sur les Sciences Sociales*, *22*(2), 191–235. doi:10.1177/053901883022002003

Chen, C. (1999). *Information Visualization and Virtual Environments*. London: Springer Verlag. doi:10.1007/978-1-4471-3622-4

Chen, C. (2004). Searching for intellectual turning points: Progressive knowledge domain visualization. In *Proceedings of the National Academy of Sciences of the United States of America*, *101*(Suppl. 1), 5303–5310. doi:10.1073/pnas.0307513100 PMID:14724295

Chen, C. (2006). CiteSpace II: Detecting and visualizing emerging trends and transient patterns in scientific literature. *Journal of the American Society for Information Science and Technology*, *57*(3), 359–377. doi:10.1002/asi.20317

Chen, C. M., & Paul, R. J. (2001). Visualizing a knowledge domain's intellectual structure. *Computer*, *34*(3), 65–71. doi:10.1109/2.910895

Cobo, M. J., López-Herrera, A. G., Herrera-Viedma, E., & Herrera, F. (2011). An approach for detecting, quantifying, and visualizing the evolution of a research field: A practical application to the fuzzy sets theory field. *Journal of Informetrics*, *5*(1), 146–166. doi:10.1016/j.joi.2010.10.002

Cook, D. J., & Holder, L. B. (2006). *Mining graph data*. Hoboken, NJ: John Wiley. doi:10.1002/0470073047

Couclelis, H. (1998). Worlds of Information: The Geographic Metaphor in the Visualization of Complex Information. *Cartography and Geographic Information Science*, *25*(4), 209–220. doi:10.1559/152304098782383034

Coulter, N., Monarch, I., & Konda, S. (1998). Software engineering as seen through its research literature: A study in co-word analysis. *Journal of the American Society for Information Science*, *49*(13), 1206–1223. doi:10.1002/(SICI)1097-4571(1998)49:13<1206::AID-ASI7>3.0.CO;2-F

De Bellis, N. (2009). *Bibliometrics and Citation Analysis: From the science citation index to Cybermetrics*. Maryland: Scarecrow Press.

Ding, Y., Chowdhury, G., & Foo, S. (1999). Mapping the intellectual structure of information retrieval studies: An author co-citation analysis; 1987-1997. *Journal of Information Science*, *41*(6), 391–407.

Garfield, E. (1994). Scientography: Mapping the tracks of science. *Current Contents: Social & Behavioural Sciences*, *7*(45), 5–10.

Garfield, E., & Small, H. (1989). *Identifying the changing frontiers of science*. The S. Neaman Press. Retrieved from http://www.garfield.library.upenn.edu/papers/362/362.html

Morris, S., & Van DerVeer Martens, B. (2008). Mapping research specialties. *Annual Review of Information Science & Technology*, *42*(1), 213–295. doi:10.1002/aris.2008.1440420113

Noyons, E. (2001). Bibliometric mapping of science in a science policy context. *Scientometrics*, *50*(1), 83–98. doi:10.1023/A:1005694202977

Noyons, E. C. M., Moed, H. F., & Luwel, M. (1999). Combining mapping and citation analysis and science mapping. *Scientometrics*, *46*(3), 591–604. doi:10.1007/BF02459614

Noyons, E. C. M., & van Raan, A. F. J. (1998). Monitoring scientific developments from a dynamic perspective: Self-organized structuring to map neural network research. *Journal of the American Society for Information Science*, *49*(1), 68–81.

Rosvall, M., & Bergstrom, C. T. (2010). Mapping change in large networks. *PLoS One*, *5*(1), e8694. doi:10.1371/journal.pone.0008694 PMID:20111700

Salton, G., & McGill, M. J. (1983). *Introduction to modern information retrieval*. New York: McGraw-Hill.

Salvador, M. R., & Lopez-Martinez, R. E. (2000). Cognitive structure of research: Scientometric mapping in sintered materials. *Research Evaluation*, *9*(3), 189–200. doi:10.3152/147154400781777214

Shneiderman, B. (1992). Tree visualization with tree maps: A 2-d space filling approach. *ACM Transactions on Graphics*, *11*(1), 92–99. doi:10.1145/102377.115768

Chapter 10
Scientograph-Based Visualization of Computer Forensics Research Literature

Jahir Hussain Haneefa Mohamed
Waljat College of Applied Sciences, Oman

ABSTRACT

Studying the communication pattern of scientific information produced as a result of academic and industrial research using mapping techniques is part of scientometric studies. In this chapter, the bibliographic data related to computer forensics retrieved from SCOPUS is subjected to mapping and visualizing. The results showed that the computer forensics literature grows exponentially. USA ranks first among the countries contributing to the literature growth. The University of South Australia is top ranking institution and K.K.R. Choo is the top-ranking author. Conference proceedings are the dominant form of communication flow. The scientograph of keywords reveals six clusters and the keywords with high frequency of occurrences are computer crime, security of data, network security, electronic crime countermeasures, digital forensic, internet, cryptography, cloud computing, cyber security, malware, and intrusion detection.

INTRODUCTION

Scientometrics is the quantitative study of scientific communication (Leydesdorff, 2001). It requires the use of a multitude of sophisticated techniques including citation analysis, social network analysis and other quantitative techniques for mapping and measurement of relationships and flows between people, groups, organizations, research papers, computers and other knowledge entities (Niazi & Hussain, 2011). Domain visualization is a relatively newer research front. The idea is to use information visualization to represent large amount of data in research fronts (Chen, Paul & O'Keefe, 2001). This allows the viewer to look at large corpus and deeper insights based on high level view of the map (Card, Mackinlay & Shneiderman, 1999). Garfield (1994) used bibliographic data to create longitudinal maps which are called scientographs, though not widely used.

DOI: 10.4018/978-1-5225-3457-0.ch010

Information and Communication Technologies (ICT) are among the fastest growing technologies around the globe. Information Technology (IT) and Information Technology Enabled Services (ITES) industry has grown out of proportion throughout the world. It has influenced every sphere of life and hence we come across a bunch of knowledge domains attached with the letter '*e*', which means electronic. Thus, we have e-banking, e-commerce, e-payment, e-libraries, etc. This in turn has led to e-crimes happening everywhere and a new field of computer forensics came into existence.

Though there are many areas such as, network forensics, data forensics, information forensics, cyber forensics, digital forensics each a specific field in its own right the term computer forensics is being used as a synonymous term for all these.All these fields are closely associated with each other as the keyword analysis in this reveals.

REVIEW OF LITERATURE

Numerous scientometric and bibliometric studies have been conducted various domains of knowledge by various professionals of field as well as knowledge managers. A comparative scientometric study on Indian and Chinese computer science research has been done by Suresh Kumar and Garg (2005). Heilig and VoB (2014) conducted a scientometric study of cloud computing literature. However, no scientometric study has been done on computer forensics except a single journal study on "Digital Investigation" by Jeysekar and Saravanan (2014). Hence this study was conducted by the authors to fill the felt need.

MATERIALS AND METHOD

The data for the study is downloaded from the bibliographic database SCOPUS. SCOPUS is an international multidisciplinary indexing and abstracting database. It covers more than 15,000 international peer reviewed journals in Science and Technology and more than 500 international conference and seminar proceedings. It is a product of *Elsevier* and is subscription based.The data retrieved is further cleaned for analysis and then analyzed using *Microsoft-Excel*. The computer algorithm *VOSviewer* and *Microsoft-Word* picture tools are used for visualization.

RESULTS AND DISCUSSION

The results obtained from the study are offered under the sub-heading (1) Year-wise contribution; (2) Country-wise contribution; (3) Top contributing institutions; (4) Author-wise contributions; (5) Journal-wise contributions; (6) Document type; (7) Subject-wise contributions; and (8) Scientographs of keywords.

1. Year-Wise Contribution

The year-wise contribution of bibliography is tabulated and presented in Table 1. A total of 4978 bibliographic records on computer forensics are indexed in SCOPUS.

Table 1. Year-wise Contribution

Year	No of Records	Percentage	Cumulative Growth	Percentage
1989	1	0.02	1	0.02
1990	1	0.02	2	0.04
1998	1	0.02	3	0.06
1999	0	0	3	0.06
2000	3	0.06	6	0.12
2001	5	0.10	11	0.22
2002	10	0.20	21	0.42
2003	30	0.60	51	1.02
2004	24	0.48	75	1.51
2005	70	1.41	145	2.91
2006	72	1.45	217	4.36
2007	130	2.61	347	6.97
2008	103	2.07	450	9.04
2009	152	3.05	602	12.09
2010	242	4.86	844	16.95
2011	308	6.19	1152	23.14
2012	478	9.60	1630	32.74
2013	558	11.21	2188	43.95
2014	799	16.05	2987	60.00
2015	822	16.51	3809	76.52
2016	1169	23.48	4978	100
Total	4978	99.99		

The first available record belongs to the year 1989. During the period 1991 to 1997 there are no records. The cumulative total of records is less than one percent till 2002. In the year 2013 the growth trend became double digit mark. It is also found that only about forty four percent of the total literature is published till 2013 whereas rest (56%) of the literature during the last three years of study i.e., between 2014 and 2016. About quarter of the total literature is published during the period 2016. This growth is remarkable when compared to the growth scientific literature other domains of knowledge. The possible reasons for this may be the availability of personal computing facilities to individuals, growth in internet connectivity, and the increasing use of digitized form of information transfer.

The year-wise production of computer forensic research literature is represented as a bar diagram in Figure 1 to enhance visualization. This graph reveals that negligible amount of literature till the year 2005 and small quantity between 2005 and 2010. After 2010 there is a sudden growth and in 2016 there is a surge.

The cumulative growth of literature is plotted as a graph in Figure 2. The trend-line drawn on the graph fits exactly the exponential growth pattern as visualized. The exponential trend-line showed the value $Y=0.2608e^{0.3527x}$.

Figure 1. Year-wise growth of literature

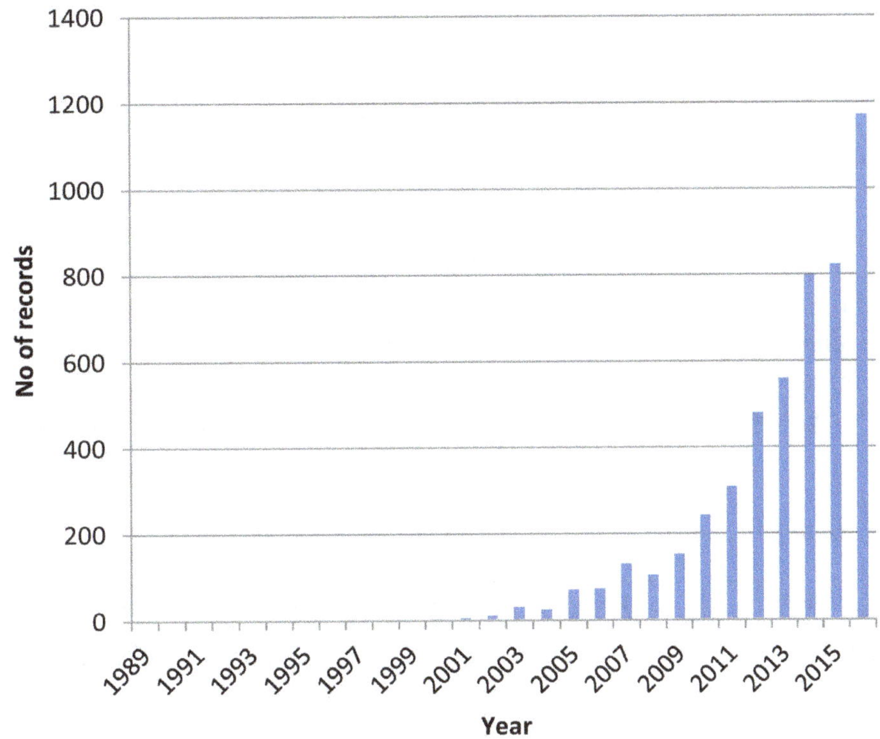

Figure 2. Cumulative growth of literature

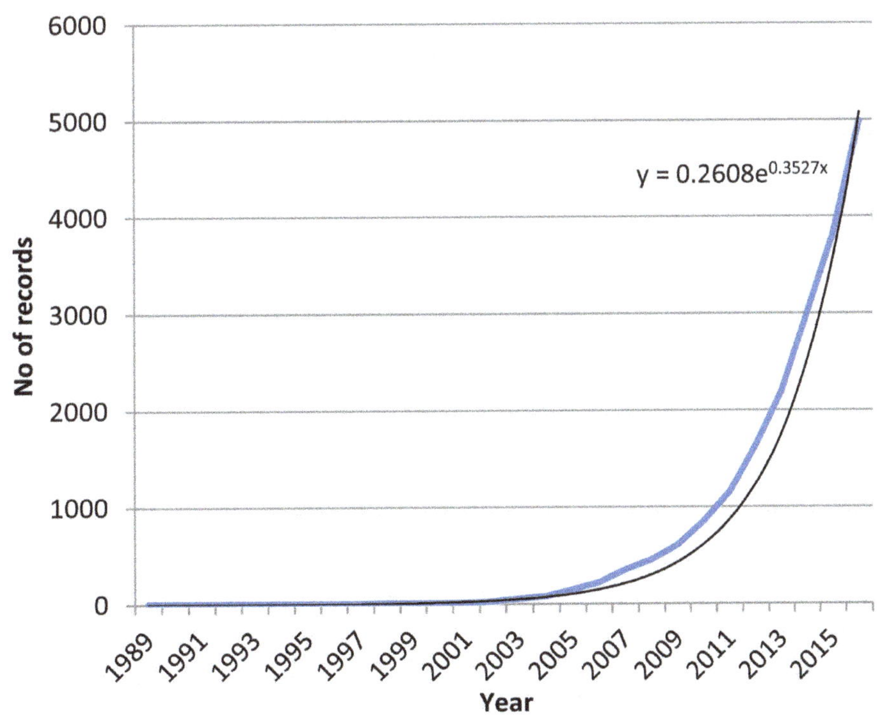

2. Country-Wise Contribution

The countries contributing more than one percent of the total literature produced are listed in Table 2, according to the number of documents produced. United States has contributed 1443 which is about twenty nine percent of the total and ranks first. United Kingdom comes second with about ten percent contributions. Interestingly, the fast growing economic and technological super-powers, China and India too have contributed about ten and nine percent each. They are ranked at third and fourth places respectively. Totally twenty three countries have contributed more than one percent of the output and their contributions together constitute about one hundred and three percent.

The countries contributing more than ten papers were visualized using the computer algorithm VOS-viewer and the resultant image in presented in Figure 3. A total of twelve clusters, which are characterized by different colors are observed in the image. The top contributing twenty-three countries' co-authorship occurrences are also listed in Table 2 and they are also ranked based on the co-authorship occurrences.

Table 2. Top contributing countries

Sl. no.	Country	Documents	Percent	Rank	Co-Authorship	Rank
1	United States	1443	28.99	1	465	1
2	United Kingdom	522	10.49	2	235	3
3	China	484	9.72	3	308	2
4	India	446	8.96	4	73	10
5	Australia	337	6.77	5	222	4
6	Malaysia	193	3.88	6	105	6
7	Canada	188	3.78	7	113	5
8	South Korea	182	3.66	8	73	11
9	Germany	175	3.52	9	88	7
10	Undefined	153	3.07	10		
11	Italy	143	2.87	11	79	8
12	Taiwan	129	2.59	12	46	14
13	South Africa	111	2.23	13	20	22
14	Japan	90	1.81	14	46	15
15	France	85	1.71	15	61	13
16	Spain	65	1.31	16	35	18
17	Saudi Arabia	62	1.25	17	75	9
18	Ireland	61	1.23	18	34	19
19	Greece	59	1.19	19	32	20
20	Netherlands	56	1.12	20	29	21
21	Singapore	55	1.10	21	64	12
22	Sweden	54	1.08	22	36	17
23	United Arab Emirates	53	1.06	23	38	16
	Total	5146	103.37		2277	

Figure 3. Scientograph of countries contributing more than ten papers

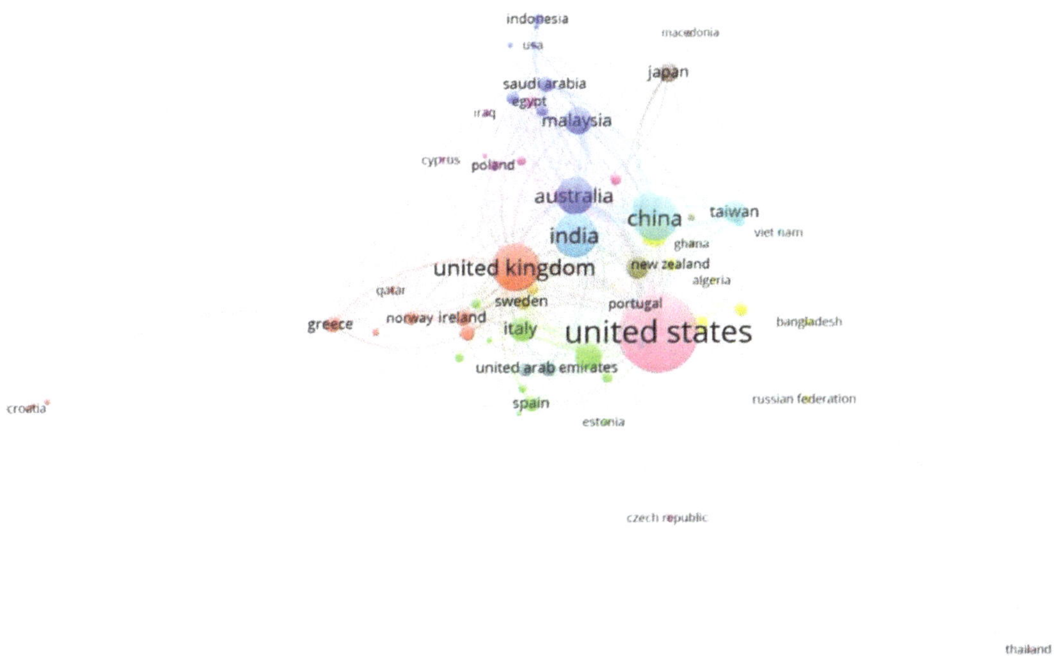

The United States is the number one country this too. However, the United Kingdom has come number three and China number two. India though ranked number four in terms of total output is ranked number ten in terms of co-authorship, which means India's international collaborative research is dismal. A cursory glance of the top contributing countries reveals a grand mix of super-powers, developing countries, big countries, small countries, highly populous countries, and also less populous countries. Hence it is inferred that computer forensic research work is going on all around the world though the quantum may vary. It is also because of the large scale developments in ICT taking place all over the world.

The cluster analysis visualized in the figure shows twelve clusters, which are as follows. Cluster # 1 consists of United Kingdom, Ireland, Greece, Netherlands, Norway, Slovenia, Croatia, Qatar, Bosnia and Herzegovina, and Serbia. Cluster # 2 consists of Germany, Italy, Spain, Brazil, Romania, Switzerland, Belgium, Estonia, Mexico and Sri Lanka. Cluster # 3 consists of Australia, Malaysia, Saudi Arabia, Pakistan, Iran, Indonesia, USA, Oman and Morocco. Canada, France New Zealand, Tunisia, Russian Federation, Algeria, Bangladesh and Lebanon constitute cluster # 4. Singapore, Poland, Egypt, Israel, Iraq, Cyprus, and Ukraine form cluster # 5. China, South Korea, Taiwan, Hong Kong, and Vietnam belong to cluster # 6. Countries belonging to cluster # 7 are India, Portugal, Finland, and Luxembourg. Three countries, namely, Sweden, Austria and Denmark constitute cluster # 8. South Africa, Nigeria, and Ghana are part of cluster number 9. The Czech Republic and Thailand belong to cluster number 10 whereas Japan and Macedonia are part of cluster # 11. Cluster # 12 consists of United Arab Emirates, and Turkey.

3. Top Contributing Institutions

The institutions that have contributed a minimum of twenty documents are listed according to their total productivity and also ranked accordingly and are presented in Table 3. This ranked list reveals that a total of 30 institutions have contributed twenty or more documents.

Among this ranked list, University of South Australia tops with seventy-nine documents followed by the second ranked University of Texas at San Antonio, which has published sixty-three documents during

Table 3. Top contributing countries

Sl no	Institution	No of Records	Percentage	Rank
1	University of South Australia	79	1.59	1
2	University of Texas at San Antonio	63	1.27	2
3	Purdue University	47	0.94	3
4	Universiteit van Pretoria	47	0.94	3
5	Universite Concordia	44	0.88	5
6	Universiti Putra Malaysia	44	0.88	5
7	University College Dublin	42	0.84	7
8	Edith Cowan University, Joondalup	41	0.82	8
9	Electronics and Telecommunications Research Institute	40	0.80	9
10	Deakin University	33	0.66	10
11	Zayed University	33	0.66	10
12	Chinese Academy of Sciences	32	0.64	12
13	University of Alabama	32	0.64	12
14	University of Salford	29	0.58	14
15	King Saud University	27	0.54	15
16	Korea University	27	0.54	15
17	Central Police University Taiwan	25	0.50	17
18	Rochester Institute of Technology	25	0.50	17
19	University of New South Wales UNSW Australia	25	0.50	17
20	George Mason University	24	0.48	20
21	Xidian University	24	0.48	20
22	University of Arizona	23	0.46	22
23	Chongqing University	22	0.44	23
24	University of Malaya	22	0.44	23
25	University of Melbourne	22	0.44	23
26	Liverpool John Moores University	21	0.42	26
27	University at Buffalo, State University of New York	21	0.42	26
28	University of Alabama at Birmingham	21	0.42	26
29	Beijing University of Posts and Telecommunications	20	0.40	29
30	Centre for Development of Advanced Computing India	20	0.40	29
	Total	975	19.52	

the study period. Rank number three is shared by the Purdue University and Universiteit van Pretoria, each with forty-seven documents to their credit. However, only the top ranking two institutions alone have contributed more than one percent of the total documents and all the thirty institutions together have contributed roughly twenty percent of the total. Hence it can be assumed there is more number of institutions engaged in computer forensics research but their contributions are negligible.

4. Author-Wise Contribution

Authors who have contributed more than ten documents are ranked according to their contribution and are listed in Table 4, along with the number of their contributions.Further, their co-authorships found-out from the network analysis and also the co-authorship rank has been listed in the table.

The authorship analysis revealed that there are thirty-four authors who have contributed a minimum of ten documents. Among them Choo, KKR who has contributed sixty-three documents ranks number one. Debbabi, M and Venter, HS follow him with thirty-five and thirty-four documents respectively. The total documents contributed by these thirty-four authors together are just five hundred and thirty, which is about eleven percent of the total documents published.

Further, it is found from the cluster analysis that Choo, KKR has thirty-five co-authorships and hence ranks first with respect to co-authorship too. However, the second ranked M Debbabi is found to be ranked as number seven with seventeen co-authorships and third ranked Venter, HS has no co-authorship relationship. The author named He, H who is ranked nineteenth with just eleven documents has twenty-seven co-authorship network and hence ranked second. All the above facts lead us to infer that number of researchers in this knowledge domain is more but the literature contributed by them is very less.

The co-authorship network map as visualized in VOSviewer is illustrated in Figure 4.

Cluster map analysis done using VOSviewer with the threshold value of ten revealed that sixty three authors have authorship connections. Cluster number one consists of fifteen authors. They are Zhang X, Liu J, Zhang J, Li X, Xiao Y, Liu Y, Ma J, Li Z, Wang X, Li Q, Yang Y, Liao X, Li H, Li B, and Chen Y. Cluster number two consists of eight authors, namely, Wang Y, Liu X, Yu S, Xiang Y, Liu S, Xu S, Xiao D, and Bertino E. Cluster number three too has eight authors, namely, Zhang Y, Wang W, Liu H, Chen H, Yu W, Zhang H, Li W, and Chen Z. Six authors namely, Dehghantanha A, Mahmod R, Conti M, Udzir NI, Li F, and Abdullah MT constitute cluster number four. Baggili I, Breitinger F, Marrington A, Kechadi T, and Busch C belong to author cluster number five. Chu HC, Wang SJ, Park JH, Lee S, and Kim J together constitute cluster number six. Cluster number seven consists of Debbabi M, Wang L, Iqbal F, and Bou-Harb E. Cluster number eight has four members consisting of Sun Y, He H, Zhu Y, and Yan J while ChooKKR, Slay J, Martini B, and Glisson WB belong to cluster number nine. James JI, and Gladyshev P are cluster number ten authors whereas Liu L, and Hong W are cluster number eleven authors.

5. Source Journal-Wise Contribution

The source documents / journals that have contributed more than twenty documents are ranked as per the number of records and presented in Table 5.

Table 4. Most Prolific Authors

Sl. no.	Author	Documents	Percent	Rank	Co-Authorship	Rank
1	Choo, K.K.R.	63	1.27	1	35	1
2	Debbabi, M.	35	0.70	2	17	7
3	Venter, H.S.	34	0.68	3		
4	Dehghantanha, A.	30	0.60	4	22	5
5	Baggili, I.	25	0.50	5	26	3
6	Breitinger, F.	21	0.42	6	16	8
7	Mahmod, R.	16	0.32	7	13	9
8	Yang, S.J.	15	0.30	8		
9	Chu, H.C.	14	0.28	9	8	17
10	Jahankhani, H.	14	0.28	9		
11	Marrington, A.	14	0.28	9	18	6
12	Alazab, M.	13	0.26	12		
13	Wang, S.J.	13	0.26	12	3	20
14	Al-Nemrat, A.	12	0.24	14		
15	Gladyshev, P.	12	0.24	14	11	10
16	Kechadi, T.	12	0.24	14	1	22
17	Rogers, M.	12	0.24	14		
18	Xiao, Y.	12	0.24	14	9	13
19	Blasch, E.	11	0.22	19		
20	Glisson, W.B.	11	0.22	19	4	19
21	He, H.	11	0.22	19	27	2
22	Bertino, E.	10	0.20	22	3	21
23	Bou-Harb, E.	10	0.20	22	8	18
24	Chen, H.	10	0.20	22	9	14
25	Creese, S.	10	0.20	22		
26	Dija, S.	10	0.20	22		
27	Geethakumari, G.	10	0.20	22		
28	Gritzalis, S.	10	0.20	22		
29	James, J.I.	10	0.20	22	10	12
30	Martini, B.	10	0.20	22	11	11
31	Slay, J.	10	0.20	22	1	23
32	Udzir, N.I.	10	0.20	22	9	15
33	Yan, J.	10	0.20	22	23	4
34	Yu, W.	10	0.20	22	9	16
	Total	530	10.65			

Figure 4. Scientograph of co-authorship cluster

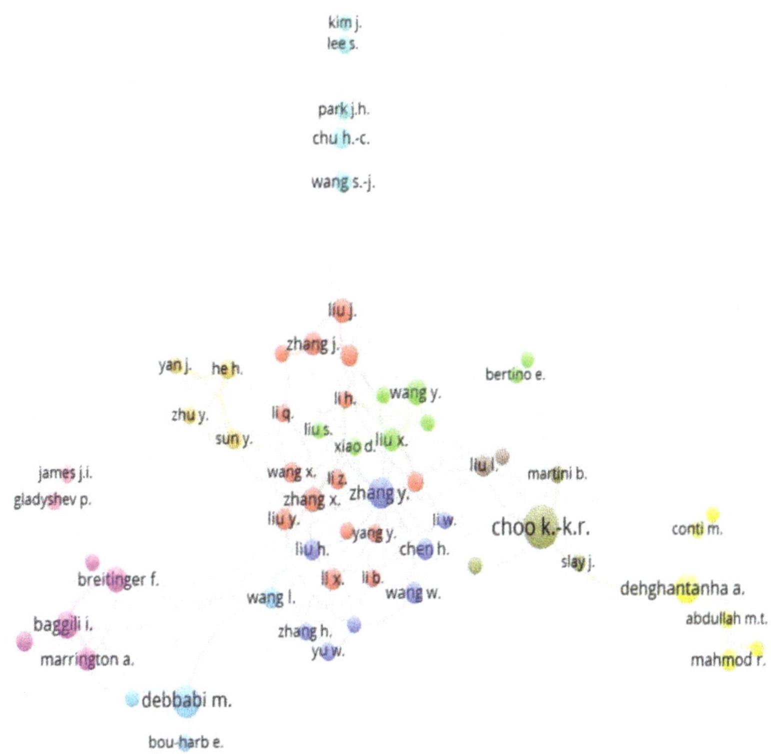

Table 5. Top Source Journals

Sl No	Source Title	Records	Percent	Rank
1	Lecture Notes In Computer Science Including Subseries Lecture Notes In Artificial Intelligence & Lecture Notes In Bioinformatics	197	3.96	1
2	Digital Investigation	118	2.37	2
3	IEEE Transactions On Information Forensics & Security	78	1.57	3
4	IFIP Advances In Information & Communication Technology	64	1.29	4
5	Communications In Computer & Information Science	54	1.08	5
6	Security & Communication Networks	39	0.78	6
7	Computers & Security	36	0.72	7
8	Advances In Intelligent Systems & Computing	35	0.70	8
9	Procedia Computer Science	30	0.60	9
10	Proceedings Of SPIE The International Society For Optical Engineering	29	0.58	10
11	Lecture Notes Of The Institute For Computer Sciences Social Informatics & Telecommunications Engineering	27	0.54	11
12	International Journal Of Electronic Security & Digital Forensics	26	0.52	12
13	Lecture Notes In Electrical Engineering	26	0.52	12
14	Lecture Notes Of The Institute For Computer Sciences Social Informatics & Telecommunications Engineering Lnicst	26	0.52	12
15	International Journal Of Digital Crime & Forensics	20	0.40	15
16	Studies In Computational Intelligence	20	0.40	15
	Total	825	16.57	

It is found that a total of sixteen journals have contributed twenty or more documents. Their contributions put together is eight hundred and twenty five documents, which is about seventeen percent of the total. The top contributor is “Lecture Notes In Computer Science Including Subseries Lecture Notes In Artificial Intelligence & Lecture Notes In Bioinformatics” followed by “Digital Investigation” and “IEEE Transactions On Information Forensics & Security”.

6. Document Type

The document type is tabulated and presented in Table 6 and also presented as a pie diagram in Figure 5 for easy visualization.

Conference papers ranks first and journal articles come second unlike other subjects where journal articles dominate as research output. This may be either due to more conference being held regarding information security / data information or non-availability journals exclusively for computer forensics except “Digital Forensics”.

7. Subject-Wise Contribution

The subject-wise contribution records obtained from SCOPUS is given in Table 7. This table reveals that computer science ranks first followed by the distant engineering. Social sciences take the third place.

8. Scientograph of Keywords

The keywords given by the authors or the journals provide us a window to peep through and understand where the research area is proceeding and also anticipate its growth. The keywords available in the SCOPUS database are analyzed and a keyword map as visualized in VOSViewer is given in Figure 6.

Table 6. Document type

Sl No	Document Type	No Of Records	Rank
1	Conference Paper	2490	1
2	Article	1721	2
3	Book Chapter	329	3
4	Book	194	4
5	Review	166	5
6	Conference Review	41	6
7	Editorial	13	7
8	Article in Press	12	8
9	Note	4	9
10	Letter	4	10
11	Short Survey	3	11
12	Erratum	1	12
	Grand Total	4978	

Figure 5. Document type

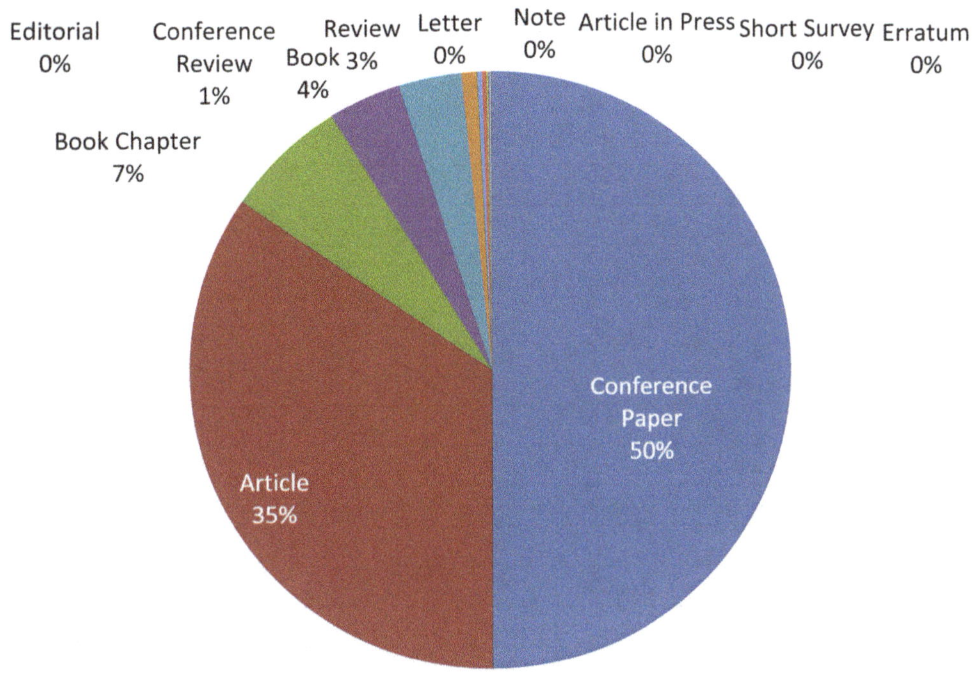

Cluster number one consists of fifty-nine terms including cloud computing, security, data privacy, digital storage, mobile security, distributed computer systems, surveys, access control, complex networks, virtual reality, big data, embedded systems, smart power grids, ubiquitous computing, and mobile telecommunication systems. Cluster number two has forty-eight words which include security of data, cyber security, security systems, information systems, cyber-attacks, information management, risk assessment, information science, semantics, data acquisition, critical infrastructures, insider threat, decision making and scada systems. Terms such as computer crime, malware, intrusion detection, social networking (online), data mining, personal computing, intrusion detection systems, network forensics, internet protocols, visualization, denial-of-service attack, computer networks, world wide web, behavioral research, malwares, mercury (metal), computer viruses, websites, electronic mail, malicious activities, computer system firewalls, phising, botnets, etc. totaling forty-eight in number are found in cluster number three. Fifty terms are found in cluster number four. Some of them are digital forensic, computer forensics, digital evidence, cyber-crimes, mobile devices, digital devices, smartphones, cyber forensics, android (operating system), telephone sets, forensic tools, file systems, mobile forensics, cyber criminals, forensic engineering, user interfaces, data structures, electronic commerce, metadata, and information retrieval. Thirty-four terms including cryptography, algorithms, artificial intelligence, biometrics, steganography, image processing, software engineering, pattern recognition, face recognition, pixels, steganalysis, digital watermarking, information hiding, and watermarking are found in cluster number five. Cluster number six has a total of twelve words including internet, human, male, female, adult, adolescent, statistics, and child.

Table 7. Subject-wise Contribution

Sl. no.	Subject area	No of records	Percent
1	Computer Science	3667	73.66
2	Engineering	1514	30.41
3	Social Sciences	921	18.50
4	Mathematics	490	9.84
5	Decision Sciences	338	6.79
6	Medicine	241	4.84
7	Psychology	188	3.78
8	Business, Management and Accounting	152	3.05
9	Health Professions	121	2.43
10	Physics and Astronomy	94	1.89
11	Arts and Humanities	82	1.65
12	Biochemistry, Genetics and Molecular Biology	80	1.61
13	Materials Science	65	1.31
14	Economics, Econometrics and Finance	52	1.04
15	Energy	44	0.88
16	Chemistry	34	0.68
17	Multidisciplinary	26	0.52
18	Environmental Science	24	0.48
19	Agricultural and Biological Sciences	17	0.34
20	Neuroscience	15	0.30
21	Nursing	13	0.26
22	Chemical Engineering	10	0.20
23	Pharmacology, Toxicology and Pharmaceutics	7	0.14
24	Dentistry	4	0.08
25	Earth and Planetary Sciences	3	0.06
26	Immunology and Microbiology	1	0.02

Some of the keywords with high frequency of occurrences are computer crime, security of data, network security, electronic crime countermeasures, crime, digital forensic, internet, cryptography, computer forensics, cloud computing, cyber security, malware, security, and intrusion detection. Keywords with high frequency of co-occurrences are computer crime, security of data, electronic crime countermeasures, network security, crime, digital forensic, internet, computer forensics, cloud computing, malware, intrusion detection, cyber security, security, digital evidence, cryptography, artificial intelligence, security systems, cyber-crimes, mobile devices, social networking (online), algorithms, data privacy, mobile security, digital storage, distributed computer systems, cybercrime, and information technology, etc.

Figure 6. Scientograph of Keywords

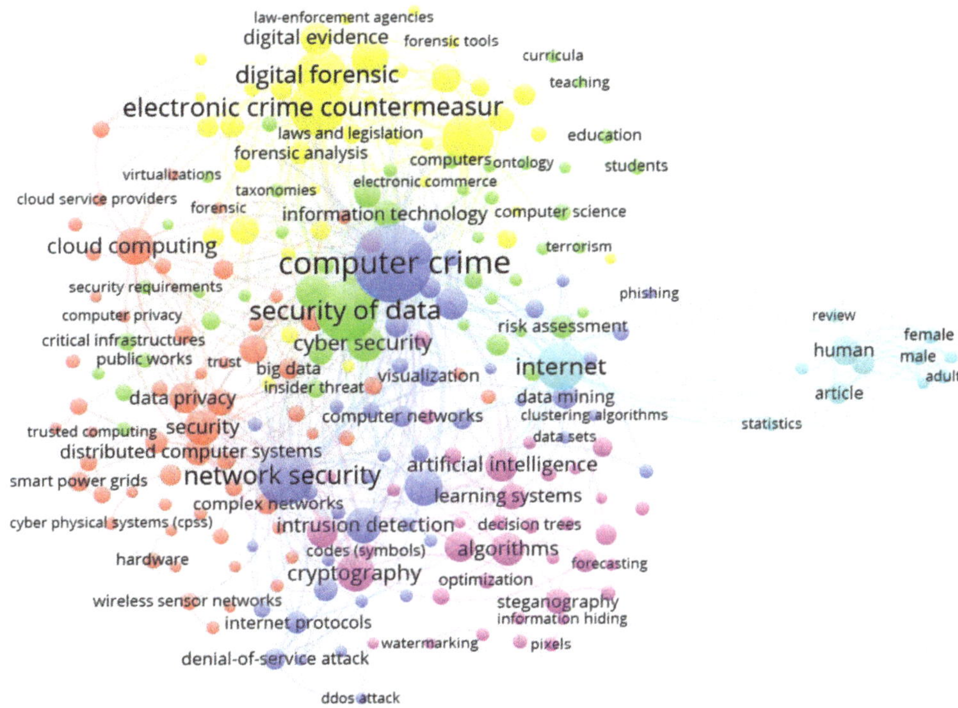

SUGGESTIONS FOR FURTHER RESEARCH

Further scope for research on computer forensic may be conducted in the following areas

1. The present study is limited to visualization and hence further full-fledged study may be conducted on computer forensics.
2. Relationship as well as difference between computer forensics and other areas like information forensic, network forensic, cyber forensic and digital forensic can be examined.
3. Comparative study may be conducted using the databases like Scopus, Web of Science and IEEE databases.

CONCLUSION

The study reveals that there is exponential growth of computer forensics literature and after the 2010 the growth is remarkable. The top contributing countries include developed nations as well as developing nations and also very small countries leading us to conclude that computer forensics research is wide spread geographically. The analysis of leading document forms shows that conference proceedings as the most preferred form of communication implying more conferences are happening in this field rather than dedicated journals. Publishing of journals in this area too can contribute to the growth of scientific communication. The keywords with high frequency of occurrences are:security of data, network security, electronic crime countermeasures, digital forensic, internet, cryptography, cloud computing, cyber security, malware, and intrusion detection. It may be presumed that these are the hot areas of research in computer forensics.

REFERENCES

Card, S., Mackinlay, J., & Shneiderman, B. (1999). *Readings in Information Visualization: Using Vision to Think*. San Francisco: Morgan Kaufmann.

Chen, C., Paul, R., & O'Keefe, B. (2001). Fitting the jigsaw of citation: Information visualization in domain analysis. *Journal of the American Society for Information Science and Technology*, *52*(4), 315–330. doi:10.1002/1532-2890(2000)9999:9999<::AID-ASI1074>3.0.CO;2-2

Garfield (1994). Scientography: Mapping the tracks of science. *Current Contents: Social and Behavioural Sciences, 7*(45), 5-10.

Heilig, L., &VoB, S. (2014). A scientometric analysis of cloud computing literature. *IEEE Transactions on Cloud Computing, 2*(3), 266-278.

Jeyasekar, J. J., & Saravanan, P. (2014). A scientometric portrait of the journal Digital Investigation. *Journal of Advances in Library and Information Science*, *3*(2), 155–162.

Kumar, S., & Garg, K. C. (2005). Scientometrics of computer science research in India and China. *Scientometrics*, *64*(2), 121–132. doi:10.1007/s11192-005-0244-9

Leydesdorff, L. (2001). *The Challenge of Scientometrics: The Development, Measurement, and Self-Organization of Scientific Communications*. Parkland: Universal-Publishers.

Niazi, M., & Hussain, A. (2011). Agent-based computing from multi-agent systems to agent-based models: A visual survey. *Scientometrics*, *89*(2), 479–486. doi:10.1007/s11192-011-0468-9

Chapter 11
The Scientific Communication Process in Homoeopathic Toxicology:
An Evaluative Study

G. Ramadhas
Noorul Islam Centre for Higher Education, India

A.S. Suman Sankar
Sarada Krishna Homoeopathic Medical College, India

N.V. Sugathan
Sarada Krishna Homoeopathic Medical College, India

ABSTRACT

The present chapter evaluates the growth of literature of Toxicology in Homoeopathy. The Scopus database is used as source for the study. Data pertaining to toxicology in homoeopathy for the period 1963 to 2017 is retrieved and analyzed. The study reveals that toxicology literature grows rapidly in the later stages than in the earlier period. The United States of America is the leading contributors at the global level and India is in second position. Among the contributors, E.J. Calabrese, Professor of Toxicology, University of Massachusetts is the most prolific author for Toxicology in Homoeopathy literature. University of Kalyani and University of Massachusetts are the leading institutions, majority of the records are published in the journal 'Homoeopathy', which is published from the USA and major contributions are in the subject domain 'medicine' (65.4 per cent). Most frequently repeated keywords in toxicology of homoeopathy literature are: homoeopathy, humans, non-human, unclassified drug and animals.

DOI: 10.4018/978-1-5225-3457-0.ch011

INTRODUCTION

The concept of homoeopathy was first introduced Dr. S. Samuel Hahnemann (1755-1843), a German based physician and a medical degree holder in Allopathy. He has produced all the symptoms of intermittent fever (malaria) but to a mild degree and without the characteristic rigors of that disease while studying the regular doses of *cinchona.* This led *Hahnemann* to an idea which was published in 1796 as Essay on a *New Principle for Ascertaining the Curative Power of Drugs*, which was followed in 1810 by his famous work *The Organon of the Healing Art* (Gevitz, 1993). Hahnemann believed that if a patient had an illness, it could be cured by giving a medicine which, if given to a healthy person, would produce similar symptoms of that same illness but to a slighter degree. The principles followed for proving is '*Similia similibus curentur*', the Latin phrase meaning "let likes be cured by likes," is the primary principle of homeopathy, called 'principle of similars'. To achieve this, Hahnemann diluted his medical preparations to such an astonishing extent that if one assumes that that the substance he employed was completely soluble, by only the fourth dilution the ratio of the medicine to the solution would be 1:100 000 000. Homeopathic drugs are prepared by a process called homeopathic dilution in which the substance is diluted repeatedly in alcohol or distilled water.

Followers of Hahnemann founded the first homeopathic medical school in the United States in the late 1800's is American Institute of Homeopathy (Loudon, 2006). It gained recognition because of its success in treating the many disease epidemics rampant at the time - including scarlet fever, typhoid, cholera and yellow fever. Homeopathy became spectacularly popular in the United States and Europe in the 1800s and its strongest advocates included European royalty, American entrepreneurs, literary giants, and religious leaders. But at the time that it was gaining widespread popularity, it became the object of deep-seated animosity and vigilant opposition from establishment medicine.

The differences between orthodox medicine and homeopathy could hardly be more vivid. Homeopathic treatment always began with a long consultation, lasting at least an hour, in which all aspects of the patient's illness and life were discussed and the appropriate treatment chosen. In contrast, orthodox medicine was immersed in the belief that advances in understanding disease could only come from a detailed correlation of symptoms and signs of the sick patient on the ward, and the findings at autopsy: clinico-pathological correlation.

Hahnemann showed no interest in detailed pathology, and none in conventional diagnosis and treatment. He was only interested in the principles of homeopathic medicine which he used to name the illness. Classical homeopathy was therefore seen by its supporters as an attractively safe system, simple, easy to understand, and centred on the patient as a whole and not on pathological lesions. The fundamental principle of homoeopathy is based on the capability to ascertain the disease producing power of drugs where it proves to be better antidote for diseases with similar symptoms. The homeopathic pharmacodynamics is acquiring knowledge of drugs is through homoeopathic drug proving, laboratory investigations and experiments through carefully conducted experiments, based on the concept that the poisonous substances when used in minute quantity produce final symptoms.

TOXICOLOGY IN HOMOEOPATHY

According to 'Macmillan Encyclopaedia', defines

Toxicology is the study of poisons, which deals with their chemical nature and origin, the pathological changes they induce, their recognition in a poisoned patient, and the determination of specific antidotes.

Encyclopaedia of Scientific Terms edited by E.C. Graham, defines toxicology as "...science of poisons, embracing the physical and chemical history of all the known poisonous substances, as well as the methods of testing them, their action on the living body, and the post mortem results they occasion..." Thus, Toxicology is the area of Science which deals with origin, properties of poisons and their reactions on cells and tissues.

Toxicology is the science of the poisons, narcotics or toxic substance. Toxic effect of drugs classified as pharmacological, pathological or genotoxic and their incidence and seriousness are related, at least over some range, to the concentration of the toxic chemical in the body. The information collected through proving, toxicological data and experimental data constitute the complete and true *materia medica*. The sum total effect of the drug is *pharmacodynamics*.

The toxicological literature of the substances which could be used as medicine in homoeopathy helps in understanding the basic information necessary for collective curative information of those drugs. The toxicological information available in literature other than homoeopathy lacks fullness and minuteness as to know a poison is reactive or irritant is vague. As also the knowledge of morbid anatomy helps us identify the organs as tissues the poison influences, the classification of the poisonous effect into cathartics, dilutents as expectorants help us further which receive later attempt.

The information toxicological/ Forensic literature is varied in homoeopathy from pharmacodynamic point of view but as poisoning. Homoeopathy requires to know poisonous active for the purpose of urging there as curative agent. But where it is applied in forensic used it relates to diagnostics, antidotes from legal point of view.

RATIONALE OF THE STUDY

Evaluation of the growth of literature is essential to ascertain the latest developments in the field as well its relations with other disciplines. Statistical techniques are applied to interpret the growth of literature in terms of year wise growth trends, the extent of collaboration at institutional level and national level, trends of interdisciplinary research and so on. These are calculated using quantitative data. These types of analysis are generally scientometric analysis. Scientometric is the study of quantitative aspects of science as a discipline or economic activity (Tague-Sutcliffe, 1992). Several attempts are made to assess the scientific growth of literature which is more significant to medicine by the information scientist to alert the health professionals and the authorities concerned at various level. The qualitative evaluation of the literature practically poses problem and hence majority of the studies are based on quantitative data downloaded from the database or the print version of sources. Homoeopathy as a Complementary and Alternative system of Medicine, is widely used at the global level to cure all sort of diseases. Toxicology is the science of poisons, narcotics or toxic substance, it is commonly used both in allopathic and homoeopathic practice. The information collected through proving, toxicological data and experimental data constitute the complete and true *materia medica*, which is a major subject in homoeopathy. Thus, the toxicological information plays key role in homeopathy, which is the significant of the study. Several attempts have been made earlier to study the scientometrics aspects of toxicological literature in different field. Some of the representative studies are: bibliometric study of core journals in toxicology (Devi,

2013), laser research in India (Ashok & Gang, 1992); bibliometrical analysis of global literature on cervical cancer (Jeyshankar & Vellaichamy, 2015); scientometric analysis of autism research output by Jeyshankar, & Vellaichamy in 2016. A few bibliometric study namely on toxicology research productivity by Zyoud,et.al.(2014) in Middle Eastern Arab countries and on toxicology publications of Iran and Turkey (Delirrad; Reshidi & Karimi, 2012). Similarly, a single bibilometric study on homoeopathy research during the period 1991 to 2003(Chiu& Ho, 2005). But a closer analysis of available studies shows that no systematic study has been conducted on growth of literature of toxicology in homoeopathy. Hence, the present chapter is intended to evaluate the growth of literature of toxicology in homoeopathy and it is titled as "An Evaluation of Growth of Literature of Toxicology in Homoeopathy".

SCOPE OF THE STUDY

The present study is intended to evaluate the growth of literature of toxicology in homeopathy. It is based on the data retrieved from the Scopus database only, and the period of coverage is limited to 1963 to 2017. The data pertains to toxicology in homoeopathy alone are downloaded from the database using the search term 'toxicology' and 'homoeopathy' are some limitations of the study. The total number of downloaded records is 1197. Using of additional key terms related to toxicology and various specialties of homoeopathy may yield more number of records. Moreover, the present chapter analyses the growth of literature of toxicology in homoeopathy based only on quantitative indicators, the qualitative analysis may explore some valuable features of growth of literature.

REVIEW OF LITERATURE

Several attempts are made by the information scientist to measure the growth of scientific literature. The approach followed by them is quantitative measurement of data based on the bibliographical data retrieved from the databases namely *Scopus*, *Web of Science*, *Pub Med. Medline*, and so on.

Tabatabaei-Malazy; Ramezani ; Atlsai ; Larijani and Abdollahi (2016) analyzed research productivity in field of antioxidative hypoglycemic herbal medicine and diabetes. Data are retrieved using search terms "type 2 diabetes", "antioxidant", "herb", "phytotherapy", "ethnopharmacology", "Chinese medicine", "traditional medicine", from the Scopus web databases until January 2015, and limited to human. 468 studies were related to this topic in human. The analysis revealed that number of publications in the field showed an increasing trend. Majority of the published papers were original articles (71%) and the most productive year was 2013. Top subject areas were medicine followed by drug. The first productive country was the US. The documents were cited totally 10724 times with average citation/article 22.91, and h-index 55. The highest cited article was a systematic review study, and top source was "Journal of Ethnopharmacology". The highest international collaboration was with the US.

Zarafshan; Mohammadi; Motevalian; Abolhassani; Khaleghi & Sharifi (2016) conducted a scientometric study by investigating published papers of Iranian researchers on autistic disorders. The trends of research in the field of autism among Iranian researchers have the same patterns as the global trends. However, some important topics, such as prevalence, causes, services, and life span issues and the infrastructures have been studied less and should be considered in future studies. Moreover, it was revealed that we are faced with shortness of sources for research in this field, and the policy makers should consider this problem.

Meera and Sahu (2014) carried out a study to depict the research performance of University College of Medical Science (UCMS) in different areas or subfields of medical and health sciences. This study is based on a bibliometric analysis of scientific research output. The data of UCMS' research output is collected from the SCOPUS database by using different searching techniques. Some bibliometric indicators such as authorship pattern, degree of collaboration, author productivity, rank distribution etc. have been used to illustrate the research performance of researchers. The study revealed that the highest contribution of 25.6 percent of total publications is made by three authorship collaboration. The degree of collaboration is 0.92, which means most of the research works are collaborative works. USA is the most preferred country by the researchers for research collaboration. This study will assist in understanding the research pattern of UCMS in the field of medical science.

Devi (2013) identified Fifty-nine journals form core journals in the field of Toxicology with a total citation of 6922. The journal 'Toxicology' was found to be ranked high. U.S. was found to be the most productive country with 47.45 per cent journals. And the period 1970's and 1980's witnessed the origin of more number of journals in Toxicology. Vellaichamy and Jeyshankar (2016) revealed in their scientometric analysis of 20 years literature on Ebola virus and demonstrated the level of research, contributions by various journals and contributing countries etc in this field.

In 2014, Zyoud, et. al. analysed the research production originating from 13 middle Eastern Arab countries with article published in toxicology journals. The study shows some evidences of toxicology research activity in toxicology journals in the Arab world. Similarly, Delirrad; Rashadi & Karimi (2012) reveals that top Iranian toxicologists had a better performance than their Turkish counterparts but overall authorship pattern was well-distributed among Turkish researchers and organization.

However, the single bibiliometric study on homoeopathy research shows that the UK, the US, and Germany produced 71% of the total output, while European countries as a whole also contributed 65% of the total share of independent publications. More document types and languages, and fewer pages have appeared in homeopathy research. 3.5% of papers were cited more than 10 times in three years after publication, and 60% were never cited. Small-group collaboration was a popular method as co-authorship. The top 3 ranking countries of publication were the UK, the US, and Germany. The US dominated citation followed by the UK, and then Germany (Chiu & Ho, 2005). But no systematic study has been conducted on growth of literature of toxicology in homoeopathy.

METHODOLOGY

Data pertaining to toxicology in Homoeopathy literature is retrieved from *Scopus* database for a period from 1963 to 2017. Scopus database is preferred over the other databases due to wide content coverage. Scopus database is considered as one of the biggest abstracting and citation database. It provides access to over 50 million records from 21,000 titles, published by over 5000 publishers. For the purpose of the present study, a common search strategy is used to search data related to toxicology in Homoeopathy literature, a total of 1197 records are indexed in *Scopus* database, and when searched in *Pubmed* the retrieved data is fewer than data retrieved from *Scopus* database. Moreover, several quantitative studies on toxicology, forensic science and similar studies show that most of the studies on growth of literature are based on *Scopus* data. Thus, *Scopus* database is used as a source for the present study. The search term 'toxicology' and 'homoeopathy' are used to search from the database. The keywords appeared in title of the article, abstract, and keywords are used to retrieve data. The period is limited to from 1963 to 2017. The downloaded data are analyzed using *SCOPUS ANALYTICS. MS-Excel, VoS viewer* and *Pajek software* are used for network analysis.

ANALYSIS AND DISCUSSION

The present study is intends to determine the growth of scientific literature of toxicology in homoeopathy Literature and the retrieved 1197 records are further analysed using appropriate tools and techniques and interpreted accordingly.

1. Year Wise Contribution

Records contributed in toxicology with reference to homoeopathy literature are analysed and the details are presented in Table 1.

Table 1 shows that a total of 1197 records are published during the study period. The decade wise contribution of records of toxicology in homoeopathy literature shows that highest number of articles is published in the period between 2011 and 2016 and lowest number is published during 1971-1980 (3 articles out of 1197). It also shows that only 120 records were published till 2000 and thereafter growth of records is sudden and rapid from 2001 onwards.

This may be due to outcome of research conducted in the area related to Toxicology in Homoeopathy in the specific period, moreover, awareness among them in dissemination of research findings in various scholarly platforms. The cumulative growth of record is presented in Figure 1.

2. Country-Wise Contribution

Country-wise contribution of records in toxicology related to homoeopathy literature is summarised in Table 2. The topmost ten countries based on their contributions in toxicology of homoeopathy literature are only listed in the table.

Table 1. Year-Wise Contribution

Year	No of Records	Cumulative Growth
1963 – 1970	11	11
1971 – 1980	3	14
1981 – 1990	22	26
1991 – 2000	94	120
2001 – 2010	502	622
2011 – 2016	565	1197
Total	1197	

Figure 1. Cumulative Growth of Toxicology in Homoeopathy

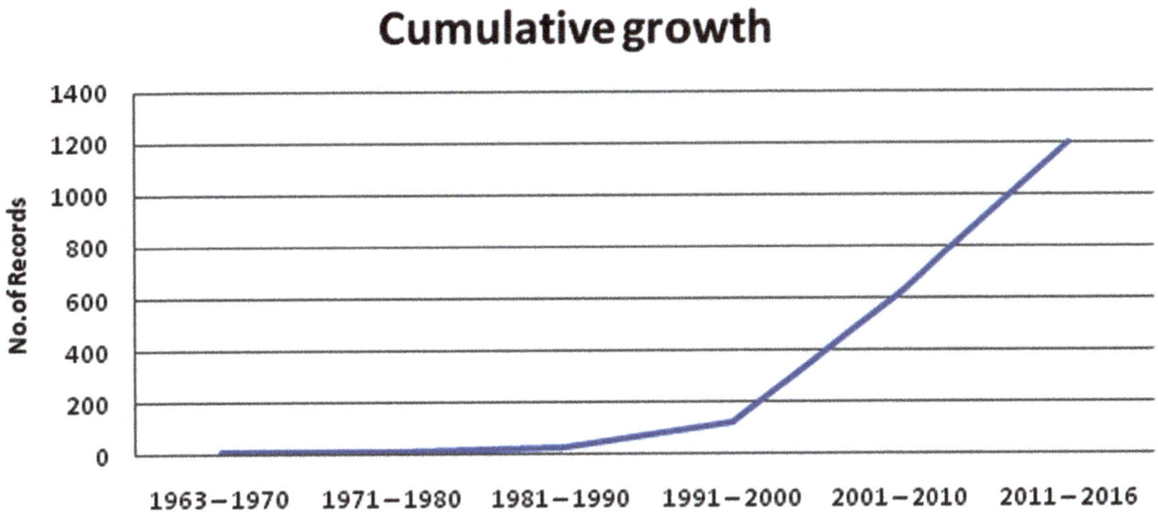

TABLE 2. Country-Wise Contribution

Rank	Country	Contribution	Per Cent
1	United States	297	24.81
2	India	202	16.88
3	United Kingdom	104	8.69
4	Germany	102	8.52
5	Italy	94	7.85
6	Brazil	68	5.68
7	France	50	4.18
8	China	37	3.09
9	Netherlands	35	2.92
10	Canada	32	2.67
10	Switzerland	32	2.67

Table-2 clearly depicts country wise ranking for toxicology in homoeopathy literature. Based on number of documents produced, major contribution for toxicology in homoeopathy is from the United States and its contribution is 24.81 per cent of total contributions. It is followed by India with 16.88 per cent. The other contributing countries are United Kingdom (8.69 per cent), Germany (8.52 per cent), Italy (7.85 per cent) Brazil (5.68 per cent), France (4.18 per cent) and China (3.09 per cent). The other countries like Netherlands, Canada and Switzerland have contributed less than three per cent of total contribution.

The first school of Homoeopathy was established in USA, the present study also reflected the same that major contribution is from the USA. However, the founder of Homoeopathy Dr. Hahnemann belongs to Germany, its contributions to homoeopathy literature are only at fourth rank. But, India is next to the USA with 16.88 per cent of contribution.

3. Collaborating Countries' Network

Research studies in toxicology related to homoeopathy are conducted in collaboration with researchers of other countries. Using *Pajek software*, collaborating countries network is created and presented in Figure 2. The collaborating countries network reveals that there are 39 vertices, which are interlinked with one or many vertices (countries). i.e. there are 39 countries which have in collaboration with other countries for 1197 records of toxicology in homoeopathy; moreover, straight line between the countries shows the collaboration and the indices on the interlinked line shows the quantum of records through collaboration.

4. Authors-Wise Contribution

The author wise contribution of toxicology in homoeopathy literature is further studied based on their contribution in order to determine the prolific authors in the field. The authors who have contributed more than 10 records for the homoeopathy literature are presented in Table 3.

Figure 2. Collaborating Countries' Network

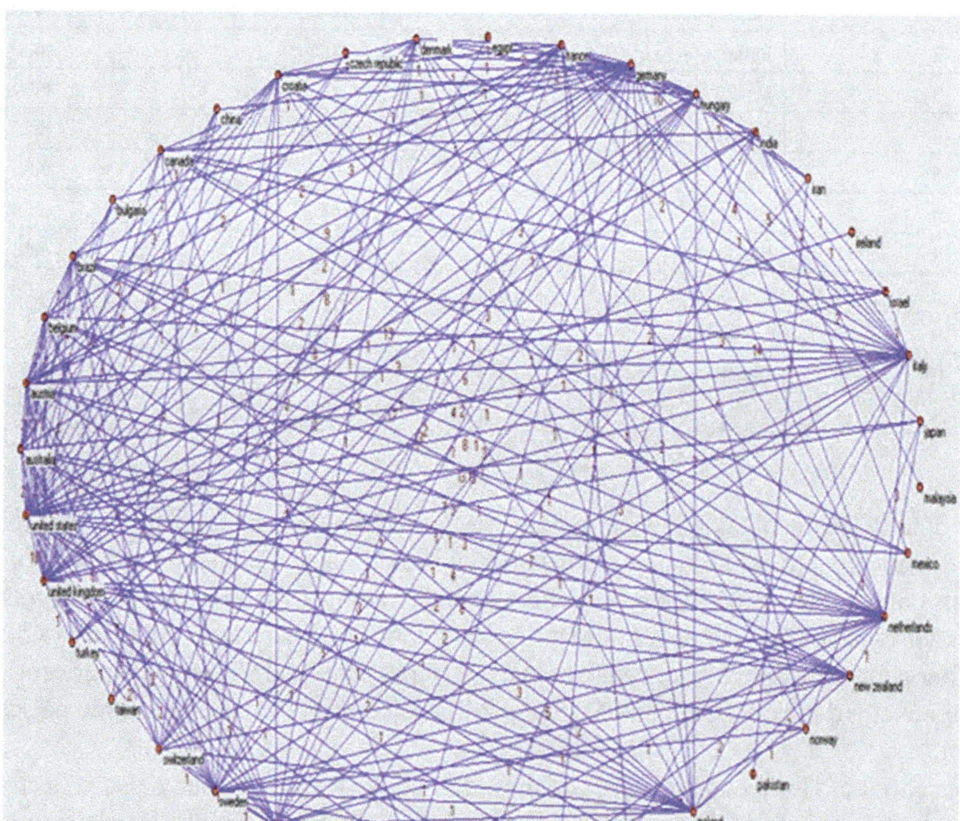

Table 3. Author-wise Contributions

Rank	Author	Contribution	%
1	Calabrese, E.J.	38	3.17
2	Khuda-Bukhsh, A.R.	37	3.09
3	Bellavite, P.	27	2.26
4	Bell, I.R.	22	1.88
5	Baumgartner, S.	18	1.5
6	Conforti, A.	17	1.42
6	Jonas, W.B.	17	1.42
8	Marzotto, M.	15	1.25
9	Endler, P.C.	14	1.17
10	Ernst, E.	13	1.09
11	Belon, P.	11	0.92
11	Biswas, S.J.	11	0.92
11	Chirumbolo, S.	11	0.92
14	Schwartz, G.E.	10	0.83

Table 3 clearly shows the author-wise contribution in toxicology of homoeopathy literature. The table shows that there are 14 authors who have contributed more than 10 records. It also shows that Calabrese, E. J., Professor of Toxicology, University of Massachusetts has contributed 38 records and secured first rank; it is 3.17 per cent of the total contribution. It is followed by Khuda-Bukshh, A.R, Professor of Zoology, University of Kalyani and Homoeopathic research secured second rank for 37 records and it is 3.09 per cent of the total contribution. Similarly, Bellivate, P., Department of Medicine, University of Italy, has contributed 27 records (2.26 per cent) and secured third rank. The contributions of others are: Bell, I.R. (1.88 per cent), Baumgartner, S (1.5 per cent), Conforti, A. (1.42 per cent) and so on.

Hence, it is concluded that Calabrese, E. J., Professor of Toxicology, University of Massachusetts is the most prolific author for toxicology in homoeopathy literature.

5. Co-Authorship Visualization

The co-authorship pattern existing among the authors in toxicology of homoeopathy literature is further studied using VOS Viewer. Co-authorship details of top co-authorship details are given in Table 4.

The co-authorship visualization is presented in Figure 3.

From the cluster analysis shows that Khuda-Bush, A.R. has 37 documents with 83 co-authorship and ranked first in co-authorship. Bellavite, P has secured second position in co-authorship for his 27 documents, which have 53 co-authorships. Similarly, third and fourth ranks are respectively by Marzotta, M for 15 documents which have 41 co-authorships and Pathak, S has 9 documents with 40 co-authorships. Rest of the authors have less than 40 co-authorships. Hence, it is interpreted that Khuda-Buhsh, A.R. is the topmost author in terms of co-authorships for toxicology in homoeopathy literature.

Table 4. Co-authorship details of Top Collaborators

label	cluster	weight<Documents>	weight<Co-authorships>
khuda-bukhsh a.r.	5	37	83
bellavite p.	1	27	53
marzotto m.	1	15	41
pathak s.	3	9	40
biswas s.j.	3	11	39
baumgartner s.	6	20	38
conforti a.	1	17	37
banerjee p.	3	8	36
boujedaini n.	5	8	32
banerjee a.	3	5	31

Figure 3. Co-authorship Visualization

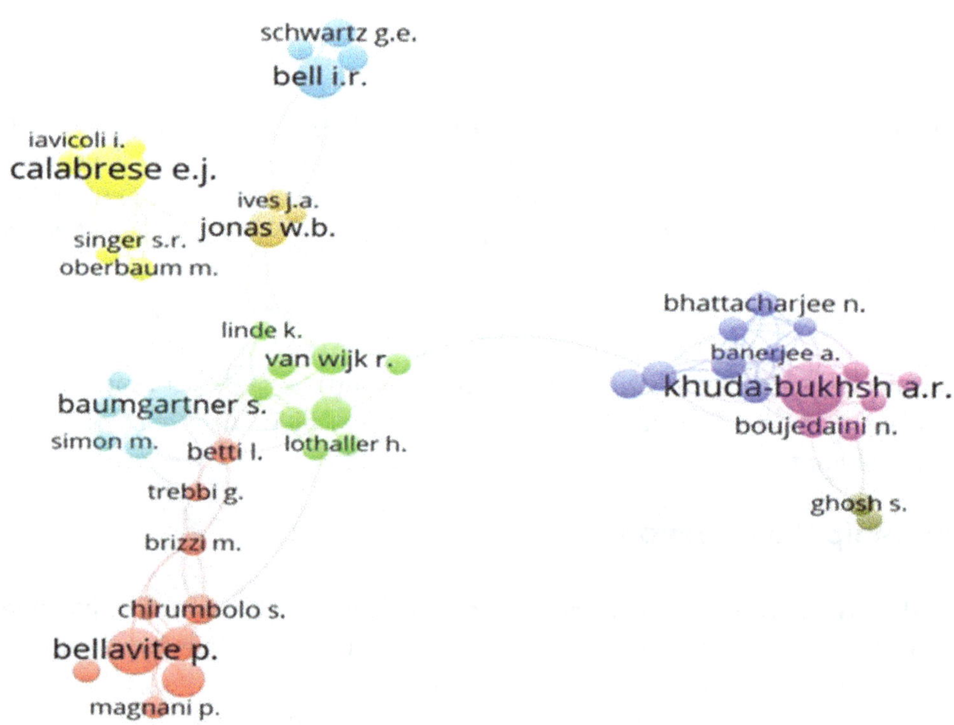

6. Institution Wise Contributions

Institution-wise contribution for toxicology in homoeopathy literature is analysed based on number of contributions made at the institutional level. The top ten most institutions based on their contributions are summarised in Table 5.

Table 5. Institution-wise Contribution

S.No	Affiliation	Contribution	%	Rank
1	University of Kalyani	40	3.34	1
2	University of Massachusetts	39	3.26	2
3	Universita degli Studi di Verona	37	3.09	3
4	Universitat Bern	22	1.84	4
5	University of Arizona	20	1.67	5
6	Universidade de Sao Paulo – USP	18	1.50	6
7	Samueli Institute	14	1.17	7
8	University of Arizona College of Medicine	13	1.09	8
9	Alma Mater Studiorum Universita di Bologna	13	1.09	8
10	Utrecht University	12	1.00	9

Table 5 informs that University of Kalyani, India is ranked first among the contributed institution for toxicology in homoeopathy literature. It has contributed 40 records out of total 1197 and its contribution is 3.34 per cent. It is followed by Univeristy of Massachusetts, USA by 3.26 per cent (39 out of 1197), Universita degli Studi di Verona by 3.09 per cent (37 out of 1197), Universitat Bern by 1.84 per cent (22 out of 1197), Univeristy of Arizona by 1.67 per cent (20 out of 1197) and Universidade de Sao Paulo – USP by 1.50 per cent. Other institutions have less than 1.5 per cent of total contributions.

Since, Kuda-Buash, A.R., Professor of Zoology, Research Centre of Homoeopathy, University of Kalyani has contributed more number of records and hence, it is interpreted that University of Kalyani, West Bengal (India) is the top most institutions, which have contributed 3.34 per cent of total contribution.

7. Institution-Wise Network

The institution wise network is created for further analysis of collaboration existing among the institution using *Pajek* software. The network is presented in Figure 4. The vertices represent the name of the institution, connections between the vertices is link reflects collaboration between the institution for published records and numerals on the links shows the strength of the collaborative works.

8. Document Type

The retrieved data consists of 1197 records of different type. Document-wise break up of data are presented in Table 6.

Table-6 clearly depicts that the retrieved documents consists of 713 article (59.57 per cent), 309 reviews (25.81 per cent), 50 book chapter (4.18 per cent), 43 books (3.59 per cent), 24 letters (2.01 per cent), 18 conference paper (1.50 per cent), 17 editorial works (1.42 per cent), 13 notes (1.09 per cent) and 10 short survey (0.84 per cent). Thus, next to the article type, number review records are published, however, the other types of records are relatively lower. This may be due to lack of such publications during the period.

Figure 4. Institution Network

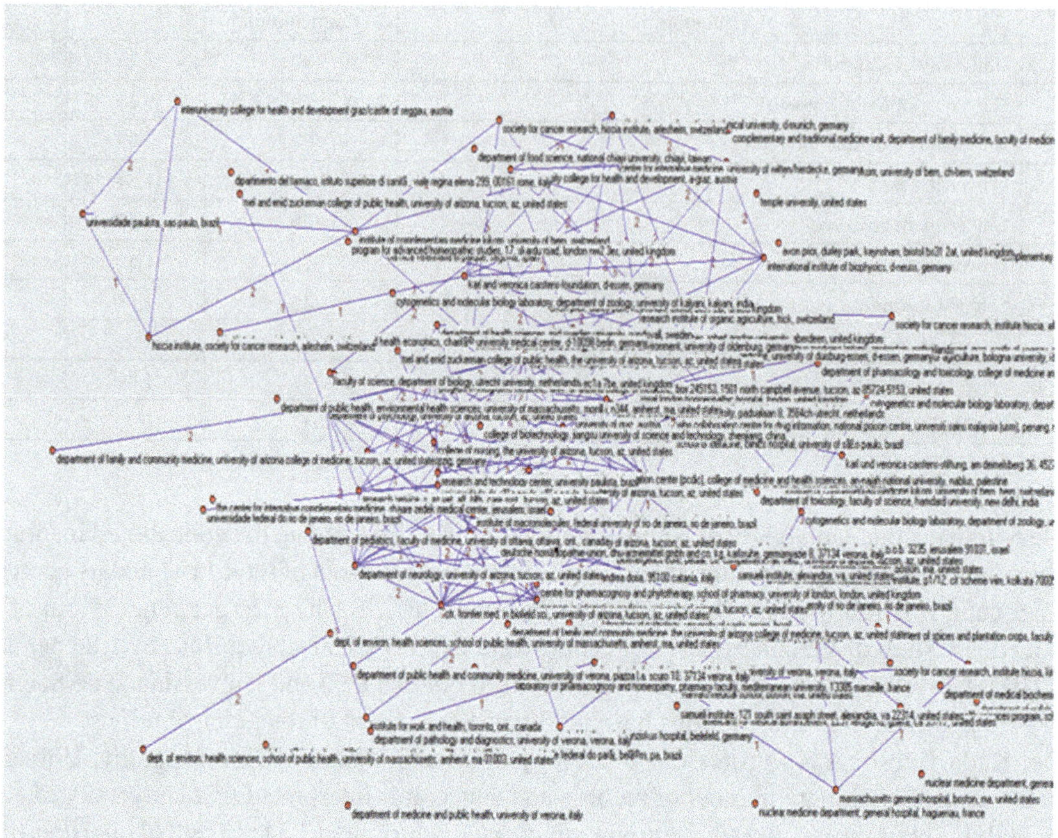

Table 6. Document Type

Document Type	Records	%
Article	713	59.57
Review	309	25.81
Book Chapter	50	4.18
Book	43	3.59
Letter	24	2.01
Conference Paper	18	1.50
Editorial	17	1.42
Note	13	1.09
Short Survey	10	0.84
	1197	100.01

Hence, it is interpreted that majority of the documents published during the period on toxicology in homoeopathy literature is of article type (59.57 per cent).

9. Distribution of Article Based on Source

The Table 6 clearly shows that 713 articles are published out of total records 1197. These articles are published in different journals. Therefore, spreading of articles over sources is summarised in Table 7. Journals having articles less than nine are not included in the table-7.

Table 7 discloses that 81 articles are published in the journal '*Homoeopathy*', it is 6.77 per cent of the total articles. However, 30 articles are published in '*Journal of Alternative & Complementary Medicine*' (2.51 per cent), 27 articles are published in '*Human & Experimental Toxicology*', 23 articles in '*Journal of Ethnopharmacology*' and 21 articles are published in '*Evidence Based Complementary & Alternative Medicine'*. Similarly, 16 articles are published in *'BMC Complementary & Alternative Medicine',* 13 articles are published in the journal 'International Journal of High Dilution Research' and 10 articles are published in '*Forschende Komplementarmedizin Und Klassische Naturheilkunde*'. The remaining journals have only less than 10 articles.

Thus, it is interpreted that more number of articles related to Toxicology on Homoeopathy are published in the journal ' *Homoeopathy*'.

10. Subject-Wise Contribution

The retrieved data are further analysed using subject domain. The details of analysis based on subject area are given in Table 8.

Table 7. Distribution of Article based on Source

Rank	Source Title	Records	%
1	Homeopathy	81	6.77
2	Journal of Alternative & Complementary Medicine	30	2.51
3	Human & Experimental Toxicology	27	2.26
4	Journal of Ethnopharmacology	23	1.92
5	Evidence Based Complementary & Alternative Medicine	21	1.75
6	BMC Complementary & Alternative Medicine	16	1.37
7	International Journal of High Dilution Research	13	1.09
8	Forschende Komplementarmedizin Und Klassische Naturheilkunde	10	0.84
9	Complementary Therapies In Medicine	9	0.75
9	Integrative Cancer Therapies	9	0.75
9	Medical Hypotheses	9	0.75
	Total	248	20.72

Table 8. Subject-wise Contribution

Rank	Subject Area	Records	%
1	Medicine	772	64.49
2	Pharmacology, Toxicology & Pharmaceutics	357	29.82
3	Biochemistry, Genetics & Molecular Biology	158	13.20
4	Agricultural & Biological Sciences	95	7.94
5	Environmental Science	90	7.52
6	Veterinary	61	5.10
7	Nursing	57	4.76
8	Chemistry	55	4.59
9	Dentistry	48	4.01
10	Chemical Engineering	33	2.76
11	Immunology & Microbiology	29	2.42
12	Health Professions	25	2.09
13	Neuroscience	24	2.01
14	Social Sciences	19	1.59
15	Engineering	18	1.50
16	Physics & Astronomy	13	1.09
17	Psychology	12	1.00
18	Materials Science	11	0.92
19	Computer Science	9	0.75
20	Arts & Humanities	8	0.69
21	Multidisciplinary	5	0.42
22	Mathematics	4	0.33
23	Earth & Planetary Sciences	3	0.25
24	Energy	1	0.08
	Total	1907	159

Table 8 reveals that 772 records are comes under the subject domain 'medicine', it is the 64.69 per cent of total. Some of the other prominent subjects are given below: Pharmacology, Toxicology & Pharmaceutics (29.28 per cent), Biochemistry, Genetics, & Molecular Biology (13.20 per cent), Agricultural and Biological Sciences (7.94 per cent), Environmental Science (7.52 per cent), Veterinary (5.10 per cent), Nursing (4.76 per cent), Chemistry (4.59 per cent) and Dentistry (4.01 per cent). The remaining subject domains have less than four per cent of the contribution.

Subject-wise contribution of records shows that *Medicine* subject is the topmost and ranked first with 46.49 per cent of total contributions and the second position is *Pharmacology, Toxicology & Pharmaceutics*. The subject-wise contribution is depicted in Figure 5.

Figure 5. Subject-wise Contribution

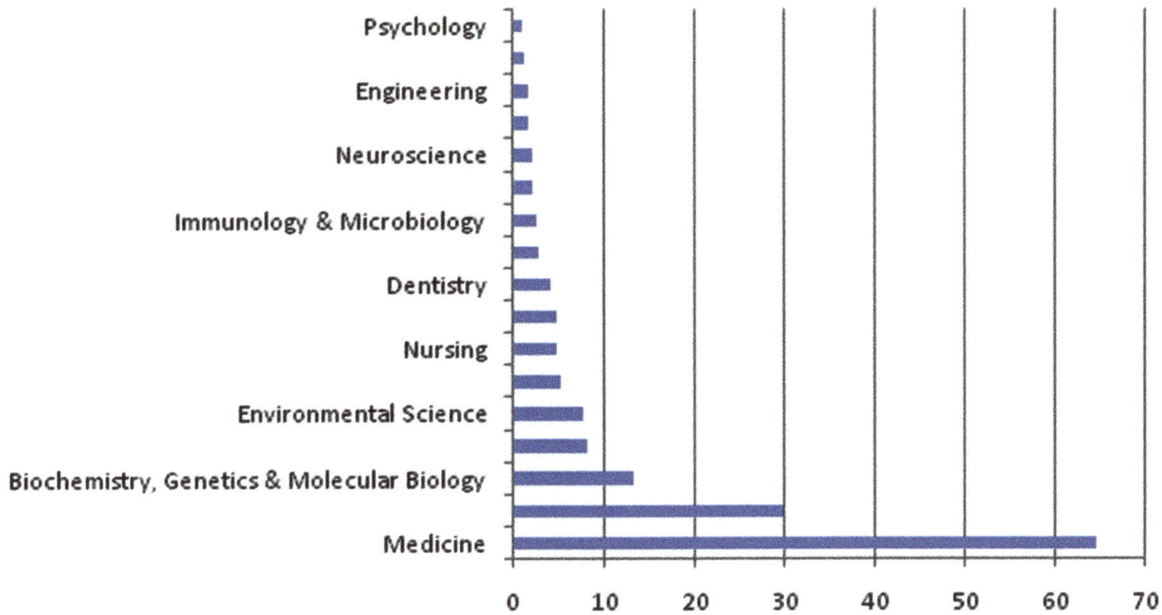

11. Keyword Visualization

Keyword Visualization is carried out using VOS Viewer, all those keywords which are occurred more than 50 times are presented in Figure 6.

Keyword visualization shows the occurrences of keywords in 1197 records are presented in three clusters. Cluster 1 consists of 22 terms including acupuncture, adult, alternatives medicine, clinical traits, complementary therapies, drug efficacy, drug safety, evidence based medicines, female, herbaceous agent, herbal medicine, human, humans, major clinical study, medicinal plant, middle aged, phytotherphy, placebo, review, traditional medicine, treatment outcome, and unindexed drugs. Twenty-one terms such as animal, animal experiment, animal model, animal tissues, animal, antineoplastic activity, antioxidant activity, article, controlled study, drug effect, drug mechanism, invitro study, male, mice, mouse, nonhuman, plant extract, rat and unclassified drug are included in cluster two.

Cluster two consist of 11 terms namely dilution, dose response, dose response relationship, drug, homoeopathic agent, homoeopathy, hormesis, materia medica, medical research, priority journal, risk assessment and toxicology.

Some of the keywords with high frequency of occurrences on toxicology in homoeopathy literature are homoeopathy, human, humans, non-human, unclassified drug and animals after omitting trivial common terms.

Figure 6. Keyword Visualization

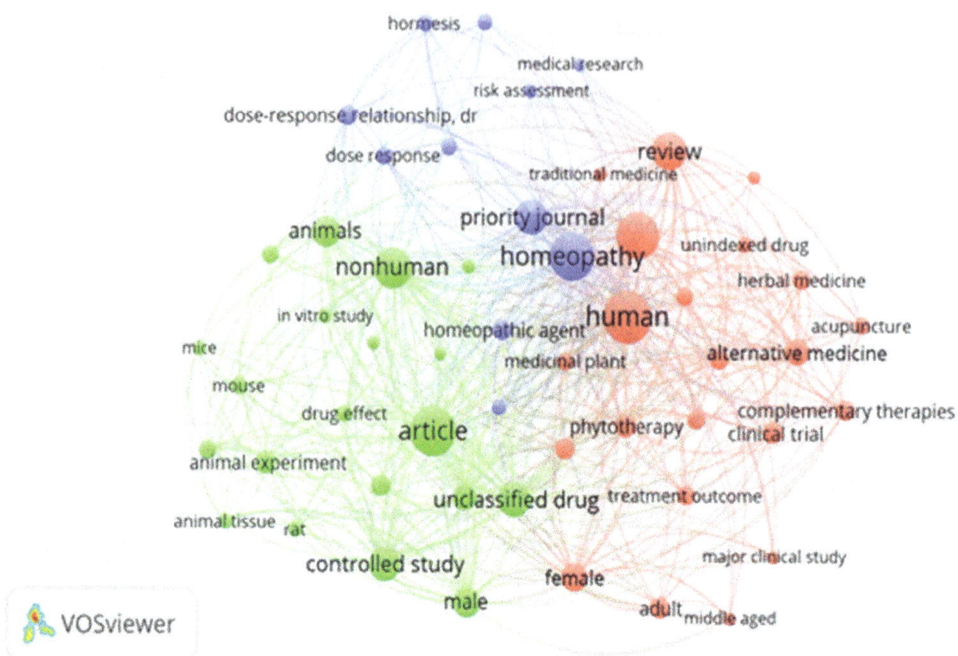

CONCLUSION

The study on growth of literature of toxicology in homeopathy reveals that toxicology literature grows rapidly in the later stage than the earlier period; growth of literature is remarkably high during the last two decades. The United States of America is the leading contributors at the global level and India is in the second position. Among the contributors, Calabrese, E. J., Professor of Toxicology, University of Massachusetts is the most prolific author for Toxicology in Homoeopathy literature. University of Kalyani and University of Massachusetts are the leading institutions which have contributed for toxicology in homoeopathy literature. Also, 60 per cent of the identified records belong to article type. Moreover, majority of the records are published in the journal '*Homoeopathy*', which is published from the USA and major contributions are in the subject domain '*medicine*' (65.4 per cent). Most frequently repeated keywords in toxicology of homoeopathy literature are: homoeopathy, human, humans, non-human, unclassified drug and animals.

REFERENCES

Ashok, J., & Gang, K. S. (1992). Laser research in India: Scientific study and model projection. *Scientometrics*, *23*(3), 395–415. doi:10.1007/BF02029806

Chiu, W.-T., & Ho, Y.-S. (2005). Bibliometric analysis of homoeopathy research during the period of 1991 to 2003. *Scientometrics*, *63*(1), 3–23. doi:10.1007/s11192-005-0201-7

Delirrad, M., Rashidi, A., & Karimi, S. (2012). A bibiliometric analysis of Toxicology publications of Iran and Turkey in ISI web of science. *Iranian Journal of Toxicology*, *6*(19), 735–745.

Devi, B. M. (2013). Bibliometric study of core journals in toxicology. *E-Science World*, *5*(3-4), 13–23.

Encyclopedia Americana (International ed.). (1981). Grolier Inc.

Fulder, S. (1996). *The Handbook of Alternative and Complementary Medicine*. London: Hodder and Stoughton.

Gevitz, N. (1993). Unorthodox medical theories. In W. F. Bynum & R. Porter (Eds.), *Companion Encyclopaedia of the History of Medicine*. London: Routledge.

Graham, E. C. (1993). Encyclopaedia of Scientific Terms (Vol. 2). New Delhi: Crest Publishing House.

Jeyshankar, R., & Vellaichamy, A. (2015). A bibliometric analysis of global literature on cervical cancer. *International Journal of Information Dissemination and Technology*, *4*(4), 298–302.

Jeyshankar, R., & Vellaichamy, A. (2016). Scientometric analysis of autism research output during 2007-2011. *SRELS Journal of Information Management*, *53*(1), 55–64. doi:10.17821/srels/2016/v53i1/86768

Loudon, I. (2006). A brief history of homoeopathy. *Journal of the Royal Society of Medicine*, *99*(12), 607–610. doi:10.1258/jrsm.99.12.607 PMID:17139061

Meera & Sahu, S.K. (2014). Research output of university college of medical science, university of Delhi: A bibliometric study. *COLLNET Journal of Scientometrics and Information Management, 8*(2), 401-418.

Tabatabaei-Malazy, O., Ramezani, A., Atlsai, R., Larijani, B., & Abdollahi, M. (2016). Scientometric study of academic publications on antioxidative herbal medicines in type 2 diabetes mellitus. *Journal of Diabetes and Metabolic Disorders*, *15*(1), 48. Retrieved 13 October 2017 from doi:10.1186/s40200-016-0273-3 PMID:27785446

Tague-Sutcliffe, J. (1992). An introduction to informetrics. *Information Processing & Management*, *28*(1), 13. doi:10.1016/0306-4573(92)90087-G

Zarafshan, H., Mohammadi, M. R., Motevalian, S. A., Abolhassani, F., Khaleghi, A., & Sharifi, V. (2016). Autism research in Iran: A scientometric study. *Iranian Journal of Psychiatry and Behavioral Sciences*, *11*(2), e7350. doi:10.5812/ijpbs.7350

Zyoud, S. H., Al-Jabi, S. W., Swelleh, W. M., & Awany, R. (2014). A bibilometric analysi of toxicology research productivity in Middle Eastern Arab countries during a 10 year period (2003-2012). *Health Research Policy and Systems*, *12*(4). PMID:24443999

Chapter 12
A Scientometric Profile of Vikram Sarabhai Space Center (VSSC) Based on Scopus Database

M. Vijayakumar
St. Xavier's College of Engineering, India

S. Lawyed Stephen
Noorul Islam Center for Higher Education, India

A. Lawrence Mary
Loyola Institute of Technology and Science, India

ABSTRACT

This chapter creates the scientometric profile of Vikram Sarabhai Space Centre (VSSC) for the period from 1973 to 2016. The scientific productivity of the scientists is analyzed using scientometric indicators based on the data retrieved from the Scopus database. The analysis revealed that linear growth is prevailing. It also shows that the scientists are actively involved in disseminating their research findings in scholarly journals and G.V. Rao secured first position among the top-ranking authors. The Indian Institute of Science is the leading collaborating institution with VSSC and the United States is topmost collaborating country. Most of the scholarly communications of VSSC are published in the 'Journal of Sound & Vibration' and a majority of the records are of an article type. Most of the articles are published in the inter-disciplinary subject of Engineering. The study also found that the total number of citations received by the documents published from VSSC are 17395 and out of the 1783 documents published and 1385 documents received citations.

DOI: 10.4018/978-1-5225-3457-0.ch012

INTRODUCTION

Scientometrics is one the technique of measuring information using quantitative and qualitative indicators. It is analogous to bibilometrics and infrometrics, it uses mathematical and statistical techniques for the evaluation and predicting the advancement of science. The scientometrics may be defined as the application of those quantitative methods which are dealing with the analysis of science viewed as an information process (Nalimov & Mulcheno,1969). Scientometrics is a technique of assessing scientific productivity of an individual, institution, subject or nation. Scientific literature produced by these entities reflects their scientific activity. Hence scientometric analysis is being increasingly used to evaluate the research performance of researchers, research institutions and the research trends in various disciplines. Mathematical and statistical tools are used to conduct these studies. Though bibliometric and scientometric techniques are found to be more or less similar, the emphasis of scientometric studies are the quantitative aspects of generation, propagation and use of scientific information, in order to contribute to understanding the mechanism of scientific research (Bormann, 2007).

BIOGRAPHY OF VIKRAM SARABHAI

Vikram Sarabhai was a pioneer of Indian Scientists and regarded as the Father of Indian Space Programme. He was born on 12th August 1919 at Ahmadabad, Gujarat and completed his college education at Cambridge University. He also worked as research scholar at Indian Institute of Science, Bangalore under the great Sir. C.V. Raman. He completed his Ph.D in 1947 from Cambridge University. His areas of interest are Solar Physics and cosmic rays. He was responsible for the establishment of Physical Research Laboratory in November 1947 at Ahmedabad. He played a major role in setting-up of a number of observation stations across the country and his continuous research on solar interplanetary physics identified the avenues for space research and space science. This had led him to form Indian National Committee for Space Research and he became chairman of the committee. Vikram Sarabhai set up the first Rocket Launching station at Thumba near Thiruvananthapuram, Kerala on the Arabian Coast as it is close to the Equator. The first rocket was launched from the launching centre on 21st November 1963. He received Shanti Swaroop Bhavnagar Memorial Award in1962 and Padma Bhushan Award 1966 among the many such in his illustrious career.

VIKRAM SARABHAI SPACE CENTRE

After the incorporation of the Indian National Committee for Space Research (INCOSPR) in 1962, its first act was establishment of Thumba Equatorial Rocket Launching Station (TERLS) at Thumba in Thiruvananthapurm in 1968. It is the major space research centre of Indian Space Research Organisation (ISRO), actively involved on rocket and space vehicle for India's Satellite programme (www.isro.org). The Thumba Equatorial Rocket Launching Station was renamed in honour Vikram Sarabhai as Vikram Sarabhai Space Research Centre (VSSC) in 1971. Developing rounding rockets, launch vehicles and certain technologies associated with space craft are the responsibilities of VSSC. In the early 1980s, VSSC was instrumental in the development of India's Satellite Launch Vehicle program, SLV-3. This was followed in the late 1980s with the Augmented Satellite Launch Vehicle (ASLV), for launching 150 kg satellites near the earth's orbit. In the 1990s, VSSC contributed to the development of India's workhorse launch vehicle, the Polar Satellite Launch Vehicle (PSLV).

REVIEW OF LITERATURE

Number of quantitative and a few qualitative studies were already conducted in different sub-discipline and branches of knowledge to assess and to evaluate the growth of literature and its various dimensions. The previous studies based on bibilometric or scientometric or informetric analyses were reviewed. From the review, it was found that few studies were conducted to assess the research productivity of the organisation, agency and individual scientists using either bibiliometric approach or scientometric approach. The present study is intended to analyse the scientific productivity of a research institution. The similar studies conducted are, on publication productivity of University of Kerala (Gopikuttan & Aswathy, 2014); among the faculty of ManonmaniamSundaranar University of Thirumagal (2012) and University of Mysore (Kumbar; Gupta & Dhawan, 2008). Similarly, research output of India is studied by Gupta and Dhawan in 2008 and scientific output of India in crop science (Tripathi & Garg, 2014); research output of Central Tuber Research Centre (Sudhier Pillai & Priyalekshmi, 2013); and research articles published by the scientists of CECRI (Jeyasankar, Babu & Rajendran, 2011). Scientific productivity of individual author Vikram Sarabhai (Kademani; Kalyane & Kumar, 2000) and publication pattern of technologist in VSSC (Kabir & Rajendran,).

Gopikuttan and Aswathy (2014) evaluated the research productivity of University of Kerala based on the data collected from Web of Science over a period of thirteen years from 2000 to 2012. The parameters such as form-wise, year-wise, subject-wise classification of published papers, most productive authors and the most preferred journals, etc. were considered for the study. The impact factor and the citation received were also analysed. The study revealed that Chemistry is the subject which produces more number of papers while the multi-authorship also possesses a lead role in this subject. Indian journals were the most preferred journals to publish the articles which were followed by UK. Collaborative Coefficient varies from subject to subject. The result also showed that the research productivity of the University of Kerala is much recognised at international level.

Thirumagal (2012) analysed the scientific publications generated by the Manonmaniam Sundaranar University, Tirunelveli, India using the data collected from the Web of Science. The analysis highlights yearly output of research publications. This study focused on publishing trend, impact factor, authorship pattern, types of articles, institutional collaboration of authors, affiliated institutions of authors, countries of contributing authors and individual author's research productivity.

Kumbar; Gupta & Dhawan (2008) described the growth, contribution and impact of research carried out by the scientists of University of Mysore in science and technology. The study indicated the patterns of communications of university scientists and studies the extent of concentration and scattering of their research output in different journals. It analysed the strong and weak areas of university research, their growth rate and impact in terms of average citations received. Also studies the output and impact of research under different existing subject departments of the university. Analyses the international collaborative share of research output at the overall level as well as across various subjects, indicating also the major countries involved in international collaboration.

Gupta and Dawan (2008) analysed India's publications output in three major international multidisciplinary databases, as indexed during 1981-2005. It reports on India's comparative strength in world's science and technology (S&T) output, its growth and decline, its strong and weak subject areas of research, media of communication, its collaborative profile and quality of S&T output, institutional productivity and quality, and dynamics of Indian research at institutional and sectoral levels. The study also provides suggestions for improving the quantity and quality of research S&T in India.

Trpathi & Garg (2014) analysed scientific output of India in the discipline of crop sciences as reflected by the coverage of scientific output in three different databases i.e. SCOPUS, CAB Abstracts and ISA (Indian Science Abstracts) during 2008-2010. The study revealed that highest number of papers was published on rice and wheat crop. Agricultural universities and institutions under the aegis of Indian Council of Agricultural Research (ICAR) were most productive institutions. Most of the papers were published in Indian journals with low impact factor. Environment and Ecology, Indian Journal of Agricultural Sciences and Research on Crops were the most preferred journals used by the Indian scientists. The major research is focused on 'genetics and plant breeding' followed by 'soil, climate and environmental aspects' and 'agronomic aspects'. The authorship pattern reveals that co-authored papers accounted for 72% of total output.

Sudhier Pillai and Priyalekshmi (2013) analysed the bibliographic details of 1076 research articles obtained from the annual reports of Central Tuber Crops Research Institute (CTCRI) and it was found that the highest number of 169 papers was published in the year 2006 and the average number of publications per year was 97.82. Most of the contributions were multi-authored (87.68%). The degree of collaboration of scientists of CTCRI was 0.87 and most of the articles published by the scientists were in the foreign journals (51.89%). Journal of Root Cropspublished by Indian Society of Root Crops tops the list with the highest number of articles 125 (39.30%).

Jeyshankar; Babu & Rajendran (2011) analyses bibliographical details of 1282 research articles published by the scientists of CECRI during the period 2000-2009. It is found that 2009 was the most productive year with 194 articles (15.13%) published in the year. Collaborative research was dominant with the highest degree of collaboration being 0.98, in the year 2005. Further, the study investigated authorship pattern, co-authorship pattern, highly prolific authors and highly preferred journals by the scientists of CECRI.

Kademani; Kalyane & Kumar (2000) analysed the citations to the publication of Vikram Sarabhai using Science Citation Index 1944-1971. The citation analysis clearly reveals in quantitative terms, the impact of contribution of Vikram Sarabhai in the inter-disciplinary domain such as Astronomy, Physics, Geology, Geography, General and extra-disciplinary subjects like Veterinary science, Computers and Telecommunication. The high rate of citations to his papers in journals of international scope and journal of high impact factor is clear indication of their high quality.

The available literature shows the present study is unique and a valuable contribution to the science of science.

NEED AND SIGNIFICANCE OF THE STUDY

The research output of the institution is very often measured using scientometric techniques. It is more reliable method for the identification of new scientific and technological knowledge. The growth of literature has become a major concern for the scientists, scholars and library professionals as they have to keep themselves abreast of the new advances in their subjects and to take policy decisions of the government on Science and Technology. The scientific productivity is normally measured in terms of their scientific output which are in the form of research papers, reports, books and technical output in the form of patent filed or accepted. The institutional productivity is measured in terms oftheir research output or scientific productivity. The Vikram Sarabhai Space Center (VSSC) is a unit of Indian Space Research Organisation (ISRO), actively involved in rocket launching technology in which the scientists are expected to communicate their scholarly findings in varied format in addition to development of models and patents. Since the reputation of such institution solely depend on their scientific contribution. Very few attempts were made previously to evaluate the scientific productivity of scientist or technologist in VSSC. Some of the representative studies on scientific productivity of individual author Vikram Sarabhai (Kademani; Kalyane & Kumar, 2000) and publication pattern of technologist in VSSC (Kabir & Rajendran, 2016). Scanning of available literature shows no systematic study has been conducted on the scientometric profiles of VSSC at institutional level. The authors felt the need to fill this gap and undertook the study to analyse the scientometric profile of VSSC based on the Scopus database.

STATEMENT OF THE PROBLEM

On the basis of the importance of the study, the investigators reviewed all available related studies and found that no systematic study has been carried out to analyse the scientometric profile of the Vikram Sarabhai Space Centre (VSSC) based on research output of scientists working in VSSC. Hence, the present study is entitled as "A Scientometric Profile of VIkram Sarabhai Space Centre (VSSC) based on *Scopus* Database".

OBJECTIVES OF THE STUDY

The specific objectives framed for this study are:

1. To determine the year-wise contribution and its trends in scientific productivity of the VSSC.
2. To identify top ranking authors, top collaborating institutions, and collaborating countries with VSSC.
3. To visualize the cluster maps of co-authors and collaborating countries.
4. To identify top ranking journals and type of records
5. To analyse the contribution of VSSC subject-wise and also to identify articles that attracted highest citations.

SCOPE AND LIMITATIONS OF THE STUDY

The scope of the present chapter is to analyse the scientometric profiles of Vikram Sarabhai Space centre (VSSC) based on the research output of scientists, which are indexed in *SCOPUS* database for the period 1973-2016. The content coverage of the selected database, time lags between publishing and indexing, and usage of more quantitative indicators are the limitations of the present chapter.

METHODOLOGY

The data for the study are downloaded from the bibliographic database *SCOPUS* using the search term '*VSSC*"; and '*Vikram Sarabhai Space Centre*'. The retrieved data are further filtered based on year between 1973 and 2016, since VSSC was established in 1972. The content coverage is more in *SCOPUS* than other databases and hence *SCOPUS* is preferred as source for the present study. A total of 1783 records are retrieved from the *SCOPUS* database. The scientometric profile of the VSSC is analysed based on qualitative and quantitative indicators. Scopus analytics tool, Microsoft tools and *VOSviewer* are used for this study.

ANALYSIS AND INTERPRETATION OF DATA

1. Year-Wise Contribution

The year-wise contribution of documents published by the scientists of VSSC is presented in Table 1. A total of 1873 documents are indexed in the *SCOPUS* database as scholarly contribution for the study period.

Table 1. Year-wise Contribution

R	No of Records	Per cent	Cumulative Growth	Per cent
1973	2	0.11	2	0.11
1974	16	0.90	18	1.01
1975	17	0.95	35	1.96
1976	29	1.63	64	3.59
1977	18	1.01	82	4.60
1978	25	1.40	107	6.00
1979	21	1.18	128	7.18
1980	12	0.67	140	7.85
1981	19	1.07	159	8.92
1982	13	0.73	172	9.65
1983	15	0.84	187	10.49
1984	26	1.46	213	11.95
1985	21	1.18	234	13.12
1986	39	2.19	273	15.31
1987	38	2.13	311	17.44
1988	34	1.91	345	19.35
1989	43	2.41	388	21.76
1990	41	2.30	429	24.06
1991	45	2.52	474	26.58

continued on following page

Table 1. Continued

R	No of Records	Per cent	Cumulative Growth	Per cent
1992	40	2.24	514	28.83
1993	32	1.79	546	30.62
1994	32	1.79	578	32.42
1995	28	1.57	606	33.99
1996	35	1.96	641	35.95
1997	39	2.19	680	38.14
1998	43	2.41	723	40.55
1999	44	2.47	767	43.02
2000	49	2.75	816	45.77
2001	43	2.41	859	48.18
2002	50	2.80	909	50.98
2003	59	3.31	968	54.29
2004	57	3.20	1025	57.49
2005	52	2.92	1077	60.40
2006	56	3.14	1133	63.54
2007	57	3.20	1190	66.74
2008	48	2.69	1238	69.43
2009	63	3.53	1301	72.97
2010	63	3.53	1364	76.50
2011	67	3.76	1431	80.26
2012	80	4.49	1511	84.74
2013	66	3.70	1577	88.45
2014	51	2.86	1628	91.31
2015	88	4.94	1716	96.24
2016	67	3.76	1783	100
	1783	100	1783	100

Table 1 show that two documents are indexed in 1973 as scholarly literature contribution of VSSC. It also reveals that the year-wise growth contribution is not uniform and high fluctuations are observed between the years. Fifty per cent of the total contribution is after 2002. The cumulative growth of contribution of VSSC scientists is only at linear level. The fluctuation in contribution may be due to specialized works undertaken by VSSC. The cumulative growth of contribution of the scientists in VSSC is presented in the Figure 1. Figure 1 clearly shows the prevailing trend of scientific publication productivity of scientists in VSSC is linear.

Figure 1. Cumulative Growth of Contribution

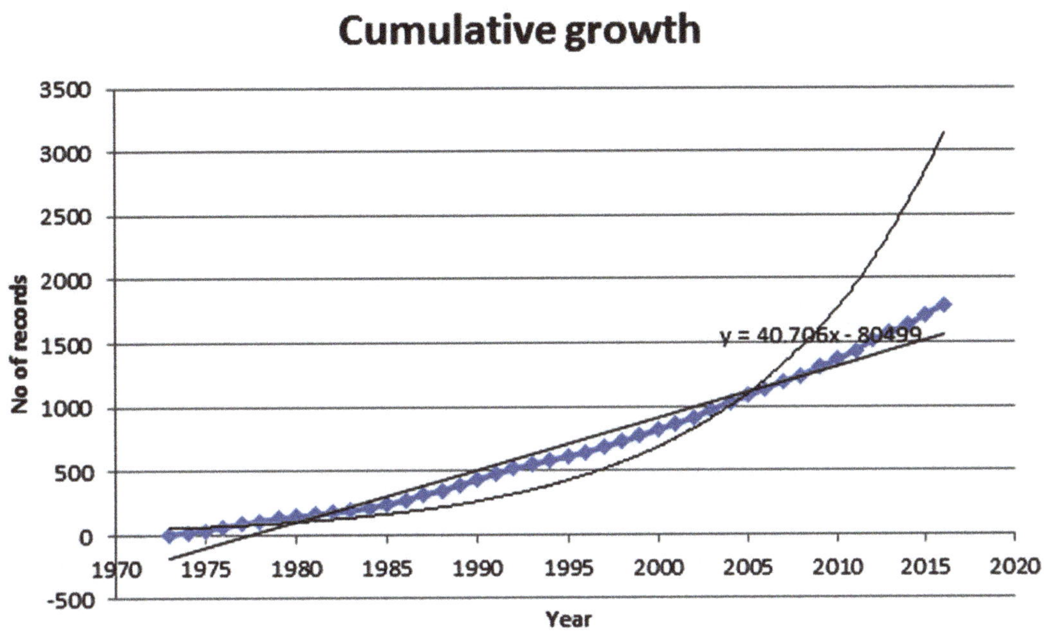

2. Top Ranking Authors

Another aspect of scientometric profile of the research institution is its author-wise research contribution. Based on the number of contribution made by the scientists of VSSC, a list of top ranking author is prepared. The scientists who have more than 50 contributions are given in Table 2.

The top-ranking list of authors reveals that there are 16 authors in the list. Among the top-ranking list, Rao, G.V. is in the first position with 128 contributions. His contribution is 7.18 per cent of the total contribution. The second and third position are respectively secured by Nageswara Rao, B. and Ninan, K.N with 113 (6.34 per cent) and 101 (5.66 percent) documents. In the top ranked list, Sreekumar, K. is in the last position (16thPosition) with 50 documents (2.80 per cent). Hence, it is interpreted that scientists are actively involved is disseminating their research findings in scholarly journals and Rao, G.V secured first position among the top-ranking authors.

The top-ranking authors for their contribution to the scientific communication are graphically presented in the bar chart in Figure 2.

3. Cluster Density Map of Co-Authors

Majority of the projects in VSSC are joint venture, involved by a group of scientists who are experts in the field and hence their scientific productivity is also linked with more than one author having some exceptional cases. Co-authors having more than 10 contributions are given in Table 3.

Table 3 shows there are nine clusters of co-authors. Cluster 1 consists of three authors namely Jha, A.K.; NarayanaMurty, S.V.S.; and Sreekumar, K. who have 77, 62 and 50 documents respectively have co-authorship links of 175, 133 and 105 numbers respectively. Similarly, in cluster-3, there are three authors Sinha, P.P.; Gupta, R.K.; and Pant, B. have respectively 76, 57 and 55 documents and with 171,

Table 2. Top Ranking Authors

Rank	Author	Contribution	Per cent
1	Rao, G.V.	128	7.18
2	NageswaraRao, B.	113	6.34
3	Ninan, K.N.	101	5.66
4	VenkateswaraRao, G.	92	5.16
5	Moorthy, K.K.	79	4.43
6	Sinha, P.P.	76	4.26
7	Jha, A.K.	74	4.15
8	NarayanaMurty, S.V.S.	72	4.04
9	Bhardwaj, A.	71	3.98
10	Pant, T.K.	58	3.25
10	Raju, K.K.	58	3.25
12	Gupta, R.K.	57	3.20
13	Mehta, R.C.	56	3.14
14	Pant, B.	55	3.08
15	Singh, G.	53	2.97
16	Sreekumar, K.	50	2.80
	Total	1193	66.90

Figure 2. Top Ranking Authors

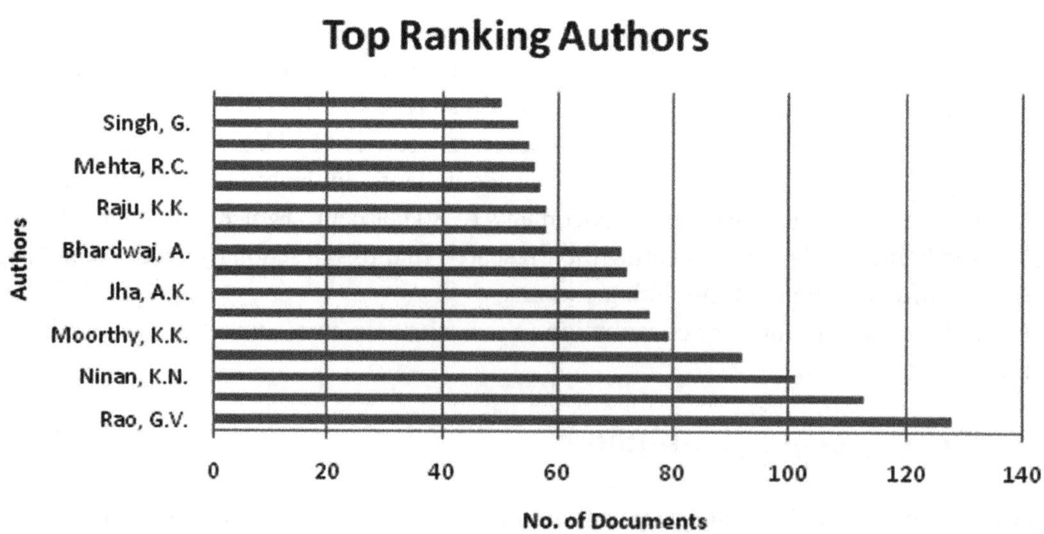

133 and 148 co-authorships. Cluster-4 consists of two authors namely Pant, T.K.; and Sridharan, R. respectively have 58 and 51 documents with 154 and 145 co-authorships. Ninan, K. N. who is in cluster-5 has 101 documents with 102 co-authorship; Cluster-6 constitutes two authors namely Nageswara Rao, B.;and Rao, B.N. have respectively 113 and 75 documents with 121 and 41 co-authorships and Cluster-7

Table 3. Co-authors with more than ten contributions

Author	Cluster	Weight<Documents>	Weight<Co-authorships>
Rao, G.V.	7	128	116
Nageswara Rao, B.	6	113	121
Ninan, K.N.	5	101	102
Venkateswara Rao, G.	7	92	90
Moorthy, K.K.	8	79	98
Sinha, P.P.	3	76	171
Rao, B.N.	6	75	41
Jha, A.K.	1	74	175
Narayana Murty, S.V.S.	1	72	133
Bhardwaj, A.	9	71	54
Pant, T.K.	4	58	154
Raju, K.K.	7	58	58
Gupta, R.K.	3	57	133
Pant, B.	3	55	148
Singh, G.	7	53	80
Sridharan, R.	4	51	145
Sreekumar, K.	1	50	105

is shared by four authors namely Rao, G.V.; Venkateswara Rao, B.; Singh, G. and Raju, K.K.. Cluster - 8 and cluster - 9 have two author search. It is observed that authors who are having more documents to their credits have comparatively lesser number of co-authorships.

Further, cluster density map of co-authors is prepared based on having more than ten contributions using *VOSviewer*. The cluster density view of the map of co-authors is given in Figure 3. The density of co-authorship is visible for the five clusters.

4. Top Ranking Collaborating Institution

The total document retrieved from the database during the study period consists of 1783 documents from the VSSC. These 1783 documents are further analyzed based on the collaborating institution. Either the scientists of VSSC or the scholars of other reputed and premier institutions having collaboration with VSSC in projects publish their scholarly communication in journals. The collaborating institution with VSSC are analysed and top ranking of collaborating institutions are listed out in Table 4.

Table 4 clearly shows that Indian Institute of Science and three Indian Institutes of Technology occupy the first five position of top-most collaborating institution with the only exception of University of Kerala, which secured 4thposition with 43 documents. Indian Institute of Science is top in the list top ranking collaborating institution with 101 documents, Indian Institute of Technology (IIT), Madras is in second position with 59 records; IIT, Kanpur in third place with 52 documents; and IIT, Bombay fifth place with 34 records. Other significant collaborating institutions are: Physical research Laboratory India, Cochin Institute of Science and Technology, National Atmospheric Research Laboratory, Chittoor; NASA Marshall Space Flight Center and Andhra University.

Figure 3. Cluster Density Map of Co-authors

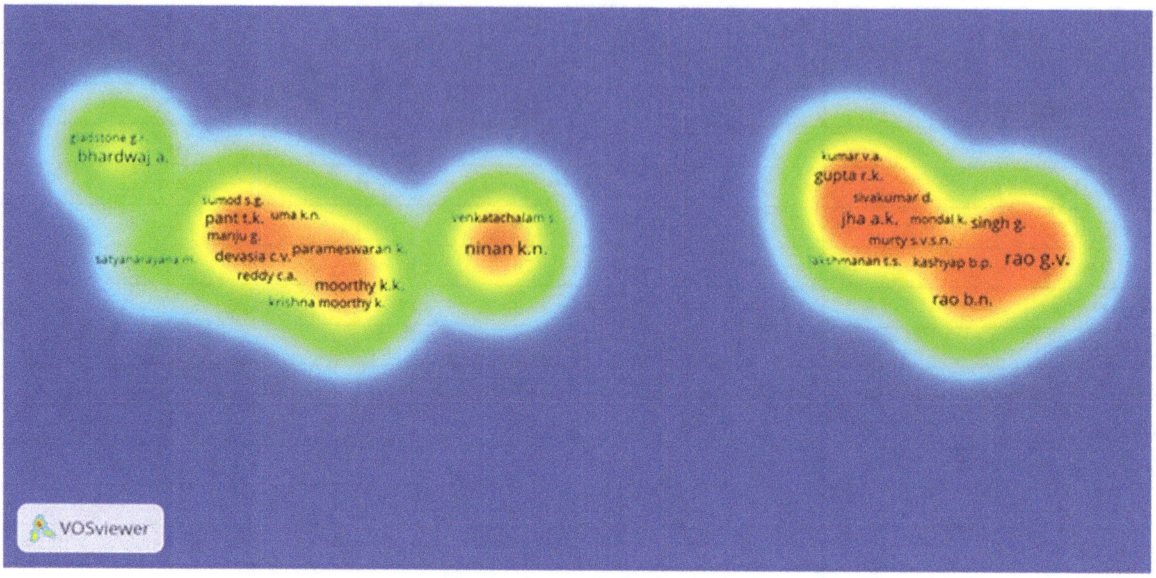

**For a more accurate representation see the electronic version.*

Table 4. Top Ranking Collaborating Institution

Rank	Institution	No of records	Per cent
1	Indian Institute of Science	101	5.66
2	Indian Institute of Technology, Madras	59	3.31
3	Indian Institute of Technology, Kanpur	52	2.92
4	University of Kerala	43	2.41
5	Indian Institute of Technology, Bombay	34	1.91
6	Physical Research Laboratory India	34	1.91
7	Cochin University of Science and Technology	23	1.29
8	Indian Institute of Technology Banaras Hindu University	22	1.23
9	National Atmospheric Research Laboratory, Chittoor	21	1.18
10	NASA Marshall Space Flight Center	20	1.12
10	Andhra University	20	1.12

University of Kerala's association with VSSC can be attributed to the geographical proximity. However, the closer analysis of collaboration shows that VSSC has links with geographically distant institution within India and also with NASA. Hence it is inferred that with the latest developments in ICT Technologies, collaborative research without geographical barriers is always possible. The pictorial representation of top ranking collaborating institutions is visually represented in figure-4.

Figure 4. Top Ranking Collaborating Institutions

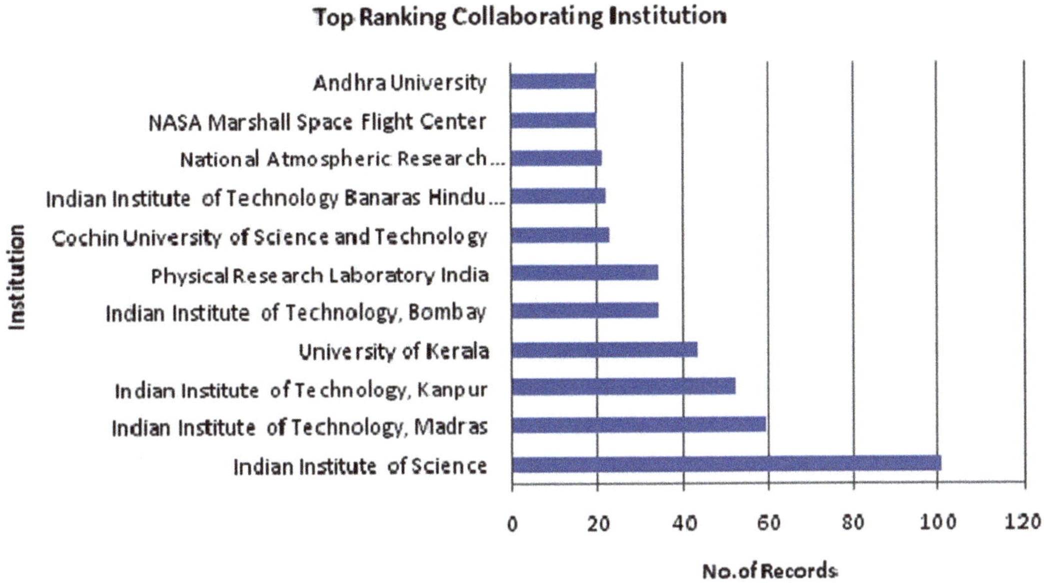

5. Top Ranking Collaborating Countries

The extent of research collaboration may not be limited to a country but it goes beyond at the global level. The degree of collaborating countries with VSSC is analysed and the details are summarised in Table 5.

All the countries which have more than five research outputs in collaboration with VSSC are listed in Table 5. Further, it reveals that the United States is first among the collaborating countries with 69 records (3.87 per cent of total contribution) and Switzerland is in 12th position with 5 documents. The second rank is shared by Japan and United Kingdom having 28 records each.

Table 5. Top Ranking Collaborating Countries

Rank	Country	Contribution	Per cent
1	United States	69	3.87
2	Japan	28	1.57
3	United Kingdom	28	1.57
4	Germany	13	0.73
5	France	12	0.67
6	Canada	8	0.45
6	Spain	8	0.45
8	Brazil	7	0.39
8	South Korea	7	0.39
10	Sweden	6	0.34
11	Russian Federation	5	0.28
11	Switzerland	5	0.28

Hence, it is interpreted that the research output of the VSSC is in collaboration with other countries at global level and the United States is in the first of top of collaborating countries.

Collaborating countries have more than five records are further visualized using VOSviewer and the same is presented in figure-5.

6. Top Ranking Journals

Based on the number of contributions made by the scientists in leading journals by the Scientist of VSSC, a list of ranking journals is prepared. The details of top 10 ranking journals preferred by the scientist of VSSC are presented in Table 6.

It is clear from the Table 6 that more records were published in the journal of '*Journal of Sound & Vibration*' with 117 records and ranked first among top 10 journals. However, it is only 6.56 per cent of the total output. Similarly, '*Computer & Structure*' secured second position; it has 4.60 per cent of the total records (82 records), whereas, '*Advances in Space Research*' is in the last place of top most 10 journals with 1.79 per cent (32 out of 1783).

Thus, it is interpreted that the choices of publication among the scientists of VSSC is highly diffused over number of wide range of journals rather than concentrated in a few core journals. Majority of the articles are centred on Engineering and allied disciplines, first rank among the top ten most journal is '*Journal of Sound & Vibration*'.

Figure 5. Visualization of Collaborating Countries

Table 6. Top Ranking Journals

Rank	Journal Title	No of records	Per cent
1	Journal of Sound & Vibration	117	6.56
2	Computers & Structures	82	4.60
3	Materials Science Forum	56	3.14
4	Journal of Applied Polymer Science	51	2.86
5	Journal of Atmospheric & Solar Terrestrial Physics	46	2.58
6	AnnalesGeophysicae	38	2.13
7	Journal of Geophysical Research Space Physics	36	2.02
8	Engineering Failure Analysis	34	1.91
8	Geophysical Research Letters	34	1.91
10	Advances In Space Research	32	1.79
	Total	526	29.5

7. Type of Documents

The scientific productivity of the research institution like VSSC is reflected in the form of scholarly communications such as articles in journals, conference proceedings, reviews, editorial comments, short communication and so on. The contributions of VSSC for the study period are further classified based on its type and the details are given in Table 7.

Table 7 reveals that majority of the records (78.80 per cent) i.e. 1405 record out of 1783 record are article type and they are published in various journals and hence article is ranked first. It is followed by conference proceeding by 13.01 per cent (232 out f 1783). The scientists of VSSC have also contributed their views in various type of documents include editorials, note, erratum, note and books only very meagre in number.

Table 7. Type of Documents

S.No	Document Type	No of records	Per cent
1	Article	1405	78.80
2	Conference Paper	232	13.01
3	Letter	78	4.37
4	Review	26	1.46
5	Editorial	16	0.90
6	Note	9	0.50
7	Book Chapter	6	0.34
8	Article in Press	6	0.34
9	Erratum	4	0.22
10	Book	1	0.06
Total		1783	100

8. Subject-Wise Contribution

The retrieved documents from the VSSC consist of 1783 records, which are categorised on the basis of subject. The subject-wise categorisation shows that Engineering has more number of articles (53.73 per cent) and it is followed by Materials Science (30.51 per cent), Earth & Planetary Sciences (26.47 per cent), Physics and Astronomy (24.22 per cent), Chemistry (10.26 per cent) and Computer Science (8.69 per cent). The remaining subjects such as mathematics, chemical engineering, Environmental Science and Energy have less number of articles. The details are summarised in Table 8.

Hence, it is interpreted that articles published from the VSSC have wider subject coverage ranging from Engineering and its various branches, related branches like Astronomy, Material Science, Chemistry, Environmental Science and Neuroscience.

9. Articles Receiving Highest Citations

The articles contributed by the VSSC, which have received more number of citations are further analyzed based on the number of citations received by individual articles. The bibiliographic details of article which have more than 100 citations are presented in Table 9. There are 13 articles which have received more than 100 citations during the study period from 1973-2016, the article which got maximum citation is

Table 8. Subject-wise Contribution

S.No	Subject Area	No of Papers	Per cent
1	Engineering	958	53.73
2	Materials Science	544	30.51
3	Earth & Planetary Sciences	472	26.47
4	Physics & Astronomy	432	24.22
5	Chemistry	183	10.26
6	Computer Science	155	8.69
7	Mathematics	81	4.54
8	Chemical Engineering	76	4.26
9	Environmental Science	60	3.37
10	Energy	48	2.69
11	Agricultural & Biological Sciences	22	1.23
12	Multidisciplinary	17	0.95
13	Medicine	10	0.56
14	Business, Management & Accounting	8	0.49
15	Biochemistry, Genetics & Molecular Biology	7	0.39
16	Decision Sciences	5	0.28
17	Social Sciences	3	0.17
18	Pharmacology, Toxicology & Pharmaceutics	2	0.11
19	Neuroscience	1	0.06
20	Undefined	5	0.28

261, is authored by Satheesh, S.K. & Krishna Moorthy, K. and it is published in the journal Atmospheric Environment in 2005. It is followed by Sreejith, S. et. al. and Madhusudanan, P. M.; Krishna, K &Ninan, K.N have citations respectively 193 and 183 citations. The table also shows that the article published in 1974 in Journal of Sound and Vibration by SadasivaRao, Y.V.K. &Nakra, B.C secured 127 citations.

Moreover, total citations received by the documents published from VSSC is 17395, among 1783 documents retrieved during the period, 1385 documents got citation and the remaining 398 documents have no citation so far. The average citations per paper are 9.76 and h-index of the institution is 55 based on Scopus data.

Thus, it is interpreted that VSSC contributed high quality articles right from the time of its inception and these articles received a good number of citations.

10. Keyword Analysis

Keywords used in the retrieved documents are further analyzed using VOSviewer. The indexing keywords used in the retrieved documents for more than 10 times are analyzed and clustered together in five clusters. The details are presented in Figure 6.

Table 9. Articles of Highest Citations Received

Authors	Title	Year	Source Title	Cited by
Satheesh S.K. & Krishna Moorthy K.	Radiative effects of natural aerosols: A review	2005	Atmospheric Environment	261
Sreejith S., et al.	Squaraine dyes: A mine of molecular materials	2008	Journal of Materials Chemistry	193
Madhusudanan P.M., Krishnan K. & Ninan K.N.	New approximation for the p(x) function in the evaluation of non-isothermal kinetic data	1986	Thermochimica Acta	183
Carol P., Sreejith S.& Ajayaghosh A.	Ratiometric and near-infrared molecular probes for the detection and imaging of zinc ions	2007	Chemistry - An Asian Journal	177
Babu S.S., Satheesh S.K.&Moorthy K.K.	Aerosol radiative forcing due to enhanced black carbon at an urban site in India	2002	Geophysical Research Letters	153
Reghunadhan Nair C.P., Mathew D.&Ninan K.N.	Cyanate ester resins, recent developments	2001	Advances in Polymer Science	136
SadasivaRao Y.V.K.&Nakra B.C.	Vibrations of unsymmetrical sandwich beams and plates with viscoelastic cores	1974	Journal of Sound and Vibration	127
Moorthy K.K., et al.	Integrated Campaign for Aerosols, gases and Radiation Budget (ICARB): An overview	2008	Journal of Earth System Science	117
Gladstone G.R., et al.	A pulsating auroral X-ray hot spot on Jupiter	2002	Nature	110
Krishna Moorthy K., Satheesh S.K. & Krishna Murthy B.V.	Investigations of marine aerosols over the tropical Indian Ocean	1997	Journal of Geophysical Research Atmospheres	109
Babu S.S. & Moorthy K.K.	Aerosol black carbon over a tropical coastal station in India	2002	Geophysical Research Letters	102
Venkateswara Rao G., Kanaka Raju K. & Raju I.S.	Finite element formulation for the large amplitude free vibrations of beams and orthotropic circular plates	1976	Computers and Structures	101
Aneesh Mohamed A.S., Gopinath A.& Baiju M.R.	A simple space vector PWM generation scheme for any general n-level inverter	2009	IEEE Transactions on Industrial Electronics	100

Figure 6. Keyword Analysis

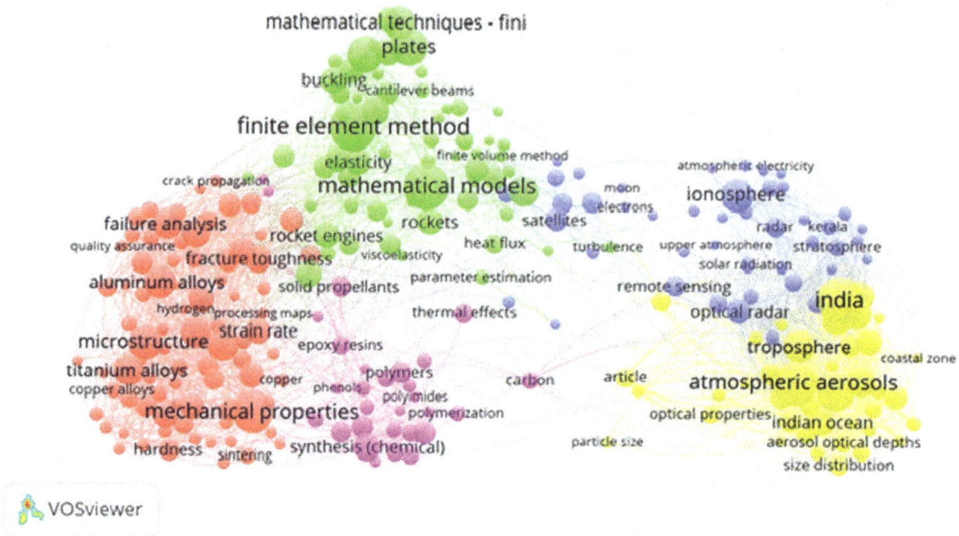

The cluster one consists of 16 terms namely crack propagation, failure analysis, quality assurance, fracture toughness, aluminum alloys, hydrogen, processing maps, strain rate, micro structure, titanium alloys, copper alloys, copper, mechanical properties, hardness and sintering.

In cluster two, the following 14 keywords included: mathematical techniques - finite, mathematical models, plates, buckling, cantilever beams, finite element method, elasticity, finite volume method, rockets, rocket engines, viscoelasticity, parameter estimation, heat flux, and solid propellants.

The cluster three also has 14 keywords. They are ionosphere, atmospheric electricity, moon, electrons, turbulence, satellites, radar, Kerala, upper atmosphere, stratosphere, solar radiation, remote sensing and optimal radar.

Cluster number four have 10 keywords such as synthesis (chemical), polymerization, polymides, polymers, phenols, carbon, thermal effects, and epoxy resins.

The fifth cluster has 9 keywords like atmospheric aerosols, troposphere, coastal zone, India, optical properties, Indian ocean, aerosol optical depths, size distribution and particle size.

FURTHER RESEARCH

The present chapter analysed the scientometric profiles of Vikram Sarabhai Space Research Centre (VSSC) based on the documents retrieved from the Scopus database and hence a similar study may be conducted based on the data retrieved from other databases like *Web of Science* (*WoS*). A comparative study on scientometric profile of similar research institution like other rocket launching centers of ISRO (Indian Space Research Organisation) may be conducted.

CONCLUSION

The Vikram Sarabhai Space Centre (VSSC) is a premier research institution involved in space research and rocket launching. Its research output or scientific productivity is to be measured occasionally to assess its present status and future course of action. The present study analysed the scientific productivity of the VSSC in terms of scientometric profiles using both quantitative and few qualitative indicators. The existing trend of growth of scientific contribution of scientists in VSSC is linear. Scientists are actively involved is disseminating their research findings in scholarly journals and Rao, G.V secured first position among the top-ranking authors. Moreover, cluster density map of co-authorship shows that authors who are having more documents to their credit have comparatively lesser number of co-authorships. Indian Institute of Science is the topmost collaborating institutions of VSSC and the United States is topmost collaborating country. Most of the scholarly communication of VSSC scientists are published in '*Journal of Sound & Vibration*' and majority of the records are of article type. Most of the articles published in Engineering related discipline. The study also found that total citations received the documents published from VSSC is 17395, among 1783 documents retrieved during the period, 1385 documents got citation and the remaining 398 documents have no citation. The average citations per paper are 9.76 and h-index of the institution is 55 based on Scopus data the outcome of the study is useful.

REFERENCES

Bhaskaran, C. (2013). Research productivity of Alagappa University during 1999-2011: A bibliometric study. *DESIDOC Journal of Library and Information Technology*, *33*(3), 236–242. doi:10.14429/djlit.33.3.4609

Bormann, L. D. (2007). What do we know about the h index? *Journal of the American Society for Information Science and Technology*, *58*(9), 1381–1385. doi:10.1002/asi.20609

Gopikuttan, A., & Aswathy, S. (2014). Publication productivity of university of Kerala: A scientometric view. *DESIDOC Journal of Library and Information Technology*, *34*(2), 131–139. doi:10.14429/djlit.34.4280

Gupta, B. M., & Dhawan, S. M. (2008). A scientometric analysis of S & T publications output by India during 1985-2002. *DESIDOC Journal of Library and Information &. Información Tecnológica*, *28*(2), 73–85.

Jeyshankar, R., Babu, B. R., & Rajendran, P. (2011). Research output of CSIR-central electro chemical research institute (CECRI): A study. *Annals Library Information Studies*, *58*(4), 301–306.

Kademani, B. S.; Kalyane, V. L. & Kumar, V. (2000). Scientometric portrait of Vikram Ambalal Sarabhai: a citation analysis. *SRELS Journal of Information Management*, 37(2).

Kumbar, M. (2009). Growth and impact of research output of University of Mysore, 1996-2006: A case study. *Annals Library Information Studies*, *55*(3), 185–195.

Matthews, A. (2013). Physics publication productivity in South African universities. *Scientometrics*, *95*(1), 69–86. doi:10.1007/s11192-012-0842-2

Nalimov, V. V., & Mulchenko, M. (1969). *Scientometrics: A study of science as information process. Macoow*: Nauka.

SudhierPillai, K.G. & Priyalekshmi, V. (2013). Research publication trend among the scientists of Central Tuber Crops Research Institute (CTCRI), Thiruvananathapuram: A scientometric studies. *Annals of Library and Information Studies*, *60*(1), 7–14.

Thirumagal, A. (2012). Scientific publications of Manonmaniam Sundaranar university, Tirunelveli: Scientometric analysis. *Library Philosophy and Practice*.

Tripathi, H. K., & Garg, K. C. (2014). Scientometrics of Indian crop science research as reflected by the coverage in Scopus, CABI and ISA databases during 2008-2010. *Annals of Library and Information Studies*, *61*(1), 41–48.

Chapter 13
Research Contributions on Nephrology During 2010-2015:
A Scientometric Approach

Radhakrishnan Natarajan
Periyar University, India

ABSTRACT

Nephrology is the adult and pediatric study of the kidneys and associated diseases. This chapter deals with the research productivity on Nephrology. Since the study is based on scientometric approach, the data has been retrieved from Web of Science database. The objectives of the study are: to identify the growth rate of research productivity, to trace the types of documents in Nephrology, to know the authorship pattern and language wise classification, and to find out the institution and geographical area wise distributions. Software packages such as Histcite and VOSviewer have effectively been used to analyze the data. In addition, various scientometric tools were also applied. As far as the contribution is concerned, the USA leads other parts of the world in Nephrology. English language dominates the globe in terms of publications.

INTRODUCTION

In 1969, Vassily V. Nalimov and Z. M. Mulchenko coined the term 'naukometriya', the Russian equivalent to the word scientometrics (Nalimov & Mulchenko, 1969). The word has been used for the study of all aspects of scientific literature of science and technology. The term had gained a wide recognition by the foundation (1978) of the journal *Scientometrics* by Tibor Braun, Hungary. Most of scientometric studies were indistinguishable from bibliometrics. The immediate and tangible output of science and technology is literature into a public domain in the form of papers, patents, etc. The focus of bibliometrics, despite many wide-ambit definitions, has always been preponderantly on the literature of Science and scholarship. There are many aspects for scientometricians to measure and analyze than the literature output; e.g., the practices of researchers, the socio-organizational structures, research and development management, the role of science and technology in the national economy and governmental policies.

DOI: 10.4018/978-1-5225-3457-0.ch013

NEPHROLOGY

Nephrology is the adult and pediatric study of the kidneys and associated diseases. The Nephrologists deal with the diagnosis and management of kidney related diseases. Kidneys are vital for maintaining normal fluid and electrolyte balance in a human body. The following is the list of kidney disorders.

- Fluid and electrolyte disorders
- Acid-base disorders
- Kidney stones
- Glomerular diseases
- Tubulointerstitial diseases
- Mineral metabolism
- Acute kidney disease
- Acute renal failure
- Chronic kidney diseases
- Chronic renal failure
- End stage renal disease and dialysis

Physicians need to be aware of medications and clinical pharmacology, high blood pressure diabetes management and its complications, epidemiology of diseases and infections as well as nutritional management for prevention and treatment of kidney related diseases.

RELATED STUDIES

Nadim Akhtar khan (2015) examined the research productivity of Government Medical College, Jammu and provided a bibliometric profile of biomedical publications of institute. Bibliometric indicators were employed to assess the trends and volume of research published. Publications by researchers and practitioners of Government Medical College, Jammu (1973 to 2011) are indexed in Scopus database. Results showed that the Government Medical College, Jammu was accelerating in terms of output. A total of 514 publications were contributed by the Institute during the study period. Highly productive departments are general medicine, pharmacology, surgery and pathology with 97 (18.87%), 69 (13.42%), 46 (8.95%) and 39(7.59%) publications respectively. The majority of publications were contributed by multi-authors (470, 91.44%). The most highly cited departments are Pediatrics, Endocrinology and Dermatology with average citations of 4.28%, 3.5% and 3.38% respectively. In addition, the study has focused on research for improving the health of individuals in the vicinity.

Mulla (2012) has described the bibliometric analysis of 998 articles on Information Science and Scientometrics (ISS) that appeared in different journals during the period 2005-2009. The study revealed that, most of the researchers have preferred to publish their research results in journals; as such 91.98% of articles were published in journals. More number (32.97%) of articles was published in 2009. The authorship trend showed that, out of 1,703 authors who contributed a total of 998 articles, out of which more number of (376, 40.96%) articles was contributed by two authors (25 papers each). The degree of collaboration was 0.78, and India could have contributed more documents i.e., 83.99% of the total publications. It examined year-wise distribution of articles, distribution of types of documents, length of the papers, authorship pattern, degree of collaboration among authors, degree of collaboration among co-authors, degree of collaboration among different category of authors, rank wise collaborators, institution wise distribution of articles, country wise and journal wise distribution of articles.

Raja and Balasubramani (2011) have carried out Plasmodium Falciparum research contributions in India which was measured using Histcite software and other tools. The results showed that the growth of Indian literature in Plasmodium Falciparum deposition and make the quantitative assessment of the research in terms of year-wise research output, geographical distribution, nature of collaboration, characteristics of highly productive institutions and the channel of communications / media used.

Dutt; Bharvi; Nikam & Khaiser (2014) have undertaken a study on collaboration in Solar Cell research in India as reflected on publications indexed in Web of Science for a period of 20 years (1991-2010). Half of the total output got emerged out of domestic and international collaborations. Among the prolific institutions, National Physical Laboratory, Delhi (CSIR) had the highest publications through collaborative research. Indian researchers had joined with academics in 31countries. However, South Korea, Japan, USA, Germany, England, France and Greece were dominant in collaborating the research partners. Various bibliometric indicators have been applied to examine the collaborative research activities.

Velmurugan and Radhakrishnan (2015) have identified the quantitative research output of Webology journal for the period 2007-2013. It is observed that the maximum number of research output was done by solo authors, whereas the minimum quantity of research articles was published by multiple authors. Maximum number of scholarly articles was published in 2008 whereas the minimum number of articles was found in 2009. The maximum articles were contributed by Indian authors. i.e. 18.99% and followed by Australia and Iran with 12.64% occupying second rank respectively. The maximum research output (50.64%) was done by multiple authors and the minimum research output (49.36%) was contributed by single author. The degree of collaboration in Webology journal is 0.506 (2007 to 2013).

OBJECTIVES OF THE STUDY

The major objectives of the present study are:

1. To identify the growth rate of research productivity on Nephrology.
2. To trace the types of documents in Nephrology.
3. To know the authorship pattern and language wise classification.
4. To find out the institution and geographical area wise distributions.

METHODOLOGY

The data were collected in October 2015 from the Web of Science Core Collection database using the following keywords. TOPIC: Nephrology, refined by: COUNTRIES/TERRITORIES: (INDIA) and Time span: 2010-2015 and Indexes: SCI-EXPANDED, SSCI, A&HCI, ESCI. The retrieved data were exported to MS Excel spreadsheet to analyze statistically and tabulate the data accordingly.

LIMITATIONS

This study covers a period of five years from 2010 to 2015 (both the years inclusive). All the records have been downloaded completely from the Web of Science online database.

SCIENTOMETRIC TOOLS USED

The following scientometrics tools namely Collaborative Index (CI) and Degree of Collaboration (DC) were applied for analysis. Simple percentage analysis is used to interpret data.

Degree of Collaboration

The degree of collaboration is defined as the ratio of the number of collaborative research papers to the total number of research papers in the discipline during a certain period of time. The formula suggested by Subramanyam is used. It is expressed as

$$DC = \frac{Nm}{Nm + Ns}$$

where, DC - Indicates the degree of collaboration in a discipline, Nm - indicates the number of multi-authored research papers in the discipline published during a year, Ns - indicates the number of single authored papers in the discipline published during the same year. Using this formula, the degree of collaboration is determined.

As the result, the degree of collaboration in the study (Nephrology) is 0.96 which showed the collaborations of multiple authors.

SOFTWARE PACKAGES USED

To retrieve the data from the database and for analysis the following software tools have been used. They are: *MS-Excel, HistCite* and *VOSviewer.*

ANALYSIS

The retrieved data are analyzed and interpreted according to the objectives. The details of analysis and interpretation of data presented as follows.

1. Year-Wise Growth on Nephrology

Table 1 shows that out of 2,622 research papers, the highest number of papers i.e. 510 (19.5%) with 1,707 total global citations and 185 total local citations got ranked first in 2013. The lowest number, i.e. 382

(14.6%) of research output with 3,098 total global citation score and the total local citation score is 365 in 2010. It is identified that the average number of research articles was 437 and also standard deviation and co-efficient variance have been calculated during the period of study.

2. Document-Wise Distribution

The Table 2 indicates the types of literature output in the forms of research Article, Review, Editorial Material, Meeting Abstract, Letter, Article; Proceedings Paper News Item, Biographical-Item, Correction, Article; Book Chapter and Review; Book Chapter which were identified. The result showed that out of 2,622, a majority of papers (69.5%) from journal articles with 9,151 global citations has been placed in first position and followed by 304 (11.6%) reviews, 230 (8.8%) editorial material, 165 (6.3%) meeting

Table 1. Year-wise Growth of Literature on Nephrology

S. No	Year	Recs	%age	TLCS	TGCS
1	2010	382	14.6	365	3098
2	2011	400	15.3	411	2879
3	2012	473	18.0	306	3304
4	2013	510	19.5	185	1797
5	2014	466	17.8	88	789
6	2015	391	14.8	23	126
Total		**2622**	**100**	**1378**	**11993**
Mean		437		229.67	1998.83
Standard Deviation		52.869		156.07	1318.95
C.V		0.121		0.67	0.65

Figure 1. Year-wise growth of literature on Nephrology

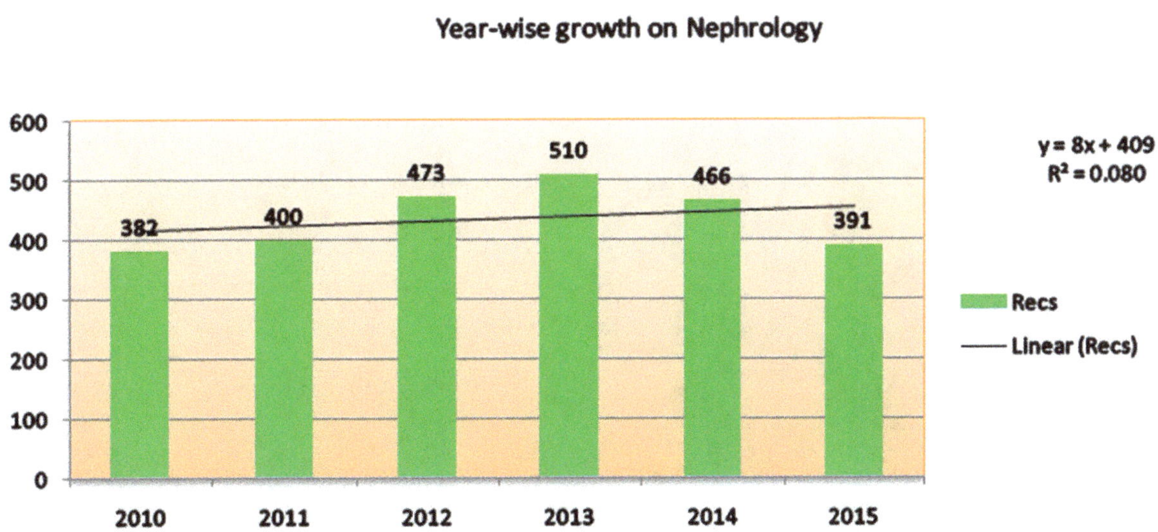

Table 2. Document- Wise Distribution

S. No	Document Type	Recs	%age	TLCS	TGCS
1	Article	1823	69.5	1099	9151
2	Review	304	11.6	144	2234
3	Editorial material	230	8.8	103	461
4	Meeting abstract	165	6.3	0	1
5	Letter	41	1.6	18	31
6	Article; proceedings paper	25	1.0	4	70
7	News Item	11	0.4	10	14
8	Biographical-Item	9	0.3	0	2
9	Correction	9	0.3	0	1
10	Article; book chapter	3	0.1	0	18
11	Review; book chapter	2	0.1	0	10
Total		**2622**	**100**	**1378**	**11993**

abstract, 41 (1.6%) letter, 25 (1.0) proceeding papers,11(0.4) News items, 9 (0.3) Bibliographic item, 9 (0.3) Correction, 3 (0.1) Article; book chapters, 2 (0.1) Reviews; book chapter were found during the study period. It is interesting to note that based on the global citation score; review manuscript has been placed in first position with 9,151 TGCS, and followed by articles with 2,234 TGCS which has occupied second level. It is noticed that most of the manuscript has cited in the form of articles globally. Hence there is a significant difference between journal articles and other documents.

The same is presented in the Figure 2.

Figure 2. Document- wise Distribution

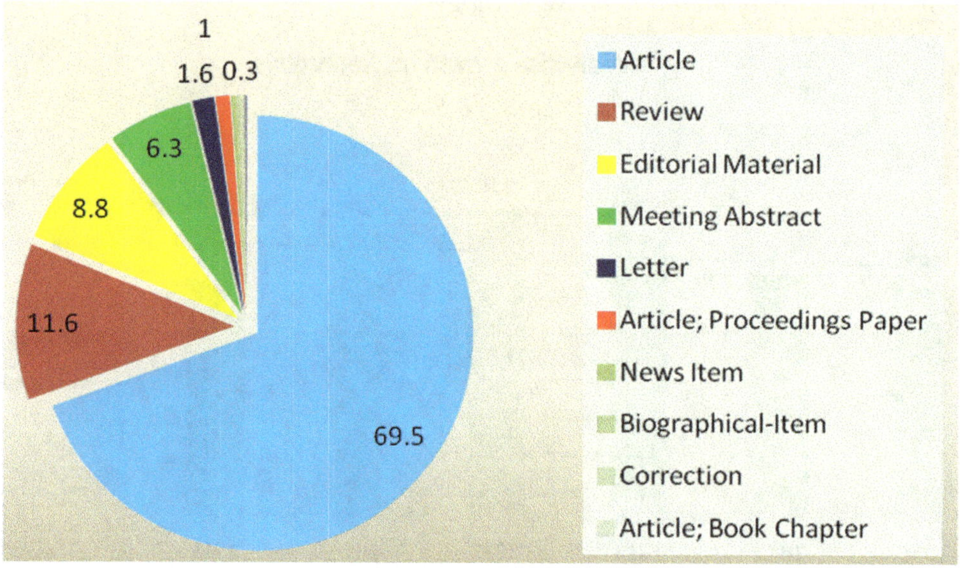

For a more accurate representation see the electronic version.

Table 3. Language- wise Distribution

S. No	Language	Recs	%age	TLCS	TGCS
1	English	2385	91.0	1333	11713
2	Spanish	120	4.6	37	228
3	French	61	2.3	4	23
4	German	30	1.1	3	17
5	Portuguese	6	0.2	0	3
6	Polish	5	0.2	1	3
7	Russian	4	0.2	0	0
8	Italian	3	0.1	0	0
9	Serbian	3	0.1	0	1
10	Korean	2	0.1	0	3
11	Turkish	2	0.1	0	2
12	Rumanian	1	0.0	0	0
Total		2622	100	1378	11993

Figure 3. Language-Wise Distribution

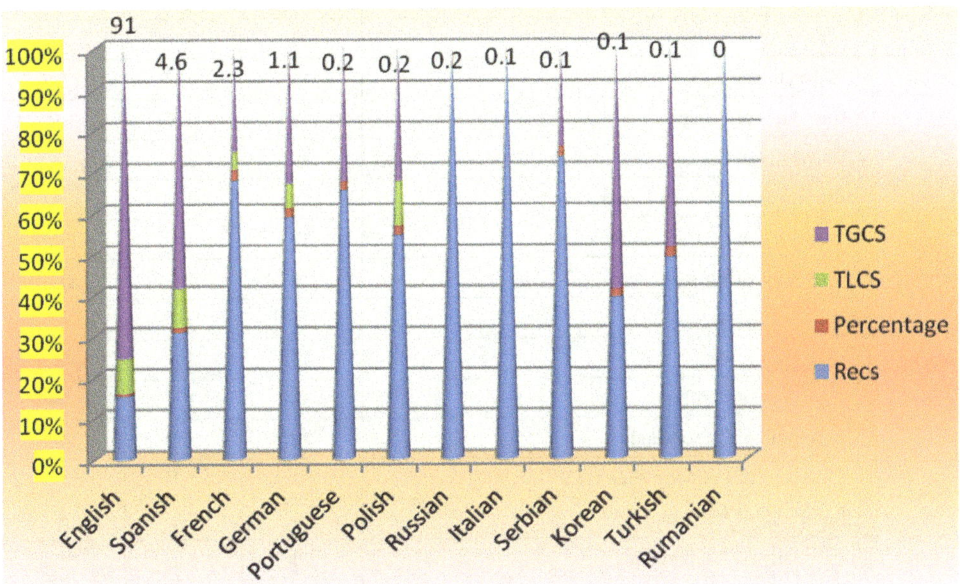

For a more accurate representation see the electronic version.

3. Language–Wise Research Output

It is identified that different languages have been used to publish research output. English, Spanish, French, German, Portuguese, Polish, Russian, Italian, Serbian, Korean, Turkish, and Rumanian have been used. Highest number of articles (91.0%) with11,713 TGCS are published in English language (predominant) and followed by Spanish articles got second position with 120 (4.6%) and the small amount of articles is written in Turkish, Korean, Serbian and Italian (each 0.1%) languages.

Language-wise distribution is presented in Figure 3.

4. Institution-Wise Distribution

Out of 380 institutions, only top 25 predominant organizations were chosen for the present study. The highest number of literature output (3.4.%) with 620 total global citations contributed by University of Amsterdam Research Center which has occupied the first position and followed by the same record count contributed by unknown contributors and Mayo Clinical which contributed 12 articles each with different citations such as 29 and 43 respectively.

Figure 4 shows institution-wise Distribution of records

Table 4. Institution -wise Distribution

S. No	Institution	Recs	%	TLCS	TGCS
1	Unknown	90	3.4	4	15
2	University of Toronto	61	2.3	61	453
3	Harvard University	46	1.8	69	393
4	University of Sydney	46	1.8	20	137
5	Mayo clinical	43	1.6	29	615
6	University of Alberta	43	1.6	85	493
7	University of Calgary	39	1.5	75	394
8	University of Western Ontario	38	1.4	85	497
9	University of Calif Los Angeles	37	1.4	26	325
10	University of Amsterdam	36	1.4	28	620
11	Emory University	35	1.3	35	134
12	University of Penn	35	1.3	41	389
13	University of Carolina	33	1.3	47	172
14	University of Queensland	30	1.1	4	98
15	University of Calif San Francisco	29	1.1	28	191
16	University of Michigan	29	1.1	20	175
17	University of Washington	29	1.1	26	286
18	Duke University	28	1.1	70	229
19	Johns Hopkins University	28	1.1	8	94
20	University of British Columbia	28	1.1	28	181
21	Yale University	28	1.1	28	187
22	University of Ottawa	25	1.0	44	208
23	University of Minnesota	24	0.9	18	198
24	Northwestern University	23	0.9	11	84
25	Ohio state University	23	0.9	21	244

Figure 4. Institution wise Distribution

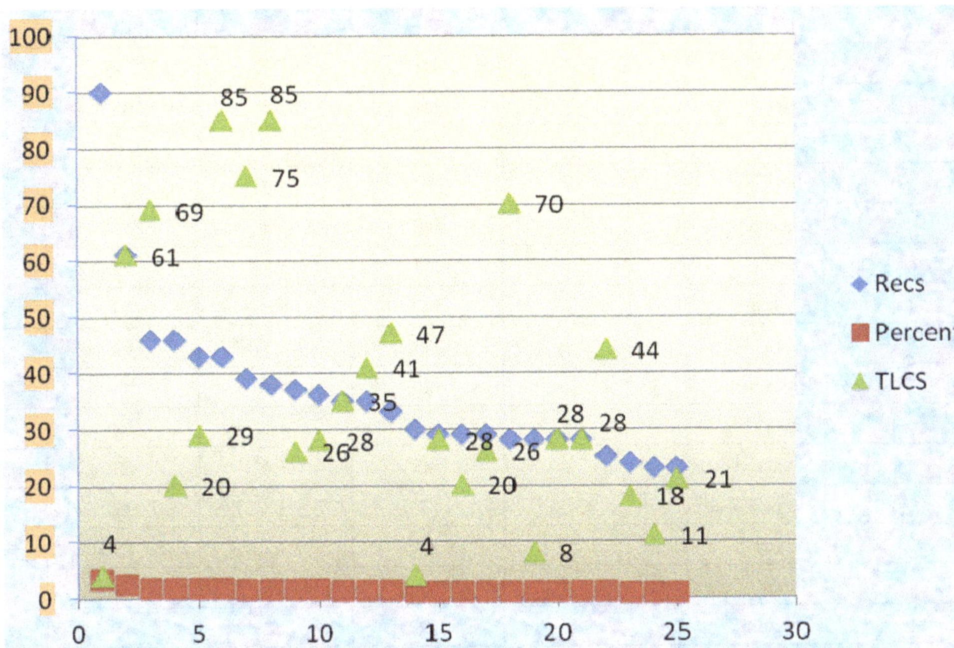

**For a more accurate representation see the electronic version.*

5. Geographical–Wise Production

It can be observed from Table 5 that 108 countries have contributed in Nephrology during the period of 6 years. USA has contributed 845 research output and the percentage rate is 32.2%. The global citation score is 5306 and has got the first place based on the record count and followed by Italy which has 216 records with 1634 global citation score and occupied the second rank, Canada got next position with 206 articles with 1912 citation score, and followed by UK which has 202 records with 1,732 citation score globally got ranked fourth. It is found that the USA has been placed first based on the citation score i.e. 5,036. Canada has been ranked in second position with 1,912 citation score and followed by UK which got placed in third position with 1,732 citation score. Based on the above analysis, there is a significant relationship between institution and country production during the period of study.

The same is presented in Figure 5.

Further country wise analysis is done by visualization using VOSviewer is presented in Figure 6.

6. Authorship Pattern

It is observed that about 90% of papers were contributed by multi authors. Out of 633 papers, the highest number of papers was published by double authors which accounts for 203 (32.07%) followed by three authored articles which lead 198 (31.28%.) 17.38% of articles was published by four authors. The authorship pattern is presented in Figure 7.

Table 5. Geographical area wise Distribution

S. No	Country	Recs	%	TLCS	TGCS
1	USA	845	32.2	736	5306
2	Italy	216	8.2	101	1634
3	Canada	206	7.9	210	1912
4	UK	202	7.7	151	1732
5	Germany	169	6.4	53	1408
6	France	151	5.8	57	955
7	Australia	147	5.6	74	802
8	Spain	143	5.5	67	836
9	Unknown	129	4.9	8	85
10	Japan	107	4.1	60	1303
11	Netherlands	90	3.4	53	1087
12	Peoples R china	87	3.3	31	711
13	Turkey	86	3.3	23	243
14	Brazil	66	2.5	34	335
15	Belgium	59	2.3	45	466
16	Poland	53	2.0	19	257
17	India	49	1.9	35	189
18	Sweden	46	1.8	29	446
19	Austria	30	1.1	25	240
20	South Korea	30	1.1	8	424
21	Denmark	29	1.1	7	222
22	Taiwan	29	1.1	13	75
23	Greece	28	1.1	22	250
24	Egypt	27	1.0	5	100
25	Argentina	26	1.0	9	42

FINDINGS

1. As far as year wise distribution is concerned, 510 research outputs (19.5%) with 1,707 total global citations and 185 total local citations got ranked first in 2013.
2. A majority of papers (69.5%) was found in journal with 9,151 global citations.
3. English, Spanish, French, German, Portuguese, Polish, Russian, Italian, Serbian, Korean, Turkish, and Rumanian have been used to publish the research papers. Highest number of articles (91.0%) with11,713 TGCS was published in English language (predominant) and followed by Spanish with 120 (4.6%)

Figure 5. Geographical area wise distribution

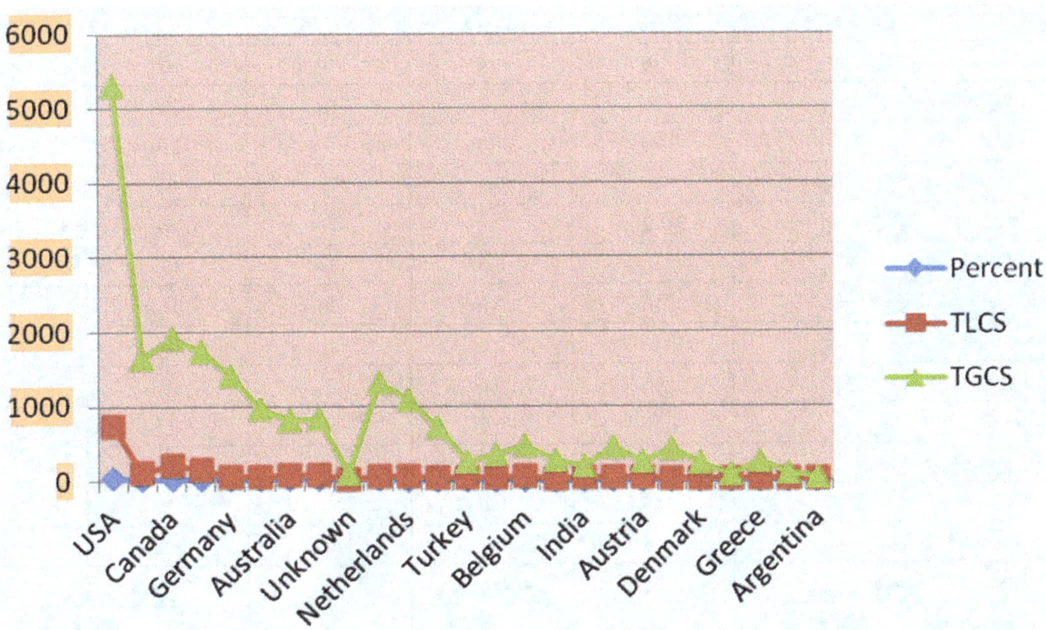

Figure 6. Country wise Map
Source: VOSviewer software used to visualize the top most countries

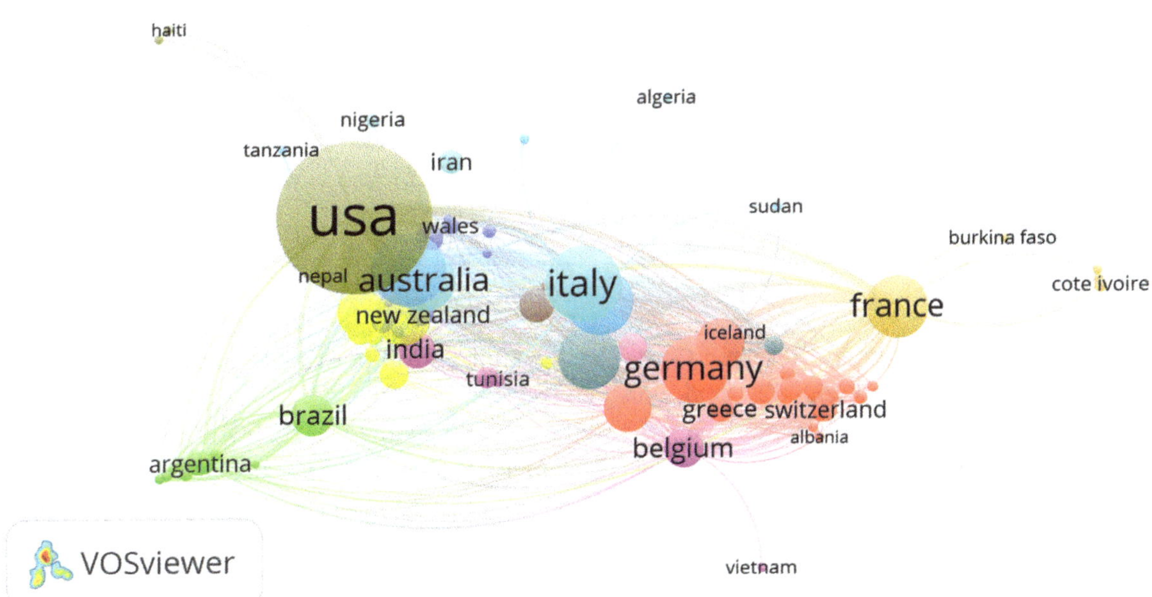

Table 6. Authorship Pattern

S. No	Author	Recs	%age	TLCS	TGCS	TLCR
1	Anonymous	36	1.4	0	0	0
2	Jager KJ	28	1.1	10	190	15
3	Zoccali C	26	1.0	9	230	15
4	Jhaveri KD	24	0.9	71	91	88
5	Craig JC	22	0.8	11	69	14
6	De Nicola L	16	0.6	6	109	12
7	Garg AX	16	0.6	30	242	12
8	Minutolo R	16	0.6	6	109	12
9	Ronco C	16	0.6	7	183	3
10	Schaefer F	16	0.6	9	77	17
11	Sinha MD	16	0.6	15	36	14
12	Conte G	15	0.6	6	108	12
13	Dekker FW	15	0.6	9	142	4
14	Hemmelgarn BR	15	0.6	48	212	11
15	Shah HH	15	0.6	36	52	75
16	Furth SL	14	0.5	17	149	12
17	Perazella MA	14	0.5	12	48	13
18	Vanholder R	14	0.5	10	88	22
19	Ferris M	13	0.5	36	85	17
20	Inward C	13	0.5	15	35	14
21	Martinez-Castelao A	13	0.5	1	32	7
22	Berns JS	12	0.5	31	98	15
23	Desai T	12	0.5	30	54	22
24	Glassock RJ	12	0.5	3	67	4
25	Levin A	12	0.5	9	111	6

4. University of Amsterdam Research Center has occupied a top position with the high number of research output (3.4%) with 620 total global citations.
5. USA has contributed 845 research papers and the percentage rate was 32.2%. Though India invest huge amount of funds for research and development, it could have got 17th position across the globe.
6. 90% of papers were contributed by multi authors. Out of 633 papers, the highest number of papers was published by double authors which account for 203 (32.07%) followed by three authored articles which has led 198 (31.28%.). 17.38% of articles were published by four authors.

Figure 7. Authorship Pattern

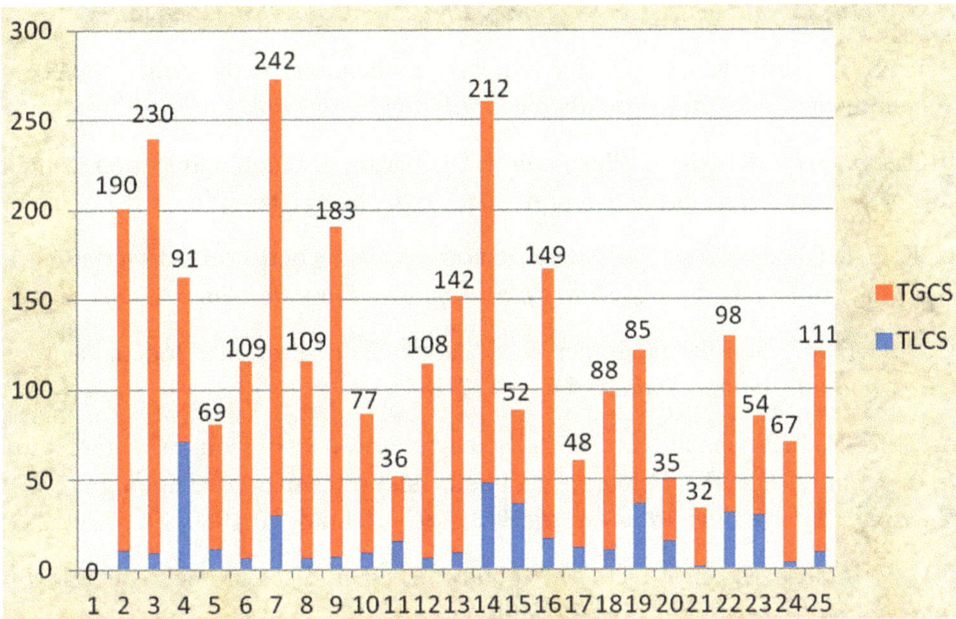

CONCLUSION

Nephrology is one of the upcoming areas of research which requires enormous funding from both government and private agencies. There are R & Ds which regularly concentrate on new diseases and medications. Researchers should also show interest in new and emerging thrust areas of health sciences. There are many agencies involve in research in India. All India Institute of Medical Sciences (AIIMS) and Indian Council of Medical Research (ICMR) have actively been involved since their inception. Still it requires a major attention on taking up of research to a greater extent.

REFERENCES

Ashokkumar, N., Radhakrishnan, N., & Mahesh, K. C. (2008). Recent trends in ophthalmology– A citation analysis. *PEARL–. Journal of Library and Information Science*, *2*(2), 13–16.

Dutt, B., & Nikam, K. (2014). Scientometrics of collaboration pattern in solar cell research in India. *Annals of Library and Information Studies*, *61*(1), 65–73.

Khan, N. A., & Ahangar, H. (2015). *Research productivity of government medical college Jammu: a Bibliometric Analysis*. Library Philosophy and Practice.

Mulla, K. R. (2012). Identifying and mapping the information science and scientometrics analysis studies in India (2005-2009): A bibliometric study. *Library Philosophy and Practice*, (June).

Radhakrishnan, N. & Velmurugan, C. (2015). Scholarly publications by the faculty members of Periyar University, Salem, India: A scientometric approach. *Library Philosophy and Practice*.

Raja, S., & Balasubramani, R. (2011). Plasmodium falciparum research publication in India: A scientometric analysis. *European Journal of Scientific Research*, *56*(3), 294–300.

Sudhierpillai, K. G. & Dileepkumar, V. (2010) Scientometric study of doctoral dissertations in biochemistry in the university of Kerala, India. *Library philosophy and practice*, June.

Sudhierpillai, K. G. (2007). Journal citation in physics doctoral dissertation of Indian institute of science. *Annals of library and Information Studies, 54*(4), 177-184.

Velmurugan, C., & Radhakrishnan, N. (2015). Scientometric observations of authorship trends and collaborative research on DESIDOC journal of library and information technology. *Collnet Journal of Scientometrics and Information Management*, *9*(2), 193–204. doi:10.1080/09737766.2015.1069957

Velmurugan, C., & Radhakrishnan, N. (2015). Webology journal: A scientometric profile. *International Journal of Information Dissemination and Technology*, *5*(2), 137.

Velmurugan, C., & Radhakrishnan, N. (2015). A scientometric analysis of research papers published on pharmacognosy as reflected in the Web of Science. *Advances in Pharmacognosy and Phytomedicine*, *1*(1), 27–40.

Chapter 14
Authorship Pattern and Collaborative Research Productivity of *Asian Journal of Dairy and Food Research* During the Year 2011 to 2015

Kavitha Elavamalai Sivasubramanian
Periyar University, India

ABSTRACT

The chapter brings out the results of a bibliometric analysis of the journal titled Asian Journal of Diary and Food Research for the period from 2011 to 2015. The data was downloaded from the journal's website. This study aims at analyzing the research output performance of co-authorship index and collaborative index. The analysis covers mainly the publication of articles, year-wise distribution of articles, degree of collaboration, institution-wise distribution, authorship pattern, etc. The results show the merits and weakness of the journal which will be helpful for its further development.

INTRODUCTION

The term bibliometrics was derived from "*biblion*" (Greek) meaning book and "*metron*" (Greek.): meaning measure; and was introduced by Pritchard in 1969. Pritchard's article "Statistical Bibliography or Bibliometrics" appeared in the December issue of the Journal of Documentation in 1969. He stated, "the term (Statistical bibliography) is clumsy, not very descriptive, and can be confused with statistics itself or bibliographies on statistics". In the same issue of Journal of Documentation appeared Robert A. Fairthorne's classic article "Empirical Hyperbolic Distributions (Bradford-Zipf-Mandelbrot) for Bibliometric Description and Prediction", in which the author used the word "bibliometric" and acknowledged Alan Pritchard as the donor of the term.

DOI: 10.4018/978-1-5225-3457-0.ch014

REVIEW OF RELATED LITERATURE

Papreja and Shukla (2016) conducted a bibliometric analysis of 539 articles published in the journal titled "Dental Clinics of North America" during 2004 to 2014 has been carried out. The paper analyzed year wise distribution of paper, average length of paper, authorship pattern, bibliographic form of documents, distributions of citations and subject trends of articles published in the journal. The result shows that the journal on an average has published 49 research articles per year and majority of the papers were contributed by two authors. Most of the articles dealt with general dentistry. The average length of the paper is 17 pages per article and journals are the most predominant form of citations.

Anil Kumar, Mallikarjun Dora and Asha Desai (2015) attempted to undertake bibliometric analysis of the research publications of Gujarat University during the ten-year period between 2004 and 2013. The data for this study was extracted from SCOPUS and included a total of 760 publications that were attributed to authors affiliated to Gujarat University. The publication data was analyzed with respect to the type of publications where 83% are journal articles. The paper also analyzed the publication trend of Gujarat University and found that from 2008 onwards there was a steady increase in the number of publications. The other aspects that were identified in the paper were the most prolific authors, collaborative authorship patterns and trends, most preferred publications, and so on.

Shanta and Betageri (2015) presented the findings of a bibliometric study one of the renowned Journal of Dairying, Foods and Home Sciences into consideration with an aim to analyze the contributions of the author and the citations received by various articles published in it. The paper covered the bibliometric aspect of year wise distribution of articles, subject wise distribution of articles and authorship of contributions. The study analyzed 34 issues containing 542 articles published in the said journal from 2003-2012. Highest number (65) of articles was published in 2013. Most of the articles 283 (52%) were 1-4 pages long.

Surwase, et al (2014) studied with the aim to analyze the global publication trends on food preservation using Scopus database for the period 1998-2012. The database contained 17511 publications on food preservation. The study analyzed the broad features of literature on food preservation focusing on year-wise distribution of publications, highly productive countries, international collaboration, activity index, highly productive institutes, methods of food preservation, preservation by food types, and channels of communication

When a single journal is studied bibliometrically, it creates a portrait of the journal, providing a description that offers an insight that is beyond the superficial. It can indicate the quality, maturity and productivity of the journal in any field, in a country or region (Anyi, Zainab & Anuar, 2009). The studies reviewed show the importance of bibliometric studies focusing on a journal publishing food science articles. Asian Journal of Dairy and Food Research is the official publication of the Agricultural Research Communication Centre. It provides a forum for the scientific community to publish their original research articles in the field of Dairy and Food Science. Hence this journal was subjected to a bibliometric study focusing on the growth of literature and collaboration.

OBJECTIVES OF THE STUDY

- To know the year wise distribution of articles.
- To know the issue & month wise distribution of articles.

- To classify the author wise distribution.
- To find out the degree of collaboration.
- To classify the institution wise distribution.
- To identify the geographical distribution of articles.
- To identify the length wise distribution.
- To know the year wise distribution of citation.
- To classify the single and multiple authored distribution of articles.
- To analysis of the relative growth rate and doubling time.
- To analysis of the co-authorship index.
- To classify collaborative index.

RESEARCH METHODOLOGY AND LIMITATION OF THE STUDY

The data was collected from the journals titled of *Asian journal of dairy and food research.* The data consists of 309 contributions made in the volume number 30 to 34 consisting of 4 issues in each volume. The analysis was done on authorship pattern, year-wise distribution, degree of collaboration, length-wise distribution of articles, geographical distribution of articles, year-wise citation of articles, and institution wise contributions. The data was also analyzed to find Subramaniam's degree of collaboration in quantitative term's. On the whole, data were subsequently examined, captured, analyzed and tabulated for making observations.

DATA ANALYSIS AND INTERPRETATION

The entire details of the data such as the article distributions of year wise, issue and month, authorship pattern, institution, geographical distribution, length, citation of distribution, and relative growth rate & doubling time were studied. In this chapter the investigator has presented the collected data, its interpretation by using a statistical calculation with help of Microsoft Excel in the form of inferences for academic study.

Table 1 shows the number of articles published from 2011-2015. It is noticed that the highest number of articles 66(21.36%) were published in the year 2015, followed by 65(21.03%) in 2012, 61(19.74%) in 2011 and 60(19.42%) in 2014. The lowest number i.e 57(18.45%) papers were published in the year 2013.

Table 1. Year wise distribution of articles

Year	Vol. No	No. of. Issue	No. of. Contribution	Percentage
2011	30	4	61	19.74
2012	31	4	65	21.03
2013	32	4	57	18.45
2014	33	4	60	19.42
2015	34	4	66	21.36
Total		20	309	100

Table 2 Shows that issue wise distribution of articles. One fifth of articles were published respective in the year of 2015 (66). In the year 2012, 65 articles were published, 61 articles were published in 2011, and 60 articles were published in the year 2014. A least number of articles (57) were published in year of 2013.

Table 3 identify that the details about the authorship pattern of articles published during the study period. Out of total 309 journals of articles, the highest number of contributions i.e. 96(31.07) have been contributed by two authors followed by 93 contributions (30.1%) by three authors. Four authors papers are 58(18.77%) in number. Forty-seven (15.21%) papers have more than five authors or more and the lowest number of contributions i.e.15 (4.85%) by single authors during the period of study.

Degree of collaboration (DC) is propounded by Subramanyam (1983). According Subramanyam $DC = \frac{\text{Nm}}{\text{Nm} + \text{Ns}}$ (where, Nm is the number of multi-author articles and Ns is the number of single author articles). Table 4 indicates degree of collaboration. The number of single author contributions is 15 articles out of 309 articles and the rest 294 articles are multiple authors' contributions. The degree of collaboration range started from 0.91 to 0.98. The average degree of collaboration range is 0.95.

Table 5 shows the Co-Authorship Index distribution of articles is highest sum in double authors (96), followed by three authors (93). Four authors have CAI of 58and more than four author papers have CAI of 47. The lowest co-authorship index is observed in single authors (15) during the period of year 2011 to 2015.

Table 2. Distribution of article issue wise

Year	Vol. No	Issue & Month Wise				Total
		March	June	September	December	
2011	30	14	16	16	15	61
2012	31	18	16	15	16	65
2013	32	14	17	13	13	57
2014	33	15	14	16	15	60
2015	34	17	17	15	17	66
Total		78	80	75	76	309
Percentage		25.24	25.89	24.27	24.6	100

Table 3. Authorship pattern

Author	Year						
	2011	2012	2013	2014	2015	Total	Percentage
Single	1	1	4	5	4	15	4.85
Two	23	19	21	15	18	96	31.07
Three	17	24	17	20	15	93	30.1
Four	11	11	8	11	17	58	18.77
Above four	9	10	7	9	12	47	15.21
Total	61	65	57	60	66	309	100

Table 4. Degree of collaboration

Year	Single Author	Multiple Author	Total	Degree of Collaboration
2011	1	60	61	0.98
2012	1	64	65	0.98
2013	4	53	57	0.92
2014	5	55	60	0.91
2015	4	62	66	0.93
Total	15	294	309	0.95

Table 5. Pattern of Co-Authorship Index

Year	Single Author		Double Authors		Three Authors		Four Authors		< Four Authors		Total
	No	CAI	No	CAI	No	CAI	No	CAI	No	CAI	
2011	1	34	23	121	17	93	11	96	9	97	**61**
2012	1	32	19	94	24	123	11	90	10	101	**65**
2013	4	145	21	118	17	99	8	75	7	80	**57**
2014	5	171	15	80	20	110	11	98	9	98	**60**
2015	4	125	18	88	15	75	17	137	12	119	**66**
Total	**15**		**96**		**93**		**58**		**47**		**309**

Table 6 shows the pattern of collaborative index. The highest value of CI is 0.34 (2012 and 2013) and the lowest value of CI is 0.3 in2015. The average collaborative index is 0.32.

Table 7 shows the details about the distribution of contributions by institution wise during the study period. Out of the total of 309 articles, more than half of the articles were contributed by University, around 19%, 12% of articles were published by research institution and technology respectively.

Table 8 shows that the geographical distribution of scientific publications. The findings reveal that out of 309 articles the maximum number of 64 (20.79%) documents were published by state of Haryana authors whereas the minimum number 1(0.32%) paper were published by state of Bihar and Meghalaya authors.

Table 6. Pattern of Collaborative Index

Year	No. of Articles	No. of Authors	CI
2011	61	187	0.33
2012	65	194	0.34
2013	57	169	0.34
2014	60	193	0.31
2015	66	223	0.3
Total	**309**	**966**	**0.32**

Table 7. Institution-wise contribution

Name of institution	No. of. Articles	Percentage
Research Institution	56	18.12
University	183	59.23
College	25	8.09
Technology	37	11.97
Others	8	2.59
Total	309	100

Table 8. Geographical and State Wise Distribution of Article

Name of the Institution	No. of Articles	Percentage
Tamilnadu	23	7.44
Uttrakhand	7	2.26
Uttra pradesh	23	7.44
Andhra pradesh	6	1.94
Maharastra	43	13.91
Karnataka	28	9.06
Haryana	64	20.79
Punjab	22	7.11
Jammu& Kashmir	6	1.94
Rajasthan	21	6.79
Bihar	1	0.32
Gujarat	14	4.53
West Bengal	5	1.61
Jharkhand	4	1.29
Mathya Pradesh	6	1.94
Assam	3	0.97
Telungana	9	2.91
Himachala pradesh	4	1.29
Chattisgarh	2	0.64
Kerala	4	1.29
Meghalaya	1	0.32
Others	13	4.20
Total	**309**	**100**

Table 9 Shows that the length wise distribution of publications. It is found that 212 articles published have 1-5 pages. Ninety-four articles have 6-10 pages. Very few articles were published with pages 11-15.

Table 10 shows the citation range distribution of publications. It is found that most (95 papers and 30.74%) of contributions are 11-15 citation range and lowest contributions are in the 0-5 citations are found in 39 (12.62%) publications.

Table 11 revealed the maximum citation of articles 1158(27.47) in 2015 and minimum citation of articles 529 (12.55) in 2011.

Table 12 explained the relative growth rate and doubling time. Relative growth rate gradually increased from 0.72 to 1.64 and at the same time doubling time decreased from 0.42 to 0.96 during the years 2011 to 2015.

Table 9. Length of the Article

No. of. page	Year					Total	Percentage
	2011	2012	2013	2014	2015		
1-5	45	51	27	41	48	212	68.60
6-10	15	14	28	19	18	94	30.42
11-15	1	-	2	-	-	3	0.97
Total	61	65	57	60	66	309	100

Table 10. Year Wise Citation of Distribution

No. of. Citation	Year					Total	Percentage
	2011	2012	2013	2014	2015		
0-5	18	11	4	4	2	39	12.62
6-10	21	22	9	13	12	77	24.91
11-15	14	19	22	24	16	95	30.74
16-20	6	7	10	12	20	55	17.79
Above 20	2	6	12	7	16	43	13.91
Total	61	65	57	60	66	309	100

Table 11. Total Citation of Distribution

Year	Citation	Percentage
2011	529	12.55
2012	726	17.24
2013	949	22.51
2014	853	20.23
2015	1158	27.47
Total	4215	100

Table 12. Status of Relative Growth Rate & Doubling Time.

Year	Total Article	Cumulative Total	RGR			
			W1	W2	W2-W1	DT
2011	61		4.11	4.83	0.72	0.96
2012	65	126	4.17	5.20	1.03	0.67
2013	57	183	4.04	5.49	1.45	0.47
2014	60	243	4.09	5.73	1.64	0.42
2015	66	309	4.18			

FINDINGS OF THE STUDY

- The majority of articles were published in 2015, whereas the lowest sums of articles were published in 2013.
- The maximum number of articles and contributions were published in the month of June issues where as the minimum number of articles were published in month of September, during the period 2011-2015.
- The highest number of contribution 96 (31.07) of articles have been contributed by two authors followed by contributions 93 (30.1%) of articles have been contributed by three authors.
- This research reveals the average degree of collaboration in, Asian journal of dairy and food research is 0.95, which clearly portray of dominance upon the multi-authored contributions.
- The Co-Authorship Index distribution of articles is highest among the double authors papers (96), followed by three authors papers (93).
- The highest value of CI observed is0.34 (2012 to 2013) and the lowest value of CI is 0.3 (2015), the average collaborative index is 0.32.
- The maximum sum of articles was contributed by multi-authors 294(0.95%), and the rest 60(0.98%) articles were contributed by single authors. It is inferred that majority of the articles have been contributed by multi-authors only.
- Out of the total articles 309, the majority of 183 (59.23%) of paper were published by universities followed by 56 (18.12%) of articles were published by research institutions.
- Out of the 309 articles, the geographical distribution of articles in the maximum number i.e., 64(20.79%) of articles were published by state of Haryana authors whereas the minimum number i.e., 1(0.32%) of articles were published by state of Bihar and Meghalaya authors.
- The research showed from the study of length-wise contributions majority of (1-5) pages, 212(68.60%) and least number (3 papers, 0.97%) of contributions are in the 11-15pages.
- The highest citations of contributions are 11-15 citation range 95(30.74%) articles were published and lowest contributions are in the 0-5 citation range 39(12.62%).
- The total distributions were analyzed for maximum citation of articles 1158(27.47%) and minimum citation of articles 529 (12.55%), in between during the period of citation, 2011 to 2015.
- Out of the result for relative growth rate and doubling time. The relative growth rate gradually increased from 0.72 to 1.64 at the same time doubling time decreased from 0.42 to o.96 in the year of 2011 to 2015.

CONCLUSION

Bibliometric studies are used to assess the scientific and systematic research distributions of energetic information, which shall be of use to the principle-makers. This study has demonstrated to be a useful tool in the assessment of research publications of food scientists. This study has highlighted the various measures such as growth of articles, authorship pattern, institution of distribution, pattern of co-authorship index, collaborative index, geographical distribution and citation wise distribution of articles which can be used to observe the journal's geographical and state-wise contributions of articles. The study will be helpful to the library professionals in data collection out of the journals and also to identify the authors. The study will be useful to procure core journals in the area of food science and technology.

REFERENCES

Anyi, K. W. U., Zainab, A. N., & Anuar, N. B. (2009). Bibliometric studies on single journals: A review. *Malaysian Journal of Library and Information Science*, *14*(1), 17–55.

Betageri, S. S. (2015). Bibliometric Analysis of contributions in the journal of dairying, foods and home sciences. *Journal of Advances in Library and Information Science*, *4*(1), 86–89.

Fairthorne, R. A. (1969). Empirical hyperbolic distributions (Bradford-Zipf-Madelbrot) for bibliometrics description and prediction. *The Journal of Documentation*, *25*(4), 319–343. doi:10.1108/eb026481

Gupta, B.M. (2012). Scientometric analysis of pakistan's S &T resource output. *Annals of library and information studies, 59*(1), 25-38.

Gupta, B. M., & Bhattcharya, S. (2004). A bibliometric approach to word & mapping the dynamics of science and technology. *DESIDOC Bulletin of Information Technology*, *24*(1), 3–8. doi:10.14429/dbit.24.1.3616

Haiai, Z. (1994). A Bibliometric study on medicine Chinese tradition in medicine database. *Scientometrics*, *31*(3), 241–250. doi:10.1007/BF02016874

Hertzel, D. H. (1987). Bibliometrics, history of development of ideas in". In A. Kent (Ed.), Encyclopedia of Library and Information Science (Vol. 42, pp. 144-219). New York: Marcell Dekker.

Jose, C. S., & Casado, E. (1983). Bibliometric analysis pf cheese literature covered by the food science and technology. *Journal of Dairy Science*, *66*(7), 1528–1535. doi:10.3168/jds.S0022-0302(83)81969-9

Kavitha, E. S., & Sivaraj, K. (2014). Bibliometric study of Ph.D. thesis in commerce. *e-library science research journal, 2*(11), 1-9.

Kumar, A. H., Dora, M., & Desai, A. (2015). A Bibliometrics profile of Gujarat university, Ahmedabad during 2004-2013. *DESIDOC Journal of Library and Information Technology*, *35*(1), 9–16. doi:10.14429/djlit.35.1.7699

Mishra, P. N., & Gowsami, N.U. (2013). Impact of grant aid project at CSIR national matellurgical laboratory, Indian: A Bibliometric study. *SRELS journal of information management, 50*(4), 387-412.

Papreja, S., & Shukla, A. (2016). Dental clinics of North America 2004-2014: A bibliometric study. *Library Herald*, *54*(1), 49–51. doi:10.5958/0976-2469.2016.00004.X

Pritchard, A. (1969). Statistical bibliography or bibliometrics. *The Journal of Documentation*, *25*(4), 348–349.

Rajinikanth, A. (2009). A bibliometrics analysis of surveying engineering literature. *Indian Journal of Information Science and Services*, *3*(1), 31–35.

Rao, G. M. (2014). *Food Science and Technology*. New Delhi: New Age Publishers.

Sethi, M. & Rao, E.S. (2005). *Food Science*. New Delhi: New Age International.

Subramanyam, K. (1983). Bibliometric study of research collaboration: A review. *Journal of Information Science*, *6*, 33–38.

Surwase, G., Mohan, L., Kademani, B. S., & Bhanumurthy, K. (2014). Research trends on food preservation: A scientometric analysis. *DESIDOC Journal of Library and Information Technology*, *34*(3), 257–264. doi:10.14429/djlit.34.5979

Thirumagal, A., & SethuKumari., N.S. (2013). Mapping of scholarly research in cloud computing: A bibliometric study. *SERLS journal of information management, 50*(5).

Van Raan, A. F. S., & Van Leeuwen, T. H. N. (2002). Assessment of the scientific basic of interdisciplinary applied research, and Application of bibliometric methods in nutrition and food research. *Research Policy*, *31*, 611–632. doi:10.1016/S0048-7333(01)00129-9

Vijay, R. K. (2005). Bibliometric study of research publication trends among, Indian food scientists and technologists. *Annals of Library and Information Studies*, *52*(3), 77–81.

Webster's new world dictionary of the American language. (1960). Cleveland: World Publishing Company.

Chapter 15
The Concepts of Academic Humility and Seniority in Scholarship:
A Critique of Research Traditions in Nigerian Universities

Floribert Patrick C. Endong
University of Calabar, Nigeria

ABSTRACT

The concept of academic humility has often been ill-defined or ill-conceived by members of the Nigerian academia. It argues that this concept has objectionably been associated – nay confused – with "academic subservience", "academic hypocrisy", "false modesty" and "yesmanism". Such misconception could partially be attributed to the prevalence of the "seniority syndrome". In effect, the seniority syndrome has often wanted that pertinence be ascribed unto a research idea or thesis not necessarily on account of the robustness of the methodology that led to its enunciation, but principally on account of the "seniority" (credentials) of the researcher who authors the thesis or idea in question. This tendency has generally stemmed from the faulty assumption that a senior researcher is theoretically more knowledgeable than his or her junior counterparts; and that the "junior researcher" must manifest unconditional reverence for his or her senior colleagues. This scenario is most often observed during Ph.D. these defenses and similar forums aimed at evaluating research in Nigerian universities.

INTRODUCTION

From conceptualization to publication/dissemination of scholarship, research is guided by a wide range of cannons and ethical principles. Some of these principles and cannons include, academic honesty, objectivism, professionalism, contribution to knowledge and/or profession and academic humility among others. These principles and traditions generally govern specific levels of the research process. Academic honesty and objectivity for instance mainly shape methodological designing; while profes-

DOI: 10.4018/978-1-5225-3457-0.ch015

sionalism guides the conceptualization process. In tandem with this, academic humility is mainly required at the publication or dissemination of research results. It is a pillar in the performance of the researcher, particularly in such forums as research conferences, workshops, seminars, thesis defense and other research-related oral exercises. Academic humility is therefore supposed to be a cardinal aspect of the researcher's personality (Kaplan, 2013; O'Leary, 2016; Straits et al., 2012; University of Otago, 2011). Adedemeji (2013) presents it as the trait of "true" and "genuine" scholars. It is an ethical value or scholarly behavior which ought to be reflected in the latter's research ideas as well as in his or her approaches to selling and defending these ideas.

However, academic humility has proven to be very elusive in nature, particularly among Nigerian researchers. This is not unconnected to the fact that the phenomenon of "humility" itself, has often been found semantically sleepery and contextually equivocal. As Worthington (2007) insightfully argues, irrespective of the context in which it is used, the concept of humility raises a number of paradoxes. Its virtuous nature is subject to personal or individual conceptions. Using the American experience, Worthington further posits that though most people think of humility as a personal psychological strength, research has demonstrated that humility is not equally valued in all people.

People say they must value humility within religious seekers; the same people do not value humility quite as much when it is shown in a close partner or close friend. Studies show that Americans are least accepting of humility in leaders, yet even here they seem divided. Some people surveyed say they want their leaders to have outspoken confidence, bordering on arrogance; others say they prefer for their leaders to have almost saintly humility. So, is humility a virtue? Well, it depends on who you ask. (Worthington, 2007, p.4)

Furthermore, the phenomenon of humility exemplifies concepts that are difficult, if not impossible to scientifically study. In Worthington's language, a growing body of research on humility is susceptible to help researchers and other professionals to be more humble. However, research suggests that there are limits to what science can tell us about humility.

In Nigeria in particular, the concept of academic humility has often been ill-defined or ill-conceived within academic and intellectual environments. This concept has objectionably been associated – nay confused – with "academic subservience", "academic hypocrisy", "false modesty" and "*yesmanism*" (the tendency of being a yes man). This could partially be attributed to the prevalence of the "seniority syndrome". In effect, the seniority syndrome has often motivated many members of the Nigerian academia to ascribe pertinence unto research ideas, a theory or a thesis, not necessarily on account of the robustness of the methodology that led to its enunciation, but principally on account of the "seniority" (academic title and years of experience) of the person who authors the thesis, theory or idea in question. This tendency has generally followed the faulty assumption/ axiom that a veteran (senior in the domain of research) is theoretically and "infallibly" more knowledgeable and more experienced than his/her junior counterparts; and that the "junior researcher" must, in all situations show or manifest unconditional respect towards his senior colleague by taking the latter's "dictation" on the research issue being raised (Agu, Omenyi & Odimegwu 2015; Balarabe 2009; Olayinka 2014).This questionable trend has visibly been in line with the African traditionalism current which among other tenets, exalts unconditional respect for elders and authorities in all social contexts.

Using both empirical understandings and conceptual constructions, this paper stresses the necessity to rethink the value of academic humility in the context of research in Nigeria. This rethinking exercise actually appears pertinent and timely as the ill-definition or misconceptions of the value of academic humility have tended to negatively affect the free generation, sharpening and argumentation of ideas (particularly by junior colleagues in research) in such contexts as thesis defenses, conferences and seminars organized in some Nigerian universities. As this paper will strongly argue and illustrate, junior colleagues in research are often made to kill their "intellectual virility" in the face of aggressive and academically arrogant "seniors" who believe in "imposing" their views on the arguable ground of their experience, and not necessarily on the robustness of their argumentation ideas and theses. In order words, academic humility has been understood by many junior researchers as "academic subservience" as well as, as being a "yes man/woman". By such an invidious system, the evaluation of research via such forums as research conferences, Ph.D. Oral defenses as well as the assessment of other postgraduate research works have, in some instances, been systemically perverted.

This paper is based on (un)official semi-structured interviews with Ph.D. and former postgraduate students as well as on interviews with a number of supervisors working with some Nigerian universities. It equally considered non-participatory observations. In the process of such observations, the researcher attended oral examinations in many faculties at three universities situated in the South-Eastern part of Nigeria. These included the Universities of Calabar, Uyo and the Cross River University of Technology (CRUTEC). The use of (un)official interviews and non-participatory observations is justified by the fact that the "seniority syndrome" (academic arrogance) is more or less a taboo in the Nigerian academia. The above-mentioned data collection methods were complemented by a critical exploitation of relevant secondary sources.

The present discourse is divided into four main parts. The first part provides a theoretical framework composed of African traditionalism (respect for elders' theory) and the constructivist theory of learning. The second part similarly offers conceptual definitions of two terms, namely academic humility and seniority. The third part shows how the concept of seniority has over the years been upheld in research in the world; while the fourth part of the paper deals with the ways in which the seniority syndrome perverts a number of research traditions in Nigerian universities.

THEORETICAL FRAMEWORK

As earlier mentioned, this paper is anchored on two theories namely African traditionalism and the constructivist theory of learning. These theories have many dimensions. However, this section will exclusively highlight the specific tenets (of the theories) that will be considered in the discourse.

African Traditionalism

African traditionalism is a socio-philosophical current which exalts African traditional values. These values include communalism, ancestral worship, respect for authorities and the elders, hospitality, solidarity, sense of sacredness of life, sense of good human relation, sense of the sacred and the sense of language and proverbs among others. In this discourse, we shall exclusively consider tenets of the theory that are related to the issues of respect for elders.

African traditionalism stresses on the unconditional respect for elders and authorities. In effect, though the African continent is not monolithic in nature, all its cultures have at least one thing in common: the unconditional reverence for elders. This reverence is manifested in various ways notably through models of greetings and interactions with elders. According to the traditionalists, when an elder dies, it is like a library went up in flames. A similar traditionalist aphorism stipulates that what an elderly man can see while seated, a young one cannot see even if he climbs up a baobab or the tallest tree in the forest. In other words, elders are naturally wise, more experienced and good educationists. By virtue of such attributes, they should be respected if not revered whatsoever the context. Respect here means the youths should avoid contradicting the elders and should content themselves with listening and obeying the old. By this tenet, the African traditionalism logic establishes a seniority system in which elders and authorities are unconditionally placed at the top of the knowledge pyramid while the youths are at the bottom.

Given the fact that traditionalism and modernism still co-habit in good number of social institutions in Africa including Nigeria, the reverence of elders and authorities tends to be subtly superimposed on interpersonal relations in professional environments. An egregious illustration is the fact that senior staffs in many services often assume the position of elders who must religiously be respected by their juniors. In the same line of thought, Ofiaja (2014) canvases manifestations of unconditional respect for elders in Nigerian politics. She notes that:

We are too deferential to elders and authority figures alike, even when it's quite clear that the people we are deferring to don't have our best interests at heart. It's why many of our political leaders can remain in office for so long. The populace won't revolt despite being treated poorly. We will complain, but we won't revolt. (para 7)

In the Nigerian academia, a similar situation is observed with unconditional respect often given to senior lecturers and professors by junior staffs. This myth is therefore extended to teaching and research domains in that, senior researchers most often assume – and are considered by their junior colleagues as having – the monopoly of knowledge. Seniority in teaching and research is thus equated to dexterity in the supervision and assessment of research works completed by their junior (notably PhD and Master students).

Constructivism

As a theory of learning, constructivism is based on observation and scientific study. As its name indicates, the theory states that people construct their own understanding and knowledge of the world based on their personal experiences (previous knowledge and understanding, preferred mode of learning and their personal motivations and feelings). In other words, the things we experience and the way we reflect on them help us construct our knowledge of the world. The constructivist theory of learning suggests that, rather than being transmitted, knowledge is actively constructed by learners. The latter learn by applying their knowledge to meaningful problems, actively building their own understanding.

In the context of teaching, supervision and evaluation of research, knowledge ceases to be like a relatively unproblematic commodity which is transmitted to the student or supervisee who is (faultily) perceived as suffering from a knowledge deficit (Shen, Wu, Achhpiliya, Bieber & Hiltz, 2004; Wellington 2010). No one (notably the student or junior researcher) is a *tabula rasa* (meaning an empty slate) or a passive recipient but an "active constructor" with immense potential to contribute to the generation of new knowledge. The Educational Broadcasting Corporation (2004) corroborates this position when it contends that the constructivist view of learning usually "means encouraging students to use active techniques (experiments, real-world problem solving) to create more knowledge and then to reflect on and talk about what they are doing and how their understanding is changing". The teacher, examiner, supervisor makes sure that he or she understands the students' pre-existing conceptions, and guides the activity to address them and then build on them.

In line with this, oral examinations should not be considered a site for a unidirectional flow of knowledge but a give-and-take business – call it a collaborative exercise. The students learn from their lecturer or examiner the same as the lecturer learns from the students. Corroborating this position, Stanford professor Scott Dick (2009) views research evaluation contexts such as Ph.D. *viva voce* examinations as a site of mutual respect and collaboration between examiner, supervisors and students. He notes that Ph.D. research work defenses provide: "A miniature model of the scientific community at its best. It fosters an 'open architecture' in which teachers and student are able to intensively collaborate for a few hours, each participant attempting to contribute ideas to advance a common body of work. What transpires is an intensive, highly charged, purposeful and personal encounter. At its best it is, for me, the university's golden hour" (para 9).

CONCEPTUAL FRAMEWORK

To ensure clarity of analyses in this discourse, it may be expedient to provide the conceptual definitions of two of the key terms featuring in the topic under study namely "academic humility" and "seniority".

Academic Humility

The concept of academic humility does not have a universally accepted definition. It is somehow derived from the word "humility" which is dominantly interpreted as a kind of virtue, though variously conceived. As noted by Austin (2014), humility is variously associated with low self-esteem or self-denigration. Nevertheless, as Archbishop of Canterbury William Temple contends, "humility does not mean thinking less of yourself than of other people, nor does it mean having a low opinion of your own gifts. It means freedom from thinking about yourself at all" (cited in Worthington, 2007).

A good number of theorists (philosophers) have, in their reflections, associated humility with forms of vices. Hume (1983) notes for instance that humility and self-denial – among other "monkish virtues" – are everywhere rejected by men of sense. They serve to no manner of purpose; neither do they advance a man's fortune in the world, nor render him a more valuable member of society. They neither qualify man for the entertainment of company, nor do they increase his power of self-enjoyment. "We observe, on the contrary, that they cross all these desirable ends; stupify the understanding and harden the heart, obscure the fancy and sour the temper. We justly, therefore, transfer them to the opposite column, and place them in the catalogue of vices" (Hume 1983, p.73-74).

It goes without saying that Hume's conceptualization of humility is glaringly arguable. Though subject to multiple interpretations, humility is dominantly viewed as – and has proven to be – a positive value. It has been demonstrated that humble people are likely to perform high in individual and team settings (Austin, 2014; Exline & Hill 2012). Humility is a three-part personality trait which includes an accurate view of the self, *teachability* and the recognition or appreciation of others' strengths. All these point to positivity. The concept equally implies the absence of arrogance, pride and narcissistic entitlement. Based on this, the least definitional illumination one may provide on the concept of academic humility is that, the term refers to a special kind of positive or scholarly attitude required of researchers. It follows the educational logic that a scholar should shy away from being arrogant – he should avoid showing off for knowledge accumulated – and rather be respectful towards his peers, whatsoever the latter's level of knowledge.

The academic humility aphorism therefore stipulates that one (the researcher) should not brag even when he has attained a high and impressive level of knowledge. He should not brag when he does not know. Rather, he should consider knowledge as being "an eternal conquest". This simply means that academic humility warrants the researcher to always seek ways to improve on his knowledge and see himself as someone who does not know everything and who must try new things. This imagination brings to mind the great British scientist and inventor Isaac Newton who, despite his life changing inventions, always considered himself to be like a little boy "playing on the sea-shore, and diverting [himself] now and then [...] whilst the great ocean of truth lay all undiscovered before [him]" (cited in Adedemeji 2013).

Academic humility equally warrants the researcher to accept the possibility that anyone can teach him a new thing. Corroborating this position Umbecco (2015) critically examines the rationale of academic humility. He notes that:

Academic humility [entails] the knowledge that anyone can teach us something. Perhaps this is because we are so clever that we succeed in having someone less skilled than us teach us something; or perhaps even someone who does not seem very clever to us has some hidden skills; or also because someone who inspires us may not inspire others. The reasons are many. The point is that we must listen with respect to anyone, without this exempting us from pronouncing our value judgments; or from the knowledge that the author's opinion is very different from ours, and that he is ideologically very distant from us. But even the sternest opponent can suggest some ideas to us. It may depend on the weather, the season, and the hour of the day. (p.143)

The contrary of academic humility is therefore academic arrogance, manifested by the tendency by a (pretentious) scholar to always think high of himself, even in context where he or she knows nothing or too little. Academic arrogance is equally manifested by the tendency for one to use his title (say of professor) to "lord it over" to junior colleagues irrespective of the level to which his or her argumentation is scientific or logical. In this paper we will use the neologism "seniority syndrome" as a synonym of academic arrogance.

Seniority

The term "seniority" is generally used in the corporate world. It refers to the length of time someone has served in a job or the number of years one has worked for a specific organization. MacLeod (1987) defines it as a system used in an industrial relations context, to designate the status of an employee in relation to other employees of the same workplace. This brings to the fore the binary opposition between senior and junior. Senior workers naturally have higher status than their junior colleagues. In some contexts, the latter are compelled to show a relative deference – if not reverence – to the former. MacLeod identifies two types of systems of seniority namely straight seniority (which is job security acquired solely through length of service), and qualified seniority (which considers other factors such as ability in conjunction with length of service).

In this paper we shall consider seniority systems employed in universities (involved in research) and research institutes particularly in the Nigerian context. These seniority systems remarkably vary across universities/research institutes and countries in terms of the nomenclature used to identify stages, grades or ranks in the teaching/research career as well as in terms of the systems of academic promotion being adopted. However, one notes that in spite of these differences in ranking and promotion systems, seniority is generally principally based on teaching experience, research experience, level of education, volume and quality of scholarly publications and ability to get research support among other factors. In line with this, professors are generally senior, compared to lecturers who are simply holders of PhD certificates. Similarly, senior lecturers are higher in rank than junior or assistant lecturers and so on. By implication/ extension, lecturers may, to an extent, be considered senior to students and research assistants. In line with this, the relationship between a Ph.D. scholar and his supervisor or examiner could be equated to a form of senior-junior relationship (University of Edinburgh 2016). As explained by Agu, Omenyi and Odimegwu (2015), the two parties have a mentor-mentee relationship. "The supervisee is expected to tap into the knowledge base of his supervisor, who should be more experienced in the field of study/ endeavor. This implies that the student should be shown by the supervisor how to write and successfully complete a dissertation" (p.565). This paper will equally consider issues like erudition, celebrity and greater visibility among peers as other indexes of seniority.

SENIORITY AND THE QUALITATIVE EVALUATION OF RESEARCH WORKS

It will be very important from the outset to illustrate how the concept of seniority subtly affects the evaluation of research. This section of our discourse speaks to this objective. In effect, as a value, seniority is variously celebrated in the international research community. This is vividly illustrated by a number of issues, one of which is the fact that the credibility of research works (scholarly articles, books, working papers among others) is often evaluated on the basis of the status or credentials of their authors or assessors (publishers). In line with this, research works that are authored by professors, Ph.D. holders and other senior researchers have sometimes been acclaimed not necessarily on the basis of merit but on ground of authors' credentials. A piece of research work authored by a professor is likely to be viewed as the work of an expert nay, the contribution "of a superhuman" (Fedler 1994). This tradition is, of course questionable as the credibility of information should, in principle, depend on the information itself and not on the status of authors and endorsers of this information. Corroborating this position, Standler (2004) reviews cases of fraudulent and misleading research results having been generated by professors and so-called *authoritative sources*. He notes that:

Some physicians and scientists have submitted fraudulent research results to respected journals, which published those fictitious results after approval by peer-review. While one hopes that such episodes are rare, it is clear that peer-review offers no protection against fraud [even] by [respected] authors. [...] It is inherently unscientific to trust an expert, instead of making a personal evaluation of the reason for the expert's conclusion (p.2)

Another factor illustrating the paramount nature of seniority in the evaluation of research is the prevalence of various myths which stipulate that only professors and Ph.D. holders can be credible researchers and assessors of research works. In tandem with this, there exist many (specialized) reviews, journals, and research platforms which are deliberately not open to contributions from junior researchers – those who are not professors or Ph.D. holders. In the same line of thought, many journals do not admit peer-reviewing by Master degree holders. Furthermore, a number of journals are open exclusively to contributions from experts and researchers perceived to be erudite and authorities in their field of specialization. Though it may be argued that such a choice is – at first sight – rationalized by the need to ensure quality research, it somehow remains clear that such a choice illustrates the arguable myth stipulating that only Ph.D. holders, professors and experts can credibly carry out or assess research works. As put by Holmes (2015), many people think a Ph.D. is an indicator that one is a qualified academic. It can be tempting to consider the degree as a chance to prove that one is already qualified for that title. In the same line of argument, Dhiman (2016) enumerates a long list of factors pleading in favor of the myth which states that undergoing a Ph.D. program and writing a thesis constitute an essential step in the (self-)training of a researcher, if not the foundation of good research. He explains that well-conceived Ph.D. programs enable researcher to have a wider view of issues and to develop critical skills, as well as broader and rigorous methodological designs which are central in research.

According to him, a Ph.D. scholar hardly limits his research within the scope of his own discipline but searches even beliefs that may not directly be related to his discipline. This cerebral activity is beneficial in two ways: first, it enables the researcher explores other beliefs which are consistent with his own observations and which will help him further explain, interpret, widen and broaden his own beliefs about issues under investigation. Second, the activity (the act of undergoing a Ph.D. program) enables the researcher explore gaps, differences and contradictions in beliefs or theories related to issues. Ph.D. programs actually sharpen the evaluative skills of researchers making them more trained than their junior colleagues (those not possessing doctoral qualifications).

Dhiman further contends that the various kinds of evaluation exercise a Ph.D. scholar goes through during the writing of his thesis help him unearth differences in beliefs for factors such as time, place, person, history, socio-economic conditions, and environment. Such exercises help him develop skills to "weigh stronger reasons against other weaker reasons, which is the precondition for creating better knowledge. But writing articles do not always follow this epistemic process. That is why a Ph.D. work is usually considered a foundation for doing good research and writing good articles". In the same line of argument, Agu, Omenyi and Odimegwu (2015) contend that completing a Ph.D. program is *sine qua non* for someone to be able to complete a research work independently and be seen as an authority in it.

It must be recognized that one's possession of a Ph.D. degree or a professorial status is evidence of research experience. These statuses automatically indicate that one has, at least initiated or completed a number of research projects (Ph.D. thesis, supervision of dissertations, journal articles and research programs among others). However, as earlier said, this does not – in itself – make one *ipso facto* a good researcher. It is unscientific to conclude that impressive credentials are a guarantee of research competence as a number of studies have reviewed poor quality and incredible research works generated by some Ph.D. holders and professors (Felder 1994; Lislie 2000; Standler 2004). Conversely, the fact that one does not have the above-mentioned statuses does not automatically mean he is "limping" in terms of research productivity or simply a novice in the practice of research. As Umbecco (2015) argues, the "best ideas may not come from the major authors". By extension, it may conversely be enthused that the best ideas may come from self-educated pundits who are neither Ph.D. holders nor professors. The plausible argument here is that, researchers' academic statuses and level of education are not perfect criteria for research evaluation.

A number of research principles subtly gives credence to the foregrounding of early theorists and major authors (which, to a reasonable extent, may be viewed as senior researchers); while relegating emergent researchers (junior ones) to the background. A good illustration of this scenario is the ensemble of research theorems which underscore the imperative for researchers to always anchor the analyses of their findings on relevant theories or stress the imperative to discuss findings in the light of revered scholarly sources. In effect, taping from authoritative sources (authors of pioneer studies and major authors) to strengthen ideas is believed to render the research report/work more credible. As noted by Maher (2010) the inclusion of such authoritative sources gives the reader or the assessor confidence that the researcher has thoroughly "done his homework" when conducting his research. It shows how the researcher's results and conclusions fit within the established work of others, besides allowing the interested reader or assessor to find sources of additional in-depth information. The research paradigms which underscore the centrality of authoritative sources subtly re-enforce or institute the "hegemony" of a category of researchers (established authors) over emergent authors (minor authors/researchers). The later must "stand on the shoulders of the former" to gain credence and credibility before their peers. All these observations, according to this author, imply a subtle form of seniority system.

HOW THE "SENIORITY SYNDROME" PERVERTS RESEARCH TRADITIONS IN NIGERIAN UNIVERSITIES

As reviewed in the preceding sections of this discourse, a good number of research principles and the internal politics/policies of research related organizations (research institutes, journals and conferences among others) have subtly encouraged various seniority systems. These systems are not always inherently antithetical to effective evaluation of research. However, problematic conceptions of seniority have tended to adversely affect research. In the Nigerian universities particularly such questionable conceptions of seniority have tended to affect research traditions in various ways.

There is a form of academic arrogance among some senior lecturers and researchers which tends to seriously affect the quality of research produced in Nigerian universities. We have chosen to use the term "seniority syndrome" in reference to this academic arrogance, as it mainly consists of viewing "presumed" senior/experienced researchers (such as [associate] professors, senior lecturers and other highly ranked lecturers) as infallible arbiters and custodians of the research technology. What we mean here is that the seniority syndrome has caused many members of the Nigerian university communities – including junior researchers – to view senior colleagues in research as "demigods" of research simply by virtue of their credentials rather than by the quality of research works they produce or author.

This syndrome has equally often spurred Nigerian junior or less informed researchers to ascribe pertinence unto a research idea, a theory or a thesis, not necessarily on account of the robustness of the methodology that led to its enunciation, but principally on account of the "seniority" (academic title and years of experience) of the person who authors the thesis, theory or idea in question. The adverse effects of the seniority syndrome on research traditions in Nigerian universities can be explored in at least three main areas: (i) the evaluation of research (thesis and seminars) produced by Ph.D. scholars and Master students, (ii) popular conceptions of academic humility among both junior and senior teaching staffs and (iii) lecturers' attitude towards the intellectual property.

The Evaluation of Research Works Produced by Ph.D. and Master Scholars

The evaluation of research works (Ph.D. and Master dissertations and theses) has in some instances, been variously perverted by adepts or perpetrators of the "seniority syndrome". One way through which this pervasion has taken place is the instauration of mechanisms which more or less accidentally intimidate candidates (junior researchers) during Ph.D. and Master oral examinations. Such mechanisms subtly stultify the free generation and defense of research ideas by candidates. A common tradition in Nigerian universities has been to conceive graduate seminars and Ph.D. theses defense as harrowing experiences during which candidates (generally perceived as less experienced) are made to listen to quasi omniscient examiners or a panel of "all knowing" professors. This politics often favors a scenario wherein the Ph.D. candidate is bombarded with all forms of both relevant and flimsy dictations from the examiners and evaluation panels. No doubt, terms and idioms such as "sitting on the firing seat" have been conceived in Nigerian university milieus to capture the often-antagonistic nature of panels that examine Ph.D. scholars during thesis defenses in Nigerian universities. The "sitting on the firing seat" neologism refers to the fact that most often, the Ph.D. candidate is viewed as a target which the examiner should, as a matter of compulsion, "fire" at will, with tough questions and "injunctions". This however, is not to insinuate that all examiners are adepts of such questionable methods of assessment.

Ph.D. these defenses in some Nigerian universities are often organized in a way that suggests that the examiner is *ipso facto* more knowledgeable than the candidate and can provide better rationalization for the outcome/findings of the research work which, curiously, happened to have (diligently) been carried out by the student. It is therefore common that, instead of offering the candidate the opportunity to freely and constructively present his ideas and arguments, some examiners tend to transform the exercise into a "magisterial lecture". In two of such Ph.D. oral examinations that the author attended, the external examiners did not even give the floor to candidates to discuss their research works. They reduced the examinations to criticisms and injunctions aimed at the candidates. It is based on this observation that Balarabe (2009) censures the seniority syndrome among examiners and supervisors working with Nigerian universities positing that: "most of our professors are so academically arrogant as if they have the monopoly of all knowledge. Nothing describes this than the adage 'an empty vessel makes the loudest noise'" (para 3).

It goes without saying that such an attitude (the tendency of viewing the postgraduate students or junior researchers *ipso facto* as kinds of tabula rasa or inexperienced scholars) is in total contradiction to the logic stipulating that the author of the research work is (in principle) the most suited person to present the details and implications of his work. Besides this, the attitude represents a serious threat to the effective evaluation of research done by Ph.D. and Master scholars, as it tends to downplay any relevant argumentation that may emanate from junior scholars on the ground that the candidate is likely a novice or "empty slate" which automatically has to learn from the more experienced – if not omniscient – examiner. This, in itself, contradicts the constructivist theory of learning which stipulates that the student is not a *tabula rasa* but an entity which can circumstantially generate interesting and well enunciated knowledge. As argued by Wellington (2010), the myth stipulating that postgraduate students inherently suffer from knowledge deficit is outdated. This is so as Knowledge has since long, ceased to be a relatively unproblematic commodity which needs to be passed to students (junior scholars) who are viewed as lacking this commodity. The act of teaching, supervising or evaluating students, "rather like filling a car with petrol, is to make good this deficit" (p.9). Therefore, fair models of evaluating research in higher education need to take a view of the postgraduate student as being an "active constructor" rather than a passive recipient.

Criticizing the tendency by examiners and assessors to view postgraduate students (junior researchers) as people who *ipso facto* suffer from knowledge deficit, Olayinka (2014) remarks that many senior colleagues in the Nigerian academia have fallen into the perception trap which makes many postgraduate students to rightly or wrongly accuse them of being braggarts who are fond of "boasting" that no student can score an "A grade" in their examination. Such academically arrogant examiners claim "C is for the intelligent student, B is for the genius and A is for God" (Olayinka, 2014, p.8). In the same line of thought, Ekundayo (2014) refers to such questionable methods of perceiving and assessing students as cancers which persistently erode and weaken the Nigerian university system as a whole. He succinctly notes that:

There are some invidious maggots, which have been ravaging standards in our universities, weakening performance and "wiping off smiles' from the faces of students, using a fine phraseology by Tanure Ojaide in Matters of the Moment. The maggots in question operate in the form of rules and manifests in certain assessment practices in the universities. Simply stated, there exist in our university system some unfair questionable and objectionable established scoring methods, invidious assessment and feedback habits and rules which harm [postgraduate] students and lecturers [supervisors] consistently. (p.11)

The undue credence given to the myth of the superhuman professors and seniors researchers has sometimes motivated faculties to appoint examiners and supervisors who are not experts in the field of specialization to which the work under examination is classified. The latter are thus made to perform in these capacities simply by virtue of their status as senior researchers or professors. This questionable paradigm stems from the faulty assumption that one simply needs to be a professor or a senior lecturer to be considered thick enough to supervise or assess research works done by Ph.D. scholars. In one of the oral examination that this researcher attended, a senior lecturer in English linguistics who had no proficiency in foreign languages was assigned to assess a Ph.D. work written and presented in the French language. In another case, a professor still in English linguistics and communication was made to examine a work done in English-French translation. The exercise turned out to be a veritable fiasco as the examiner inadvertently showed real signs of knowledge deficit in translation and ironically became the person who was examined in the process. This finding corroborates Balarabe's (2009) succinct remark that:

Quite a number of Ph.D. theses are supervised [and examined] by people who do not have any proven training, research contribution or in-depth knowledge in the area they are supervising. They are supervisors just because they have Ph.D. in that subject! I know a professor in such a situation who has supervised many Ph.D. students in Topology, Algebra, History of mathematics, and Computer sciences; all added to his laureates. A good undergraduate student of mathematics should be able to understand the impossibility of this ridiculous Academic. (para 3)

It must be underlined that what is problematic here is not the fact that a Ph.D. holder or professor is assigned to assess a junior colleague; but the fact that one's status as a professor is perceived as a valid passport which grants him or her the right to venture in domains which are outside his or her "comfort zones". It is therefore problematic to believe that professorship alone makes one abreast of the task of supervising or assessing the research work of a junior colleague who specializes in a different discipline. As posited by Awuni (2016), research has proven that the myth of superhuman professors and senior lecturers has led to bad quality assessment and supervision of Ph.D. or Master students' research. Studies show that many academic staff in the universities actually need further training to be able to supervise and assess graduate students.

Perverted Perceptions About Academic Humility

The seniority syndrome has visibly perverted popular perceptions of academic humility in some Nigerian universities. A good number of researchers – particularly junior ones – in these universities tend to view academic humility as "academic subservience" or "yesmanism" (being a yes man), particularly in the face of senior academic debaters or opponents. This, again, is usually seen in Ph.D. these defenses when a candidate is challenged by a professor or another senior member of the examination panel. The most probable reaction of candidates in such confrontational contexts has often been guided by the questionable belief that, it is inherently risky and totally contrary to the principle of academic humility to challenge a professor or senior colleague in matters of research, even if one (the junior researcher) is sure in his heart to be "on track" and to have a better idea than his senior. Challenging a professor in the context of an intellectual debate or oral examination is still viewed by many members of the Nigerian research community as pure arrogance and a lack of respect. It has even been found that such argumentative attitude by junior scholars is often taken in bad faith by some professors (Adedemeji 2013; Balarabe 2009). It is therefore not uncommon to find Ph.D. scholars who resent contradicting professors and senior colleagues even when they have better and more plausible ideas. The scenario seems to be worse particularly in examination conditions where the examiners (senior colleagues) have immense power to apply more or less subjective parameters to assess the works of Ph.D. and Master students. Agu, Omenyi and Odinmegwu's (2015) somehow share corollaries as they argue that, "though various faculties provide explicit rubric for Ph.D. research work assessment, most supervisors/assessors assess their candidates' works implicitly. This indicates [the] use of global rather than analytic forms of assessment" (p.565).

For fear of provoking the "raft" or "enmity" of the examiner; and to avoid the latter's negative assessment of their research works, some Ph.D. candidates often resolve in their hearts to avoid challenging their supervisors, assessors and examination panels (in dissertation defense) by defending positions that are contrary to those of the latter. This has been so even when these Ph.D. scholars follow a better logic than their senior colleagues. Effiong C. (one of our informant who is a former Ph.D. scholars) shares corollaries as he exposes the risks of contradicting a professor or senior lecturer in the context of an oral examination. He contends that:

It takes more than boldness to frontally confront a professor and push your idea in the context of a postgraduate oral examination. Most Nigerian professors believe in bragging and intimidating junior colleagues in such contexts [...] I experienced it during my Master thesis defense. I faced a panel led by a professor who, in the course of the examination, repeatedly asked me to discuss how my work contributes to knowledge and (the mass media) profession. About five times she asked me to discuss my contribution to knowledge. Each time she interrogated me, I gave the same answer. The professor interpreted my act (the fact that I gave the same answer each time she asked the question) as arrogance, over confidence and disrespect for the panel [of assessors]. At a point, she asked me whether I thought I was more knowledgeable than the panel of examiners. She poured reproaches and turned the whole exercise to a confrontational and harrowing experience.

Data collected through informal interviews with postgraduate students reveal that most junior researchers resent "intellectual arm wrestle" with professors even when they are confident of the scientific soundness of their ideas. Most of them think it is disrespectful even in an intellectual environment to frontally contradict a senior lecturer or a professor. The tendency to resent "intellectual divergence" with senior colleagues as a sign of academic humility is very much in line with African traditionalism. In effect, African traditionalism strongly advocates the observance of a number of core African traditional values, among which is counted the respect and reverence for elders and authorities. Professors and senior researchers are symbolically or practically viewed as elders and authorities who, according to the traditional African social systems, are infinitely wise, knowledgeable and more experienced than the junior members of the society.

Professors and senior colleagues are thus viewed as entities that must be revered by junior members of the community whatsoever the context. As reviewed by Hamzat (2013), the traditional society is structured in such a way that juniors' right to talk and contribute to any public intellectual debate is strictly censored and sometimes censured by the elders. As he puts it, the young is generally expected not to question the objectionable or unobjectionable actions of elders or leaders even when it is clear that he or she risks to be affected by such actions. If the young person tries to express his displeasure with any wrong doing and offers advice, his act is likely ignored and he himself is likely to be tagged with unpleasant names. "Irrespective of how mannered and respectful a young person faulted the wrong actions of the elder/leaders, such young person would always be accused of rudeness, arrogance and disrespect". The traditionalist perspective on senior-junior relationship is equally captured in the popular maxim which stipulates that "what an old man sees sitting down, a young cannot see even if he climbs on top of the tallest tree". A similar scenario is thus observed in research contexts where junior researchers do not always enjoy the "luxury" of challenging senior colleagues without incurring retribution.

Perverted Attitudes Towards the Intellectual Property

The seniority syndrome has equally caused some lecturers and researcher to have a questionable conception of the intellectual propriety. Under the guise of improving the chances of their works (articles prepared for publication in refereed journals) to enjoy greater credibility, some junior researchers include the names of professors and other senior lecturers as co-authors of their works irrespective of the latter's (level) of participation in the completion of the research works. By making professors co-authors of their research papers, some junior researchers hope to technically secure at least two things: (i) the endorsement of their works by experts and (ii) the eventual publication of these works in reputed journals. This is in spite of the fact that their act will mean sharing the credits/merits attached to their works with parties (professors) which may not have taken any serious part in the completion of their research projects. It could be argued that such an authorship politics is no other thing than "recruiting" fictive co-authors and endorsers. It visibly follows the popular belief that professors are custodians of knowledge and superhuman entities (Fedler, 1994). Their inclusion in any research project has the potential to favorably affect the rating of the work in question among peers and in the international research community. The act of including professors as co-authors of research works irrespective of their (level) of participation in the completion of such works is visibly contrary to the principles of intellectual propriety. In such a process, the veritable authors (unjustly – though willfully –) lose parts of the credits attached to their innovations or ideas, to professors or senior colleagues who have not contributed to their efforts.

CONCLUSION

The seniority syndrome is a form of academic arrogance and should therefore be construed as the contrary of academic humility. Its prevalence in Nigerian university communities affects research in multiple ways: it perverts research evaluation exercises such as the assessment of Ph.D. research works; it negatively affects junior and senior researchers' perception of academic humility and is at the origin of ill-conceptions of the intellectual propriety. As has been illustrated in the preceding sections of this chapter, the seniority syndrome affects the evaluation of research in the sense that it causes academic credentials rather than facts to be the measure for judging the credibility of research works and ideas. In line with this, credibility is often ascribed unto an ideas, thesis or theory, not necessarily by virtue of the robustness of the methodology that framed or drove its enunciation but rather by virtue of the seniority of the researcher who emits it. This enables Nigerian professors and senior lecturers to arguably be viewed as demi-gods and all-knowing entities even in contexts where their ideas are not scientifically tenable. The seniority syndrome equally perverts conceptions about academic humility as many view the act of being academically humble as being subservient as well as killing their intellectual virility for fear of being perceived as disrespectful.

As highlighted in this chapter, the ill-conceptions of academic humility as academic subservience and the unconditional respect for senior colleagues are in line with the African traditionalism current. This current preaches unconditional reverence for the elders. In effect, by according unconditional respect for their senior colleagues in research, junior colleagues simply replicate the tradition of always respecting elders and authority (seniors) and avoiding challenging them on matters of general interest. However, it has been argued in multiple forums that some of the tenets of African traditionalism do not have practical relevance in a scientific world, a world which in principle is driven by such values as objectivity and scientific rigor among others. True researchers and scientists are not respecters of persons but of the truth and robust methodologies liable to generate (new) knowledge and drive human development.

The problems explored in this chapter call for a number of corrective and research-friendly actions within Nigerian universities. There is need to shape the content of research (methodology) courses taught in Nigerian university in a way as to integrate activities that will enable a better sensitization of young researchers on academic humility. Research training programs in Nigerian universities should not only be geared towards, building the capacity of young researchers in research methodologies but equally facilitate the "production" of very assertive researchers who will constructively defend their ideas in all knowledge-sharing forums and this, irrespective of the credentials of their intellectual opponents.

In addition to this, Nigerian universities should elaborate clear rules and regulations that define and guide doctoral dissertation defenses. A modicum of attention should be given to the establishment of objective parameters for assessing candidates. Such parameters should be made known to the candidates prior to their thesis defenses. This will go a long to reduce incidences of intuitive or implicit assessments of Ph.D. dissertations and will psychologically prepare the candidates by making them less liable to be intimidated by assessors who appear not animated by academic humility. Universities should equally adopt serious mechanisms to redress and sanction academic arrogance among researchers and lecturers working under them. A special attention should be given to cases of manifest arrogance by assessors and examiners in oral examination situations.

REFERENCES

Adedemeji, M. (2013). What is academic humility. *The Alma Matter*. Retrieved July 29, 2016, from http://mahfouzadedimeji.com/what-is-academic-humility/

Agu, N. N., Omenyi, A. S., & Odimegwu, C. (2015). Evaluation of doctorate dissertation in Nigerian universities: do faculties provide and use explicit criteria/rubrics? *IJTIE: International Journal of Technology and Inclusive Education*, *4*(1), 565-569.

Austin, M. (2014) Humility is a trait worth having. *Psychology Today*. Retrieved 14 August, 2016. from https://www.psychologytoday.com/blog/ethics-everyone/201206/humility

Awuni, J. A. (2016). Students' perspective of effective supervision of graduate programmes in Ghana. *American Journal of Educational Research*, *4*(2), 163–169.

Balarabe, Y. (2009). PhD as minimum qualification for academic staff in Nigerian universities: a policy of self deception. *Modern Ghana*. Retrieved March 12, 2016, from http//:www.modernghana/phd-as-minimum-qualification-for-academic-staff-in-nigeria-universities.htm.

Dhiman, C. (2016). Professors must have PhDs, at least. *The Daily Star*, November 13 edition, 23-25.

Educational Broadcasting Corporation. (2004). Constructivism as a paradigm for teaching and learning. *Concept to Classroom* Retrieved March 14, 2016, from http://www.thirteen.org/edonline/concept2class/constructivism/

Ekundayo, S. (2014). Maggots in our university kola nut. *The Guardian Newspaper*, March 9 edition, p.11.

European University – Industry Network. (2007). Teaching and research synergy in the context of university – industry cooperation. Eindhoven: ZKP. Chevalier de Seyn Publishers.

Exline, J., & Hill, P. (2012, May). Humility: A consistent and robust predictor of generosity. *The Journal of Positive Psychology*, *7*(3), 208–218. doi:10.1080/17439760.2012.671348

Felder, R. M. (1994). The myth of the superhuman professor. *Journal of Engineering Education*, *82*(2), 105–110. doi:10.1002/j.2168-9830.1994.tb01087.x

Healthfield, S. M. (2016). What seniority means at work. *The Balance*. Retrieved March 7, 2016, from https://www.thebalance.com/what-seniority-means-at-work-1919372

Holmes, O. (2015). Dissertation Diaries: Lessons in academic humility. *Society for Personality and Social Psychology*, November.

Hume, D. (1983). *An enquiry into the principles of morals*. Indianapolis: Hackett.

Kaplan, S. (2013). *Scholarliness developed*. London: Oxford University Press.

Leslie, S. (2000). The conflict between research and teaching. In H. Kline & M. Alder (Eds.), *Research and development*. New York: Prentice Hall.

MacLeod, K. (1987). The seniority principle. Is it discriminatory? Kingston: IRC: Queens' University.

Mafa, O., & Mapolisa, T. (2014). Supervisors' experiences in supervising postgraduate education students' dissertations and theses at the Zimbabwe Open University (ZOU). *International Journal of Asian Social Sciences*, *2*(10), 1685–1697.

Maher, R.C. (2010). Authoritative sources. *EE417*. Retrieved online March 9, 2016 from http//:authoritative-sources/ee417.fall2010/htm

O'Leary, P. D. (2016). Surviving oral exam. Retrieved March 14, 2016, from https://www.cs.umd.edu/~oleary/gradstudy/node7.html

Ofiaja, A. (2014). Why elders deserve only earned respect. *This is Africa*. Retrieve March 14, 2016, from http://www.why-elders-deserve-only-earned-respect/thisisafrica.com

Olayinka, A. I. (2014). My stewardship. Ibadan: Office of the Deputy Vice-Chancellor (Academic, University of Ibadan.

Scott, D. (2009, June 3). After 50 years of teaching, professor sees Ph.D. oral exam as the golden hour. Stanford Report. Retrieves March 14, 2016, from http://news.stanford.edu/news/2009/june3/scott-060309.html

Shen, J., Wu, D., Achhpiliya, V., Bieber, M., & Hiltz, R. (2004). Participatory learning approach: A research agenda. *Semantic Scholar*, *1*(6), 1–13.

Smith, M. M. (2013). *Humble leadership: The research shows it's a competitive advantage. Talent management and human resources*. New York: TLNT.

Standler, R. B. (2004). Evaluating credibility of information on the internet. Retrieved January 30, 2017 from http://www.rbs0.com

Stehlik, T. (2008). *Final report: The teaching-research nexus in the division of education, arts and social sciences*. Sydney: University of Australia.

Straits, K. J. E., Bird, D. M., Tsinajinnie, E., Espinoza, J., Goodkind, J., Spencer, O., & Willging, C. et al.The Guiding Principles Workgroup. (2012). *Guiding principles for engaging in research with native American communities (Version 1)*. Washington: UNM Center for Rural and Community Behavioral Health & Albuquerque Area Southwest Tribal Epidemiology Center.

Umbecco, E. (2015). *How to write a thesis*. New York: Academic Emergence Worldpress.

University of Edinburgh. (2016). *Code of practice for supervisors and research students*. Edinburgh: University of Edinburgh.

University of Otago. (2011). *Pacific research protocols*. Otago: University of Otago.

Wellington, J. J. (2010). *Making supervision work for you: A student's guide*. London: Sage.

Worthingtong, E. L. (2007). The Paradox of Humility. *Greater good: The science of a meaningful Life*. Retrieved July 29, 2016, from http://greatergood.berkeley.edu/article/item/paradox_humility

KEY TERMS AND DEFINITIONS

Academic Humility: A scholarly attitude which consists of avoiding showing off knowledge accumulated and being respectful towards peers whatever their level of education or experience in research. Academic humility is demonstrated when a researcher has an accurate view of himself; when he recognizes the strength of others (his peers) and he accepts that anybody can, in one way or the other, teach him something.

African Traditionalism: A socio-philosophical current which celebrates African core cultural values such as communalism, solidarity and respect for elders and authorities among others. Adepts of the current preach a return to the old ways, they view the ancestral traditions as the solutions to human development.

Authoritative Sources/References: These are references that are archival, corroborated and sanctioned in nature. By archival we mean sources that are presently accessible by the general public and which are expected to remain available in future. By corroborate we mean that the information contained in the sources appears in other reliable works notably magazine journals and research monographs. By sanctioned we mean that information is from trustworthy sources notably an encyclopedia, a peer reviewed journal, a professional society magazine or any other source known for its reliance on thorough fact-checking procedures.

Intellectual Propriety (IP): Concept used in reference to the creations of the mind which may be invention, literary and artistic works, symbols, images and names used in commerce as well as designs. IP systems are established and protected by law through schemes such as copyright laws, trademarks and patents. They enable creative people and researchers to earn recognition or financial benefits for their creation, invention or discovery.

Seniority: This term refers to the length of time someone has served in a job or the number of years one has worked for a specific organization. It can equally be defined as a system used in an industrial relations context, to designate the status of an employee in relation to other employees of the same workplace.

Seniority Syndrome: A form of academic arrogance which consists in giving undue credence to the status or credentials (seniority) of a researcher.

Related References

To continue our tradition of advancing research in the area of environmental science and technologies, we have compiled a list of recommended IGI Global readings. These references will provide additional information and guidance to further enrich your knowledge and assist you with your own research and future publications.

Abayomi, K., de la Pena, V., Lall, U., & Levy, M. (2011). Quantifying sustainability: Methodology for and determinants of an environmental sustainability index. In Z. Luo (Ed.), *Green finance and sustainability: Environmentally-aware business models and technologies* (pp. 74–89). Hershey, PA: IGI Global. doi:10.4018/978-1-60960-531-5.ch004

Abdel Gelil, I. (2012). Globalization of the environmental issues: Response of the Arab region. In M. Tortora (Ed.), *Sustainable systems and energy management at the regional level: Comparative approaches* (pp. 147–165). Hershey, PA: IGI Global. doi:10.4018/978-1-61350-344-7.ch008

Adewumi, J., Ilemobade, A., & van Zyl, J. (2013). Application of a multi-criteria decision support tool in assessing the feasibility of implementing treated wastewater reuse. *International Journal of Decision Support System Technology, 5*(1), 1–23. doi:10.4018/jdsst.2013010101

Adeyemo, J., Adeyemo, F., & Otieno, F. (2012). Assessment of pollutant loads of runoff in Pretoria, South Africa. In E. Carayannis (Ed.), *Sustainable policy applications for social ecology and development* (pp. 115–127). Hershey, PA: IGI Global. doi:10.4018/978-1-4666-1586-1.ch009

Ahmad, A. F., & Panni, M. F. (2014). Green marketing strategy: A pedagogical view. In H. Kaufmann & M. Panni (Eds.), *Handbook of research on consumerism in business and marketing: Concepts and practices* (pp. 92–124). Hershey, PA: IGI Global. doi:10.4018/978-1-4666-5880-6.ch005

Ahmad, M. F., & Siang, A. Y. (2014). Modelling of hydrodynamics and sediment transport at Pantai Tok Jembal, Kuala Terengganu Mengabang Telipot, Terengganu, using MIKE 21. In O. Olanrewaju, A. Saharuddin, A. Ab Kader, & W. Wan Nik (Eds.), *Marine technology and sustainable development: Green innovations* (pp. 109–126). Hershey, PA: IGI Global. doi:10.4018/978-1-4666-4317-8.ch007

Ahmed Al-kerdawy, M. M.Ahmed Al-kerdawy. (2011). The role of environmental innovation strategy in reinforcing the impact of green managerial practices on competitive advantages of fertilizer companies in Egypt. *International Journal of Customer Relationship Marketing and Management*, *2*(1), 36–54. doi:10.4018/jcrmm.2011010103

Akkaya, C., Wolf, P., & Krcmar, H. (2011). Efficient information provision for environmental and sustainability reporting. In I. Management Association (Ed.), Green technologies: Concepts, methodologies, tools and applications (pp. 1587-1609). Hershey, PA: IGI Global. doi:10.4018/978-1-60960-472-1.ch705

Al-kerdawy, M. M. (2013). The role of environmental innovation strategy in reinforcing the impact of green managerial practices on competitive advantages of fertilizer companies in Egypt. In R. Eid (Ed.), *Managing customer trust, satisfaction, and loyalty through information communication technologies* (pp. 37–53). Hershey, PA: IGI Global. doi:10.4018/978-1-4666-3631-6.ch003

Alves de Lima, A., Carvalho dos Reis, P., Branco, J. C., Danieli, R., Osawa, C. C., Winter, E., & Santos, D. A. (2013). Scenario-patent protection compared to climate change: The case of green patents. *International Journal of Social Ecology and Sustainable Development*, *4*(3), 61–70. doi:10.4018/jsesd.2013070105

Ang, Z. (2011). The impact of electricity market and environmental regulation on carbon capture & storage (CCS) development in China. In Z. Luo (Ed.), *Green finance and sustainability: Environmentally-aware business models and technologies* (pp. 463–471). Hershey, PA: IGI Global. doi:10.4018/978-1-60960-531-5.ch024

Antonova, A. (2013). Green, sustainable, or clean: What type of IT/IS technologies will we need in the future? In P. Ordóñez de Pablos (Ed.), *Green technologies and business practices: An IT approach* (pp. 151–162). Hershey, PA: IGI Global. doi:10.4018/978-1-4666-1972-2.ch008

Antonova, A. (2014). Green, sustainable, or clean: What type of IT/IS technologies will we need in the future? In *Sustainable practices: Concepts, methodologies, tools and applications* (pp. 384–396). Hershey, PA: IGI Global. doi:10.4018/978-1-4666-4852-4.ch021

Appiah, D. O., & Kemausuor, F. (2012). Energy, environment and socio-economic development: Africa's triple challenge and options. In M. Tortora (Ed.), *Sustainable systems and energy management at the regional level: Comparative approaches* (pp. 166–182). Hershey, PA: IGI Global. doi:10.4018/978-1-61350-344-7.ch009

Appleby, M. R., Lambert, C. G., Rennie, A. E., & Buckley, A. B. (2011). An investigation into the environmental impact of product recovery methods to support sustainable manufacturing within small and medium-sized enterprises (SMEs). *International Journal of Manufacturing, Materials, and Mechanical Engineering*, *1*(2), 1–18. doi:10.4018/ijmmme.2011040101

Appleby, M. R., Lambert, C. G., Rennie, A. E., & Buckley, A. B. (2013). An investigation into the environmental impact of product recovery methods to support sustainable manufacturing within small and medium-sized enterprises (SMEs). In J. Davim (Ed.), *Dynamic methods and process advancements in mechanical, manufacturing, and materials engineering* (pp. 73–90). Hershey, PA: IGI Global. doi:10.4018/978-1-4666-1867-1.ch004

Arora, K., Kumar, A., & Sharma, S. (2012). Energy from waste: Present scenario, challenges, and future prospects towards sustainable development. In P. Olla (Ed.), *Global sustainable development and renewable energy systems* (pp. 271–296). Hershey, PA: IGI Global. doi:10.4018/978-1-4666-1625-7.ch014

Arora, K., Kumar, A., & Sharma, S. (2014). Energy from waste: present scenario, challenges, and future prospects towards sustainable development. In *Sustainable practices: Concepts, methodologies, tools and applications* (pp. 1519–1543). Hershey, PA: IGI Global. doi:10.4018/978-1-4666-4852-4.ch085

Ashraf, G. Y. (2012). A study of eco-friendly supply chain management at cement industries of Chhattisgarh. In M. Garg & S. Gupta (Eds.), *Cases on supply chain and distribution management: Issues and principles* (pp. 146–157). Hershey, PA: IGI Global. doi:10.4018/978-1-4666-0065-2.ch007

Ashraf, G. Y. (2013). A study of eco-friendly supply chain management at cement industries of Chhattisgarh. In *Supply chain management: Concepts, methodologies, tools, and applications* (pp. 823–830). Hershey, PA: IGI Global. doi:10.4018/978-1-4666-2625-6.ch048

Asiimwe, E. N., & Åke, G. (2012). E-waste management in East African community. In K. Bwalya & S. Zulu (Eds.), *Handbook of research on e-government in emerging economies: Adoption, e-participation, and legal frameworks* (pp. 307–327). Hershey, PA: IGI Global. doi:10.4018/978-1-4666-0324-0.ch015

Aspradaki, A. A. (2013). Deliberative democracy and nanotechnologies in health. *International Journal of Technoethics*, *4*(2), 1–14. doi:10.4018/jte.2013070101

Ayadi, F. S. (2013). An empirical investigation of environmental kuznets curve in Nigeria. In K. Ganesh & S. Anbuudayasankar (Eds.), *International and interdisciplinary studies in green computing* (pp. 302–310). Hershey, PA: IGI Global. doi:10.4018/978-1-4666-2646-1.ch022

Ayuk, E. T., Fonta, W. M., & Kouame, E. B. (2014). Application of quantitative methods in natural resource management in Africa: A review. In P. Schaeffer & E. Kouassi (Eds.), *Econometric methods for analyzing economic development* (pp. 205–234). Hershey, PA: IGI Global. doi:10.4018/978-1-4666-4329-1.ch013

Bachour, N. (2012). Green IT project management: Optimizing the value of green IT projects within organizations. In W. Hu & N. Kaabouch (Eds.), *Sustainable ICTs and management systems for green computing* (pp. 146–178). Hershey, PA: IGI Global. doi:10.4018/978-1-4666-1839-8.ch007

Baginetas, K. N. (2011). Sustainable management of agricultural resources and the need for stakeholder participation for the developing of appropriate sustainability indicators: The case of soil quality. In Z. Andreopoulou, B. Manos, N. Polman, & D. Viaggi (Eds.), *Agricultural and environmental informatics, governance and management: Emerging research applications* (pp. 227–261). Hershey, PA: IGI Global. doi:10.4018/978-1-60960-621-3.ch013

Baginetas, K. N. (2012). Sustainable management of agricultural resources and the need for stakeholder participation for the developing of appropriate sustainability indicators: The case of soil quality. In *Regional development: Concepts, methodologies, tools, and applications* (pp. 632–665). Hershey, PA: IGI Global. doi:10.4018/978-1-4666-0882-5.ch401

Bailis, R., & Arabshahi, N. (2011). Voluntary emissions reduction: Are we making progress? In Z. Luo (Ed.), *Green finance and sustainability: Environmentally-aware business models and technologies* (pp. 241–273). Hershey, PA: IGI Global. doi:10.4018/978-1-60960-531-5.ch014

Banerjee, S., Sing, T. Y., Chowdhury, A. R., & Anwar, H. (2013). Motivations to adopt green ICT: A tale of two organizations. *International Journal of Green Computing*, *4*(2), 1–11. doi:10.4018/jgc.2013070101

Baptiste, A. K. (2013). Local vs. expert perception of climate change: An analysis of fishers in Trinidad and Tobago. In H. Muga & K. Thomas (Eds.), *Cases on the diffusion and adoption of sustainable development practices* (pp. 44–82). Hershey, PA: IGI Global. doi:10.4018/978-1-4666-2842-7.ch003

Bassey, K., & Chigbu, P. (2013). Optimal detection and estimation of marine oil spills through coherent pluralism. *International Journal of Operations Research and Information Systems*, *4*(1), 84–111. doi:10.4018/joris.2013010105

Bata, R., & Jordão, T. C. (2011). Modeling the influences of heating fuel consumption in gaseous emissions and solid waste generation. In V. Olej, I. Obršálová, & J. Krupka (Eds.), *Environmental modeling for sustainable regional development: System approaches and advanced methods* (pp. 162–185). Hershey, PA: IGI Global. doi:10.4018/978-1-60960-156-0.ch008

Beall, A. M., & Ford, A. (2012). Reports from the field: Assessing the art and science of participatory environmental modeling. In J. Wang (Ed.), *Societal impacts on information systems development and applications* (pp. 195–213). Hershey, PA: IGI Global. doi:10.4018/978-1-4666-0927-3.ch013

Ben Brahim, H., & Duckstein, L. (2011). Descriptive methods and compromise programming for promoting agricultural reuse of treated wastewater. In H. do Prado, A. Barreto Luiz, & H. Filho (Eds.), *Computational methods for agricultural research: Advances and applications* (pp. 355–388). Hershey, PA: IGI Global. doi:10.4018/978-1-61692-871-1.ch017

Bentley, G. C., Cromley, R. G., Hanink, D. M., & Heidkamp, C. P. (2013). Forest cover change in the northeastern U.S.: A spatial assessment in the context of an environmental kuznets curve. *International Journal of Applied Geospatial Research*, *4*(3), 1–18. doi:10.4018/jagr.2013070101

Berke, M. Ö., Sütlü, E., Avcioglu, B., & Gem, E. (2013). Identification of priority areas for conservation in Lake Egirdir and Lake Kovada, Turkey. In J. Papathanasiou, B. Manos, S. Arampatzis, & R. Kenward (Eds.), *Transactional environmental support system design: Global solutions* (pp. 199–202). Hershey, PA: IGI Global. doi:10.4018/978-1-4666-2824-3.ch018

Bhattarai, N., Khanal, S., Pudasaini, P. R., Pahl, S., & Romero-Urbina, D. (2011). Citrate stabilized silver nanoparticles: Study of crystallography and surface properties. *International Journal of Nanotechnology and Molecular Computation*, *3*(3), 15–28. doi:10.4018/ijnmc.2011070102

Bier, A. (2012). A system dynamics approach to changing perceptions about thermal water quality trading markets. In J. Wang (Ed.), *Societal impacts on information systems development and applications* (pp. 182–194). Hershey, PA: IGI Global. doi:10.4018/978-1-4666-0927-3.ch012

Biller, D., & Sanchez-Triana, E. (2013). Enlisting markets in the conservation and sustainable use of biodiversity in South Asia's Sundarbans. *International Journal of Social Ecology and Sustainable Development, 4*(3), 71–86. doi:10.4018/jsesd.2013070106

Bonadiman, R. (2013). Sustainability: Brazilian perspectives and challenges after the first kioto's protocol period. *International Journal of Social Ecology and Sustainable Development, 4*(3), 52–60. doi:10.4018/jsesd.2013070104

Boote, K. J., Jones, J. W., Hoogenboom, G., & White, J. W. (2012). The role of crop systems simulation in agriculture and environment. In P. Papajorgji & F. Pinet (Eds.), *New technologies for constructing complex agricultural and environmental systems* (pp. 326–339). Hershey, PA: IGI Global. doi:10.4018/978-1-4666-0333-2.ch018

Boulton, A., Devriendt, L., Brunn, S. D., Derudder, B., & Witlox, F. (2014). City networks in cyberspace and time: Using Google hyperlinks to measure global economic and environmental crises. In *Crisis management: Concepts, methodologies, tools and applications* (pp. 1325–1345). Hershey, PA: IGI Global. doi:10.4018/978-1-4666-4707-7.ch067

Bradbury, M. (2012). The sustainable waterfront. In O. Ercoskun (Ed.), *Green and ecological technologies for urban planning: Creating smart cities* (pp. 274–292). Hershey, PA: IGI Global. doi:10.4018/978-1-61350-453-6.ch015

Bradbury, M. (2014). The sustainable waterfront. In *Sustainable practices: Concepts, methodologies, tools and applications* (pp. 1683–1700). Hershey, PA: IGI Global. doi:10.4018/978-1-4666-4852-4.ch093

Brimmo, A., & Emziane, M. (2014). Carbon nanotubes for photovoltaics. In M. Bououdina & J. Davim (Eds.), *Handbook of research on nanoscience, nanotechnology, and advanced materials* (pp. 268–311). Hershey, PA: IGI Global. doi:10.4018/978-1-4666-5824-0.ch012

Brister, E., Hane, E., & Korfmacher, K. (2011). Visualizing plant community change using historical records. *International Journal of Applied Geospatial Research, 2*(4), 1–18. doi:10.4018/jagr.2011100101

Buxton, G. (2013). Nanotechnology and polymer solar cells. In S. Anwar, H. Efstathiadis, & S. Qazi (Eds.), *Handbook of research on solar energy systems and technologies* (pp. 231–253). Hershey, PA: IGI Global. doi:10.4018/978-1-4666-1996-8.ch009

Buxton, G. (2014). Nanotechnology and polymer solar cells. In Nanotechnology: Concepts, methodologies, tools, and applications (pp. 384-405). Hershey, PA: IGI Global. doi:10.4018/978-1-4666-5125-8.ch015

Cadman, T., & Hume, M. (2012). Developing sustainable governance systems for regional sustainability programmes and 'green' business practices: The case of 'green' timber. In M. Tortora (Ed.), *Sustainable systems and energy management at the regional level: Comparative approaches* (pp. 365–382). Hershey, PA: IGI Global. doi:10.4018/978-1-61350-344-7.ch019

Cai, T. (2014). Artificial neural network for industrial and environmental research via air quality monitoring network. In Z. Sun & J. Yearwood (Eds.), *Handbook of research on demand-driven web services: Theory, technologies, and applications* (pp. 399–419). Hershey, PA: IGI Global. doi:10.4018/978-1-4666-5884-4.ch019

Cai, T. (2014). Geospatial technology-based e-government design for environmental protection and emergency response. In K. Bwalya (Ed.), *Technology development and platform enhancements for successful global e-government design* (pp. 157–184). Hershey, PA: IGI Global. doi:10.4018/978-1-4666-4900-2.ch009

Calipinar, H., & Ulas, D. (2013). Model suggestion for SMEs economic and environmental sustainable development. In N. Ndubisi & S. Nwankwo (Eds.), *Enterprise development in SMEs and entrepreneurial firms: Dynamic processes* (pp. 270–290). Hershey, PA: IGI Global. doi:10.4018/978-1-4666-2952-3.ch014

Carvajal-Escobar, Y., Mimi, Z., Khayat, S., Sulieman, S., Garces, W., & Cespedes, G. (2011). Application of methodologies for environmental flow determination in an andean and a Mediterranean basin: Two case studies of the Pance River (Colombia) and Wadi River (Palestine) basin. *International Journal of Social Ecology and Sustainable Development, 2*(4), 26–43. doi:10.4018/jsesd.2011100103

Carvajal-Escobar, Y., Mimi, Z., Khayat, S., Sulieman, S., Garces, W., & Cespedes, G. (2013). Application of methodologies for environmental flow determination in an andean and a Mediterranean basin: Two case studies of the Pance River (Colombia) and Wadi River (Palestine) basin. In E. Carayannis (Ed.), *Creating a sustainable ecology using technology-driven solutions* (pp. 296–314). Hershey, PA: IGI Global. doi:10.4018/978-1-4666-3613-2.ch020

Cascelli, E., Crestaz, E., & Tatangelo, F. (2013). Cartography and geovisualization in groundwater modelling. In G. Borruso, S. Bertazzon, A. Favretto, B. Murgante, & C. Torre (Eds.), *Geographic information analysis for sustainable development and economic planning: New technologies* (pp. 49–67). Hershey, PA: IGI Global. doi:10.4018/978-1-4666-1924-1.ch004

Ceccaroni, L., & Oliva, L. (2012). Ontologies for the design of ecosystems. In T. Podobnikar & M. Čeh (Eds.), *Universal ontology of geographic space: semantic enrichment for spatial data* (pp. 207–228). Hershey, PA: IGI Global. doi:10.4018/978-1-4666-0327-1.ch009

Charuvilayil, R. A. (2013). Industrial pollution and people's movement: A case study of Eloor Island Kerala, India. In H. Muga & K. Thomas (Eds.), *Cases on the diffusion and adoption of sustainable development practices* (pp. 312–351). Hershey, PA: IGI Global. doi:10.4018/978-1-4666-2842-7.ch012

Chen, E. T. (2011). Green information technology and virtualization in corporate environmental management information systems. In *Green technologies: Concepts, methodologies, tools and applications* (pp. 1421–1434). Hershey, PA: IGI Global. doi:10.4018/978-1-60960-472-1.ch605

Chen, H., & Bishop, I. D. (2013). Collaborative environmental knowledge management. *International Journal of E-Planning Research, 2*(1), 58–81. doi:10.4018/ijepr.2013010104

Chen, X. M. (2011). GIS and remote sensing in environmental risk assessment. In *Green technologies: Concepts, methodologies, tools and applications* (pp. 840–847). Hershey, PA: IGI Global. doi:10.4018/978-1-60960-472-1.ch415

Chen, Y. (2013). Generalize key requirements for designing IT-based system for green with considering stakeholder needs. *International Journal of Information Technologies and Systems Approach, 6*(1), 78–97. doi:10.4018/jitsa.2013010105

Chinchuluun, A., Xanthopoulos, P., Tomaino, V., & Pardalos, P. (2012). Data mining techniques in agricultural and environmental sciences. In P. Papajorgji & F. Pinet (Eds.), *New technologies for constructing complex agricultural and environmental systems* (pp. 311–325). Hershey, PA: IGI Global. doi:10.4018/978-1-4666-0333-2.ch017

Chitra, S. (2011). Adopting green ICT in business. In I. Management Association (Ed.), Green technologies: Concepts, methodologies, tools and applications (pp. 1145-1153). Hershey, PA: IGI Global. doi:10.4018/978-1-60960-472-1.ch501

Chiu, M. (2013). Gaps between valuing and purchasing green-technology products: Product and gender differences. *International Journal of Technology and Human Interaction, 8*(3), 54–68. doi:10.4018/jthi.2012070106

Cho, C. H., Patten, D. M., & Roberts, R. W. (2014). Environmental disclosures and impression management. In R. Hart (Ed.), *Communication and language analysis in the corporate world* (pp. 217–231). Hershey, PA: IGI Global. doi:10.4018/978-1-4666-4999-6.ch013

Christophoridis, C., Bizani, E., & Fytianos, K. (2011). Environmental quality monitoring, using GIS as a tool of visualization, management and decision-making: Applications emerging from the EU water framework directive EU 2000/60. In Z. Andreopoulou, B. Manos, N. Polman, & D. Viaggi (Eds.), *Agricultural and environmental informatics, governance and management: Emerging research applications* (pp. 397–424). Hershey, PA: IGI Global. doi:10.4018/978-1-60960-621-3.ch021

Christophoridis, C., Bizani, E., & Fytianos, K. (2013). Environmental quality monitoring, using GIS as a tool of visualization, management and decision-making: Applications emerging from the EU water framework directive EU 2000/60. In *Geographic information systems: Concepts, methodologies, tools, and applications* (pp. 1559–1586). Hershey, PA: IGI Global. doi:10.4018/978-1-4666-2038-4.ch094

Cincu, C., & Diacon, A. (2013). Hybrid solar cells: Materials and technology. In L. Fara & M. Yamaguchi (Eds.), *Advanced solar cell materials, technology, modeling, and simulation* (pp. 79–100). Hershey, PA: IGI Global. doi:10.4018/978-1-4666-1927-2.ch006

Cocozza, A., & Ficarella, A. (2013). Electrical resistivity measures in cohesive soils for the simulation of an integrated energy system between CCS and low-enthalpy geothermal. *International Journal of Measurement Technologies and Instrumentation Engineering, 3*(1), 48–68. doi:10.4018/ijmtie.2013010105

Cohen, E., & Zimmerman, T. D. (2012). Teaching the greenhouse effect with inquiry-based computer simulations: A WISE case study. In L. Lennex & K. Nettleton (Eds.), *Cases on inquiry through instructional technology in math and science* (pp. 551–580). Hershey, PA: IGI Global. doi:10.4018/978-1-4666-0068-3.ch020

Congedo, L., Baiocco, F., Brini, S., Liberti, L., & Munafò, M. (2013). Urban environment quality in the Italian spatial data infrastructure. In G. Borruso, S. Bertazzon, A. Favretto, B. Murgante, & C. Torre (Eds.), *Geographic information analysis for sustainable development and economic planning: New technologies* (pp. 179–192). Hershey, PA: IGI Global. doi:10.4018/978-1-4666-1924-1.ch012

Cosmi, C., Di Leo, S., Loperte, S., Pietrapertosa, F., Salvia, M., Macchiato, M., & Cuomo, V. (2011). Comprehensive energy systems analysis support tools for decision making. In *Green technologies: Concepts, methodologies, tools and applications* (pp. 493–514). Hershey, PA: IGI Global. doi:10.4018/978-1-60960-472-1.ch307

Cotton, M. (2012). Community opposition and public engagement with wind energy in the UK. In M. Tortora (Ed.), *Sustainable systems and energy management at the regional level: Comparative approaches* (pp. 310–327). Hershey, PA: IGI Global. doi:10.4018/978-1-61350-344-7.ch016

Cuccurullo, S., Francese, R., Passero, I., & Tortora, G. (2013). A 3D serious city building game on waste disposal. *International Journal of Distance Education Technologies*, *11*(4), 112–135. doi:10.4018/ijdet.2013100108

Cunha, M. D. (2011). Wastewater systems management at the regional level. In V. Olej, I. Obršálová, & J. Krupka (Eds.), *Environmental modeling for sustainable regional development: System approaches and advanced methods* (pp. 186–203). Hershey, PA: IGI Global. doi:10.4018/978-1-60960-156-0.ch009

Cunha, M. D. (2012). Wastewater systems management at the regional level. In *Regional development: Concepts, methodologies, tools, and applications* (pp. 1161–1177). Hershey, PA: IGI Global. doi:10.4018/978-1-4666-0882-5.ch607

Da Ronch, B., Di Maria, E., & Micelli, S. (2013). Clusters go green: Drivers of environmental sustainability in local networks of SMEs. *International Journal of Information Systems and Social Change*, *4*(1), 37–52. doi:10.4018/jissc.2013010103

Danahy, J., Wright, R., Mitchell, J., & Feick, R. (2013). Exploring ways to use 3D urban models to visualize multi-scalar climate change data and mitigation change models for e-planning. *International Journal of E-Planning Research*, *2*(2), 1–17. doi:10.4018/ijepr.2013040101

Dhal, S. (2013). Indigenous agricultural knowledge and innovation: A study of agricultural scientists in odisha. *International Journal of Information Systems and Social Change*, *4*(3), 57–71. doi:10.4018/jissc.2013070104

Dimitriou, D., Voskaki, A., & Sartzetaki, M. (2014). Airports environmental management: Results from the evaluation of European airports environmental plans. *International Journal of Information Systems and Supply Chain Management*, *7*(1), 1–14. doi:10.4018/IJISSCM.2014010101

Dizdaroglu, D., Yigitcanlar, T., & Dawes, L. (2011). Planning for sustainable urban futures. In *Green technologies: Concepts, methodologies, tools and applications* (pp. 1922–1932). Hershey, PA: IGI Global. doi:10.4018/978-1-60960-472-1.ch806

Djeflat, A. (2014). Harnessing knowledge for sustainable development: Challenges and opportunities for Arab countries. In A. Driouchi (Ed.), *Knowledge-based economic policy development in the Arab world* (pp. 229–244). Hershey, PA: IGI Global. doi:10.4018/978-1-4666-5210-1.ch009

Dolney, T. J. (2011). A GIS methodology for assessing the safety hazards of abandoned mine lands (AMLs): Application to the state of Pennsylvania. *International Journal of Applied Geospatial Research, 2*(3), 50–71. doi:10.4018/jagr.2011070104

Dubovski, S. (2014). Activities in oil and gas processing for avoiding or minimizing environmental impacts. In D. Matanovic, N. Gaurina-Medjimurec, & K. Simon (Eds.), *Risk analysis for prevention of hazardous situations in petroleum and natural gas engineering* (pp. 247–263). Hershey, PA: IGI Global. doi:10.4018/978-1-4666-4777-0.ch012

Dugas, D. P., DeMers, M. N., Greenlee, J. C., Whitford, W. G., & Klimaszewski-Patterson, A. (2011). Rapid evaluation of arid lands (REAL): A methodology. *International Journal of Applied Geospatial Research, 2*(3), 32–49. doi:10.4018/jagr.2011070103

Dusmanescu, D. (2013). Aspects regarding implementation of renewable energy sources in Romania up to 2050. *International Journal of Sustainable Economies Management, 2*(4), 1–21. doi:10.4018/ijsem.2013100101

Ehlinger, T., Tofan, L., Bucur, M., Enz, J., Carlson, J., & Shaker, R. (2011). Application of a participatory ex ante assessment model for environmental governance and visualizing sustainable redevelopment in Gorj County, Romania. In Z. Andreopoulou, B. Manos, N. Polman, & D. Viaggi (Eds.), *Agricultural and environmental informatics, governance and management: Emerging research applications* (pp. 61–86). Hershey, PA: IGI Global. doi:10.4018/978-1-60960-621-3.ch004

Ehlinger, T., Tofan, L., Bucur, M., Enz, J., Carlson, J., & Shaker, R. (2012). Application of a participatory ex ante assessment model for environmental governance and visualizing sustainable redevelopment in Gorj County, Romania. In *Regional development: Concepts, methodologies, tools, and applications* (pp. 743–768). Hershey, PA: IGI Global. doi:10.4018/978-1-4666-0882-5.ch407

Ekekwe, N. (2013). Nanotechnology and microelectronics: The science, trends and global diffusion. *International Journal of Nanotechnology and Molecular Computation, 3*(4), 1–23. doi:10.4018/ijnmc.2013100101

El Alouani, H., & Driouchi, A. (2014). The oil and gas sectors, renewable energy, and environmental performance in the Arab world. In A. Driouchi (Ed.), *Knowledge-based economic policy development in the Arab world* (pp. 172–228). Hershey, PA: IGI Global. doi:10.4018/978-1-4666-5210-1.ch008

El-Daoushy, F. (2011). Assessing environment-climate impacts in the Nile basin for decision-making. In *Green technologies: Concepts, methodologies, tools and applications* (pp. 694–712). Hershey, PA: IGI Global. doi:10.4018/978-1-60960-472-1.ch407

Elkarmi, F., & Abu Shikhah, N. (2012). Renewable energy technologies. In *Power system planning technologies and applications: Concepts, solutions and management* (pp. 121–142). Hershey, PA: IGI Global. doi:10.4018/978-1-4666-0173-4.ch008

Ene, C. (2013). Post-consumer waste: Challenges, trends and solutions. *International Journal of Sustainable Economies Management, 2*(3), 19–31. doi:10.4018/ijsem.2013070102

Erdoğdu, M. M., & Karaca, C. (2014). A road map for a domestic wind turbine manufacturing industry in Turkey. In B. Christiansen & M. Basilgan (Eds.), *Economic behavior, game theory, and technology in emerging markets* (pp. 57–90). Hershey, PA: IGI Global. doi:10.4018/978-1-4666-4745-9.ch005

Espiritu, J. F., & Ituarte-Villarreal, C. M. (2013). Wind farm layout optimization using a viral systems algorithm. *International Journal of Applied Evolutionary Computation, 4*(4), 27–40. doi:10.4018/ijaec.2013100102

Evangelista, P., Huge-Brodin, M., Isaksson, K., & Sweeney, E. (2013). Purchasing green transport and logistics services: Implications from the environmental sustainability attitude of 3PLs. In D. Folinas (Ed.), *Outsourcing management for supply chain operations and logistics service* (pp. 449–465). Hershey, PA: IGI Global. doi:10.4018/978-1-4666-2008-7.ch026

Evangelista, P., Huge-Brodin, M., Isaksson, K., & Sweeney, E. (2014). Purchasing green transport and logistics services: Implications from the environmental sustainability attitude of 3PLs. In *Sustainable practices: Concepts, methodologies, tools and applications* (pp. 86–102). Hershey, PA: IGI Global. doi:10.4018/978-1-4666-4852-4.ch005

Ewald, J. A., Sharp, R. J., Beja, P., & Kenward, R. (2013). Pan-European survey and database of environmental assessment factors. In J. Papathanasiou, B. Manos, S. Arampatzis, & R. Kenward (Eds.), *Transactional environmental support system design: Global solutions* (pp. 97–119). Hershey, PA: IGI Global. doi:10.4018/978-1-4666-2824-3.ch006

Fann, J., & Rakas, J. (2011). Greener transportation infrastructure: Theoretical concepts for the environmental evaluation of airports. In Z. Luo (Ed.), *Green finance and sustainability: Environmentally-aware business models and technologies* (pp. 394–421). Hershey, PA: IGI Global. doi:10.4018/978-1-60960-531-5.ch021

Fann, J., & Rakas, J. (2013). Methodology for environmental sustainability evaluation of airport development alternatives. *International Journal of Applied Logistics, 4*(4), 8–31. doi:10.4018/ijal.2013100102

Fara, L., & Yamaguchi, M. (2013). Prospects and strategy of development for advanced solar cells. In L. Fara & M. Yamaguchi (Eds.), *Advanced solar cell materials, technology, modeling, and simulation* (pp. 287–296). Hershey, PA: IGI Global. doi:10.4018/978-1-4666-1927-2.ch014

Farmani, R., Savic, D., Henriksen, H., Molina, J., Giordano, R., & Bromley, J. (2011). Evolutionary Bayesian belief networks for participatory water resources management under uncertainty. In *Green technologies: Concepts, methodologies, tools and applications* (pp. 524–539). Hershey, PA: IGI Global. doi:10.4018/978-1-60960-472-1.ch309

Fearnside, P. M. (2013). Climate change as a threat to Brazil's Amazon forest. *International Journal of Social Ecology and Sustainable Development, 4*(3), 1–12. doi:10.4018/jsesd.2013070101

Filipović, V., Roljević, S., & Bekić, B. (2014). Organic production in Serbia: The transition to green economy. In *Sustainable practices: Concepts, methodologies, tools and applications* (pp. 769–785). Hershey, PA: IGI Global. doi:10.4018/978-1-4666-4852-4.ch043

Finardi, U. (2012). Nanosciences and nanotechnologies: Evolution trajectories and disruptive features. In N. Ekekwe & N. Islam (Eds.), *Disruptive technologies, innovation and global redesign: Emerging implications* (pp. 107–126). Hershey, PA: IGI Global. doi:10.4018/978-1-4666-0134-5.ch007

Finardi, U. (2014). Nanosciences and nanotechnologies: Evolution trajectories and disruptive features. In Nanotechnology: Concepts, methodologies, tools, and applications (pp. 1-20). Hershey, PA: IGI Global. doi:10.4018/978-1-4666-5125-8.ch001

Fokaides, P. A. (2012). Towards zero energy buildings (ZEB): The role of environmental technologies. In O. Ercoskun (Ed.), *Green and ecological technologies for urban planning: Creating smart cities* (pp. 93–111). Hershey, PA: IGI Global. doi:10.4018/978-1-61350-453-6.ch006

Fokaides, P. A. (2014). Towards zero energy buildings (ZEB): The role of environmental technologies. In Sustainable practices: Concepts, methodologies, tools and applications (pp. 1742-1761). Hershey, PA: IGI Global. doi:10.4018/978-1-4666-4852-4.ch096

Gadatsch, A. (2011). Corporate environmental management information systems influence of green IT on IT management and IT controlling. In *Green technologies: Concepts, methodologies, tools and applications* (pp. 1408–1420). Hershey, PA: IGI Global. doi:10.4018/978-1-60960-472-1.ch604

Gálvez, J., Parreño, M., Pla, J., Sanchez, J., Gálvez-Llompart, M., Navarro, S., & García-Domenech, R. (2011). Application of molecular topology to the prediction of water quality indices of Alkylphenol pollutants. *International Journal of Chemoinformatics and Chemical Engineering, 1*(1), 1–11. doi:10.4018/ijcce.2011010101

Gálvez, J., Parreño, M., Pla, J., Sanchez, J., Gálvez-Llompart, M., Navarro, S., & García-Domenech, R. (2013). Application of molecular topology to the prediction of water quality indices of Alkylphenol pollutants. In A. Haghi (Ed.), *Methodologies and applications for chemoinformatics and chemical engineering* (pp. 1–10). Hershey, PA: IGI Global. doi:10.4018/978-1-4666-4010-8.ch001

Garner, N., Lischke, M. D., Siol, A., & Eilks, I. (2014). Learning about sustainability in a non-formal laboratory context for secondary level students: A module on climate change, the ozone hole, and summer smog. In K. Thomas & H. Muga (Eds.), *Handbook of research on pedagogical innovations for sustainable development* (pp. 229–244). Hershey, PA: IGI Global. doi:10.4018/978-1-4666-5856-1.ch012

Gaurina-Medjimurec, N., & Pasic, B. (2014). CO2 underground storage and wellbore integrity. In D. Matanovic, N. Gaurina-Medjimurec, & K. Simon (Eds.), *Risk analysis for prevention of hazardous situations in petroleum and natural gas engineering* (pp. 322–357). Hershey, PA: IGI Global. doi:10.4018/978-1-4666-4777-0.ch015

Ghosh, N., & Goswami, A. (2014). Biofuels and renewables: Implications for people, planet, policies, politics. In *Sustainability science for social, economic, and environmental development* (pp. 64–87). Hershey, PA: IGI Global. doi:10.4018/978-1-4666-4995-8.ch007

Ghosh, N., & Goswami, A. (2014). Biofuel sustainability and transition pathways. In *Sustainability science for social, economic, and environmental development* (pp. 88–95). Hershey, PA: IGI Global. doi:10.4018/978-1-4666-4995-8.ch008

Ghosh, N., & Goswami, A. (2014). Economics, environmental policy, trade and sustainability. In *Sustainability science for social, economic, and environmental development* (pp. 246–268). Hershey, PA: IGI Global. doi:10.4018/978-1-4666-4995-8.ch016

Ghosh, N., & Goswami, A. (2014). Energy and emission linkages from the three wheeler autorickshaws of Kolkata: An exploratory analysis of the impact on economic, environmental, social dimensions of sustainability. In *Sustainability science for social, economic, and environmental development* (pp. 221–245). Hershey, PA: IGI Global. doi:10.4018/978-1-4666-4995-8.ch015

Ghosh, N., & Goswami, A. (2014). Labour observatories for agricultural policymaking and sustainable development. In *Sustainability science for social, economic, and environmental development* (pp. 56–63). Hershey, PA: IGI Global. doi:10.4018/978-1-4666-4995-8.ch006

Ghosh, N., & Goswami, A. (2014). On value and price of environmental resources. In *Sustainability science for social, economic, and environmental development* (pp. 24–32). Hershey, PA: IGI Global. doi:10.4018/978-1-4666-4995-8.ch002

Ghosh, N., & Goswami, A. (2014). Story of live discussion in autos of Delhi: What do they say about sustainability? In *Sustainability science for social, economic, and environmental development* (pp. 216–220). Hershey, PA: IGI Global. doi:10.4018/978-1-4666-4995-8.ch014

Ghosh, N., & Goswami, A. (2014). Two first generation biofuel (biodiesel, bioethanol) and sustainability: Some other realities for India and trade patterns. In *Sustainability science for social, economic, and environmental development* (pp. 174–208). Hershey, PA: IGI Global. doi:10.4018/978-1-4666-4995-8.ch012

Ghosh, N., & Goswami, A. (2014). Valuation and market-based pricing of economic and ecosystem services of water resources. In *Sustainability science for social, economic, and environmental development* (pp. 96–132). Hershey, PA: IGI Global. doi:10.4018/978-1-4666-4995-8.ch009

Ghosh, N., & Goswami, A. (2014). Water scarcity and conflicts: Can water futures exchange in South Asia provide the answer? In *Sustainability science for social, economic, and environmental development* (pp. 147–173). Hershey, PA: IGI Global. doi:10.4018/978-1-4666-4995-8.ch011

Gil, J., Díaz, L., Granell, C., & Huerta, J. (2013). Open source based deployment of environmental data into geospatial information infrastructures. In I. Management Association (Ed.), Geographic information systems: Concepts, methodologies, tools, and applications (pp. 952-969). Hershey, PA: IGI Global. doi:10.4018/978-1-4666-2038-4.ch059

Gill, L., Hathway, E. A., Lange, E., Morgan, E., & Romano, D. (2013). Coupling real-time 3D landscape models with microclimate simulations. *International Journal of E-Planning Research, 2*(1), 1–19. doi:10.4018/ijepr.2013010101

Giuliani, G., Ray, N., Schwarzer, S., De Bono, A., Peduzzi, P., Dao, H., & Lehmann, A. et al. (2011). Sharing environmental data through GEOSS. *International Journal of Applied Geospatial Research, 2*(1), 1–17. doi:10.4018/jagr.2011010101

Giuliani, G., Ray, N., Schwarzer, S., De Bono, A., Peduzzi, P., Dao, H., & Lehmann, A. et al. (2013). Sharing environmental data through GEOSS. In *Geographic information systems: Concepts, methodologies, tools, and applications* (pp. 1260–1275). Hershey, PA: IGI Global. doi:10.4018/978-1-4666-2038-4.ch076

Godbole, N. (2011). E-waste management: Challenges and issues. In B. Unhelkar (Ed.), *Handbook of research on green ICT: Technology, business and social perspectives* (pp. 480–505). Hershey, PA: IGI Global. doi:10.4018/978-1-61692-834-6.ch035

Goel, A., Tiwary, A., & Schmidt, H. (2011). Approaches and initiatives to green IT strategy in business. In *Green technologies: Concepts, methodologies, tools and applications* (pp. 1361–1375). Hershey, PA: IGI Global. doi:10.4018/978-1-60960-472-1.ch601

Granit, J. J., King, R. M., & Noël, R. (2013). Strategic environmental assessment as a tool to develop power in transboundary water basin settings. In E. Carayannis (Ed.), *Creating a sustainable ecology using technology-driven solutions* (pp. 269–281). Hershey, PA: IGI Global. doi:10.4018/978-1-4666-3613-2.ch018

Gräuler, M., Teuteberg, F., Mahmoud, T., & Gómez, J. M. (2013). Requirements prioritization and design considerations for the next generation of corporate environmental management information systems: A foundation for innovation. *International Journal of Information Technologies and Systems Approach, 6*(1), 98–116. doi:10.4018/jitsa.2013010106

Gräuler, M., Teuteberg, F., Mahmoud, T., & Gómez, J. M. (2013). Requirements prioritization and design considerations for the next generation of corporate environmental management information systems: A foundation for innovation. *International Journal of Information Technologies and Systems Approach, 6*(1), 98–116. doi:10.4018/jitsa.2013010106

Greenlee, B., & Daim, T. (2011). Building a sustainable regional eco system for green technologies: Case of cellulosic ethanol in Oregon. In Z. Luo (Ed.), *Green finance and sustainability: Environmentally-aware business models and technologies* (pp. 535–568). Hershey, PA: IGI Global. doi:10.4018/978-1-60960-531-5.ch028

Greenlee, B., & Daim, T. (2013). Building a sustainable regional eco system for green technologies: Case of cellulosic ethanol in Oregon. In I. Management Association (Ed.), Small and medium enterprises: Concepts, methodologies, tools, and applications (pp. 993-1025). Hershey, PA: IGI Global. doi:10.4018/978-1-4666-3886-0.ch049

Grigoroudis, E., Kouikoglou, V. S., & Phillis, Y. A. (2012). Approaches for measuring sustainability. In P. Olla (Ed.), *Global sustainable development and renewable energy systems* (pp. 101–130). Hershey, PA: IGI Global. doi:10.4018/978-1-4666-1625-7.ch006

Grigoroudis, E., Kouikoglou, V. S., & Phillis, Y. A. (2014). Approaches for measuring sustainability. In *Sustainable practices: Concepts, methodologies, tools and applications* (pp. 158–184). Hershey, PA: IGI Global. doi:10.4018/978-1-4666-4852-4.ch009

Guangming, L., & Zhaofeng, A. (2013). Empirical study on the correlations of environmental pollution, human capital, and economic growth: Based on the 1990-2007 data in Guangdong China. In P. Ordóñez de Pablos (Ed.), *Green technologies and business practices: An IT approach* (pp. 128–137). Hershey, PA: IGI Global. doi:10.4018/978-1-4666-1972-2.ch006

Gupta, A. K., Chakraborty, A., Giri, S., Subramanian, V., & Chattaraj, P. (2011). Toxicity of halogen, sulfur and chlorinated aromatic compounds: A quantitative-structure-toxicity-relationship (QSTR). *International Journal of Chemoinformatics and Chemical Engineering*, *1*(1), 61–74. doi:10.4018/ijcce.2011010105

Gupta, A. K., Chakraborty, A., Giri, S., Subramanian, V., & Chattaraj, P. (2013). Toxicity of halogen, sulfur and chlorinated aromatic compounds: A quantitative-structure-toxicity-relationship (QSTR). In A. Haghi (Ed.), *Methodologies and applications for chemoinformatics and chemical engineering* (pp. 60–73). Hershey, PA: IGI Global. doi:10.4018/978-1-4666-4010-8.ch005

Habala, O., Šeleng, M., Tran, V., Šimo, B., & Hluchý, L. (2012). Mining environmental data in the ADMIRE project using new advanced methods and tools. In N. Bessis (Ed.), *Technology integration advancements in distributed systems and computing* (pp. 296–308). Hershey, PA: IGI Global. doi:10.4018/978-1-4666-0906-8.ch018

Hájek, P., & Olej, V. (2011). Air quality assessment by neural networks. In V. Olej, I. Obršálová, & J. Krupka (Eds.), *Environmental modeling for sustainable regional development: System approaches and advanced methods* (pp. 91–117). Hershey, PA: IGI Global. doi:10.4018/978-1-60960-156-0.ch005

Hall, C., Easley, R., Howard, J., & Halfhide, T. (2013). The role of authentic science research and education outreach in increasing community resilience: Case studies using informal education to address ocean acidification and healthy soils. In H. Muga & K. Thomas (Eds.), *Cases on the diffusion and adoption of sustainable development practices* (pp. 376–402). Hershey, PA: IGI Global. doi:10.4018/978-1-4666-2842-7.ch014

Hall, G. M., & Howe, J. (2013). The drivers for a sustainable chemical manufacturing industry. In *Industrial engineering: Concepts, methodologies, tools, and applications* (pp. 1659–1679). Hershey, PA: IGI Global. doi:10.4018/978-1-4666-1945-6.ch088

Hashemi, M., & O'Connell, E. (2013). Science and water policy interface: An integrated methodological framework for developing decision support systems (DSSs). In *Data mining: Concepts, methodologies, tools, and applications* (pp. 405–434). Hershey, PA: IGI Global. doi:10.4018/978-1-4666-2455-9.ch020

Heck, M., & Schmidt, G. (2013). Lot-size planning with non-linear cost functions supporting environmental sustainability. In K. Ganesh & S. Anbuudayasankar (Eds.), *International and interdisciplinary studies in green computing* (pp. 226–231). Hershey, PA: IGI Global. doi:10.4018/978-1-4666-2646-1.ch016

Herold, S., & Sawada, M. C. (2013). A review of geospatial information technology for natural disaster management in developing countries. In *Geographic information systems: Concepts, methodologies, tools, and applications* (pp. 175–215). Hershey, PA: IGI Global. doi:10.4018/978-1-4666-2038-4.ch014

Higginson, N., & Vredenburg, H. (2012). Finding the sweet spot of sustainability in the energy sector: A systems approach to managing the canadian oil sands. In M. Tortora (Ed.), *Sustainable systems and energy management at the regional level: Comparative approaches* (pp. 184–201). Hershey, PA: IGI Global. doi:10.4018/978-1-61350-344-7.ch010

Hilty, L. M. (2011). Information and communication technologies for a more sustainable world. In D. Haftor & A. Mirijamdotter (Eds.), *Information and communication technologies, society and human beings: Theory and framework (festschrift in honor of Gunilla Bradley)* (pp. 410–418). Hershey, PA: IGI Global. doi:10.4018/978-1-60960-057-0.ch033

Hin, L. T., & Subramaniam, R. (2012). Use of policy instruments to promote sustainable energy practices and implications for the environment: Experiences from Singapore. In M. Tortora (Ed.), *Sustainable systems and energy management at the regional level: Comparative approaches* (pp. 219–235). Hershey, PA: IGI Global. doi:10.4018/978-1-61350-344-7.ch012

Hrncevic, L. (2014). Petroleum industry environmental performance and risk. In D. Matanovic, N. Gaurina-Medjimurec, & K. Simon (Eds.), *Risk analysis for prevention of hazardous situations in petroleum and natural gas engineering* (pp. 358–387). Hershey, PA: IGI Global. doi:10.4018/978-1-4666-4777-0.ch016

Hsiao, S., Chen, D., Yang, C., Huang, H., Lu, Y., Huang, H., & Lin, Y. et al. (2013). Chemical-free and reusable cellular analysis: Electrochemical impedance spectroscopy with a transparent ITO culture chip. *International Journal of Technology and Human Interaction*, *8*(3), 1–9. doi:10.4018/jthi.2012070101

Hunter, J., Becker, P., Alabri, A., van Ingen, C., & Abal, E. (2011). Using ontologies to relate resource management actions to environmental monitoring data in South East Queensland. *International Journal of Agricultural and Environmental Information Systems*, *2*(1), 1–19. doi:10.4018/jaeis.2011010101

Imbrenda, V., D'Emilio, M., Lanfredi, M., Ragosta, M., & Simoniello, T. (2013). Indicators of land degradation vulnerability due to anthropic factors: Tools for an efficient planning. In G. Borruso, S. Bertazzon, A. Favretto, B. Murgante, & C. Torre (Eds.), *Geographic information analysis for sustainable development and economic planning: New technologies* (pp. 87–101). Hershey, PA: IGI Global. doi:10.4018/978-1-4666-1924-1.ch006

Imbrenda, V., D'Emilio, M., Lanfredi, M., Ragosta, M., & Simoniello, T. (2014). Indicators of land degradation vulnerability due to anthropic factors: Tools for an efficient planning. In Sustainable practices: Concepts, methodologies, tools and applications (pp. 1400-1413). Hershey, PA: IGI Global. doi:10.4018/978-1-4666-4852-4.ch078

Iojă, C., Niţă, M. R., & Stupariu, I. G. (2014). Resource conservation: Key elements in sustainable rural development. In Z. Andreopoulou, V. Samathrakis, S. Louca, & M. Vlachopoulou (Eds.), *E-innovation for sustainable development of rural resources during global economic crisis* (pp. 80–97). Hershey, PA: IGI Global. doi:10.4018/978-1-4666-4550-9.ch008

Ioja, C., Rozylowicz, L., Patroescu, M., Nita, M., & Onose, D. (2011). Agriculture and conservation in the Natura 2000 network: A sustainable development approach of the European Union. In Z. Andreopoulou, B. Manos, N. Polman, & D. Viaggi (Eds.), *Agricultural and environmental informatics, governance and management: Emerging research applications* (pp. 339–358). Hershey, PA: IGI Global. doi:10.4018/978-1-60960-621-3.ch018

Ioja, C., Rozylowicz, L., Patroescu, M., Nita, M., & Onose, D. (2013). Agriculture and conservation in the Natura 2000 network: A sustainable development approach of the European Union. In *Geographic information systems: Concepts, methodologies, tools, and applications* (pp. 1276–1296). Hershey, PA: IGI Global. doi:10.4018/978-1-4666-2038-4.ch077

Ip-Soo-Ching, J. M., & Zyngier, S. (2014). The rise of "environmental sustainability knowledge" in business strategy and entrepreneurship: An IT-enabled knowledge-based view of tourism operators. In P. Ordóñez de Pablos (Ed.), International business strategy and entrepreneurship: An information technology perspective (pp. 23-40). Hershey, PA: IGI Global. doi:10.4018/978-1-4666-4753-4.ch002

Ivask, M., Aruvee, E., & Piirimäe, K. (2013). Database of environmental decision support tools. In J. Papathanasiou, B. Manos, S. Arampatzis, & R. Kenward (Eds.), *Transactional environmental support system design: Global solutions* (pp. 70–96). Hershey, PA: IGI Global. doi:10.4018/978-1-4666-2824-3.ch005

Jacobsson, M., Linde, A., & Linderoth, H. (2011). The relation between ICT and environmental management practice in a construction company. In *Green technologies: Concepts, methodologies, tools and applications* (pp. 1099–1117). Hershey, PA: IGI Global. doi:10.4018/978-1-60960-472-1.ch430

Jafari, M. (2013). Challenges in climate change and environmental crisis: Impacts of aviation industry on human, urban and natural environments. *International Journal of Space Technology Management and Innovation*, *3*(2), 24–46. doi:10.4018/ijstmi.2013070102

Jain, H. (2011). Green ICT organizational implementations and workplace relationships. In B. Unhelkar (Ed.), *Handbook of research on green ICT: Technology, business and social perspectives* (pp. 146–168). Hershey, PA: IGI Global. doi:10.4018/978-1-61692-834-6.ch010

Jamous, N. (2013). Light-weight composite environmental performance indicators (LWC-EPI): A new approach for environmental management information systems (EMIS). *International Journal of Information Technologies and Systems Approach*, *6*(1), 20–38. doi:10.4018/jitsa.2013010102

Jan, Y., Lin, M., Shiao, K., Wei, C., Huang, L., & Sung, Q. (2013). Development of an evaluation instrument for green building literacy among college students in Taiwan. *International Journal of Technology and Human Interaction*, *8*(3), 31–45. doi:10.4018/jthi.2012070104

Jarmoszko, A., D'Onofrio, M., Lee-Partridge, J. E., & Petkova, O. (2013). Evaluating sustainability and greening methods: A conceptual model for information technology management. *International Journal of Applied Logistics*, *4*(3), 1–13. doi:10.4018/jal.2013070101

Jena, R. K. (2013). Green computing to green business. In P. Ordóñez de Pablos (Ed.), *Green technologies and business practices: An IT approach* (pp. 138–150). Hershey, PA: IGI Global. doi:10.4018/978-1-4666-1972-2.ch007

Jinturkar, A. M., & Deshmukh, S. S. (2013). Sustainable development by rural energy resources allocation in India: A fuzzy goal programming approach. *International Journal of Energy Optimization and Engineering*, *2*(1), 37–49. doi:10.4018/ijeoe.2013010103

Jirava, P., & Obršálová, I. (2011). Modeling the effects of the quality of the environment on the health of a selected population. In V. Olej, I. Obršálová, & J. Krupka (Eds.), *Environmental modeling for sustainable regional development: System approaches and advanced methods* (pp. 344–365). Hershey, PA: IGI Global. doi:10.4018/978-1-60960-156-0.ch017

Jonoski, A., & Evers, M. (2013). Sociotechnical framework for participatory flood risk management via collaborative modeling. *International Journal of Information Systems and Social Change*, *4*(2), 1–16. doi:10.4018/jissc.2013040101

Joshi, P. K., & Priyanka, N. (2011). Geo-informatics for land use and biodiversity studies. In Y. Trisurat, R. Shrestha, & R. Alkemade (Eds.), *Land use, climate change and biodiversity modeling: Perspectives and applications* (pp. 52–77). Hershey, PA: IGI Global. doi:10.4018/978-1-60960-619-0.ch003

Joshi, P. K., & Priyanka, N. (2013). Geo-informatics for land use and biodiversity studies. In I. Management Association (Ed.), Geographic information systems: Concepts, methodologies, tools, and applications (pp. 1913-1939). Hershey, PA: IGI Global. doi:10.4018/978-1-4666-2038-4.ch114

Júnior, R., Rigitano, R., & Boesten, J. (2011). Pesticide leaching models in a Brazilian agricultural field scenario. In H. do Prado, A. Barreto Luiz, & H. Filho (Eds.), *Computational methods for agricultural research: Advances and applications* (pp. 266–295). Hershey, PA: IGI Global. doi:10.4018/978-1-61692-871-1.ch013

Kader, A. S., & Olanrewaju, O. S. (2014). River transportation master plan study for environmental enhancement. In O. Olanrewaju, A. Saharuddin, A. Ab Kader, & W. Wan Nik (Eds.), *Marine technology and sustainable development: Green innovations* (pp. 178–184). Hershey, PA: IGI Global. doi:10.4018/978-1-4666-4317-8.ch011

Kamaja, C. K., Rajaperumal, M., Boukherroub, R., & Shelke, M. V. (2014). Silicon nanostructures-graphene nanocomposites: Efficient materials for energy conversion and storage. In M. Bououdina & J. Davim (Eds.), *Handbook of research on nanoscience, nanotechnology, and advanced materials* (pp. 176–195). Hershey, PA: IGI Global. doi:10.4018/978-1-4666-5824-0.ch009

Kaplan, A. (2012). "Green infrastructure" concept as an effective medium to manipulating sustainable urban development. In O. Ercoskun (Ed.), *Green and ecological technologies for urban planning: Creating smart cities* (pp. 234–254). Hershey, PA: IGI Global. doi:10.4018/978-1-61350-453-6.ch013

Kašparová, M., & Krupka, J. (2011). Air quality modeling and metamodeling approach. In V. Olej, I. Obršálová, & J. Krupka (Eds.), *Environmental modeling for sustainable regional development: System approaches and advanced methods* (pp. 144–161). Hershey, PA: IGI Global. doi:10.4018/978-1-60960-156-0.ch007

Kenward, R., Casey, N. M., Walls, S. S., Dick, J. M., Smith, R., & Turner, S. L. ... Sharp, R. J. (2013). Pan-European analysis of environmental assessment processes. In J. Papathanasiou, B. Manos, S. Arampatzis, & R. Kenward (Eds.), Transactional environmental support system design: Global solutions (pp. 120-133). Hershey, PA: IGI Global. doi:10.4018/978-1-4666-2824-3.ch007

Kokkinakis, A., & Andreopoulou, Z. (2011). E-governance and management of inland water ecosystems using time-series analysis of fishery production. In Z. Andreopoulou, B. Manos, N. Polman, & D. Viaggi (Eds.), *Agricultural and environmental informatics, governance and management: Emerging research applications* (pp. 318–338). Hershey, PA: IGI Global. doi:10.4018/978-1-60960-621-3.ch017

Kongar, E. A., & Rosentrater, K. (2013). Data envelopment analysis approach to compare the environmental efficiency of energy utilization. In K. Ganesh & S. Anbuudayasankar (Eds.), *International and interdisciplinary studies in green computing* (pp. 273–288). Hershey, PA: IGI Global. doi:10.4018/978-1-4666-2646-1.ch020

Kosaka, M., Yabutani, T., & Zhang, Q. (2014). A value co-creation model for energy-saving service business using inverters. In M. Kosaka & K. Shirahada (Eds.), *Progressive trends in knowledge and system-based science for service innovation* (pp. 292–306). Hershey, PA: IGI Global. doi:10.4018/978-1-4666-4663-6.ch016

Kram, T., & Stehfest, E. (2011). Integrated modeling of global environmental change (IMAGE). In Y. Trisurat, R. Shrestha, & R. Alkemade (Eds.), *Land use, climate change and biodiversity modeling: Perspectives and applications* (pp. 104–118). Hershey, PA: IGI Global. doi:10.4018/978-1-60960-619-0.ch005

La Greca, P., La Rosa, D., Martinico, F., & Privitera, R. (2013). Land cover analysis for evapotranspiration assessment in Catania metropolitan region. In G. Borruso, S. Bertazzon, A. Favretto, B. Murgante, & C. Torre (Eds.), *Geographic information analysis for sustainable development and economic planning: New technologies* (pp. 102–114). Hershey, PA: IGI Global. doi:10.4018/978-1-4666-1924-1.ch007

Laike, Y., & Chun, L. (2012). China-European Union trade and global warming. In E. Carayannis (Ed.), *Sustainable policy applications for social ecology and development* (pp. 18–28). Hershey, PA: IGI Global. doi:10.4018/978-1-4666-1586-1.ch003

Laing, R., Bennadji, A., & Gray, D. (2013). Traffic control and CO2 reduction: Utilisation of virtual modelling within university estates master planning. *International Journal of E-Planning Research, 2*(1), 43–57. doi:10.4018/ijepr.2013010103

Lam, J. C., & Hills, P. (2011). Promoting technological environmental innovations: What is the role of environmental regulation? In Z. Luo (Ed.), *Green finance and sustainability: Environmentally-aware business models and technologies* (pp. 56–73). Hershey, PA: IGI Global. doi:10.4018/978-1-60960-531-5.ch003

Lam, J. C., & Hills, P. (2012). Transition to low-carbon hydrogen economy in America: The role of transition management. In Z. Luo (Ed.), *Advanced analytics for green and sustainable economic development: Supply chain models and financial technologies* (pp. 92–111). Hershey, PA: IGI Global. doi:10.4018/978-1-61350-156-6.ch007

Lam, J. C., & Hills, P. (2013). Promoting technological environmental innovations: The role of environmental regulation. In Z. Luo (Ed.), *Technological solutions for modern logistics and supply chain management* (pp. 230–247). Hershey, PA: IGI Global. doi:10.4018/978-1-4666-2773-4.ch015

lba, J., Wrege, M., & Garrastazu, M. (2011). Zoning based on climate and soil for planting eucalyptus in southern region of Rio Grande do Sul State, Brazil. In H. do Prado, A. Barreto Luiz, & H. Filho (Eds.), *Computational methods for agricultural research: Advances and applications* (pp. 127-143). Hershey, PA: IGI Global. doi:10.4018/978-1-61692-871-1.ch008

Lee, S., Yigitcanlar, T., Egodawatta, P., & Goonetilleke, A. (2011). Sustainable water provision. In *Green technologies: Concepts, methodologies, tools and applications* (pp. 1768–1781). Hershey, PA: IGI Global. doi:10.4018/978-1-60960-472-1.ch714

Lee, Y. M., An, L., Liu, F., Horesh, R., Chae, Y. T., & Zhang, R. (2014). Analytics for smarter buildings. *International Journal of Business Analytics*, *1*(1), 1–15. doi:10.4018/ijban.2014010101

Leff, E. (2012). Environmental rationality: Innovation in thinking for sustainability. In F. Nobre, D. Walker, & R. Harris (Eds.), *Technological, managerial and organizational core competencies: Dynamic innovation and sustainable development* (pp. 1–17). Hershey, PA: IGI Global. doi:10.4018/978-1-61350-165-8.ch001

Leff, E. (2014). Environmental rationality: Innovation in thinking for sustainability. In *Sustainable practices: Concepts, methodologies, tools and applications* (pp. 1–17). Hershey, PA: IGI Global. doi:10.4018/978-1-4666-4852-4.ch001

Lefley, F., & Sarkis, J. (2011). A pragmatic profile approach to evaluating environmental sustainability investment decisions. In Z. Luo (Ed.), *Green finance and sustainability: Environmentally-aware business models and technologies* (pp. 321–332). Hershey, PA: IGI Global. doi:10.4018/978-1-60960-531-5.ch017

Li, H., & Zhang, X. (2012). Study on environmental tax: A case of China. In D. Ura & P. Ordóñez de Pablos (Eds.), *Advancing technologies for Asian business and economics: Information management developments* (pp. 207–219). Hershey, PA: IGI Global. doi:10.4018/978-1-4666-0276-2.ch016

Li, X., Ortiz, P., Kuczenski, B., Franklin, D., & Chong, F. T. (2012). Mitigating the environmental impact of smartphones with device reuse. In W. Hu & N. Kaabouch (Eds.), *Sustainable ICTs and management systems for green computing* (pp. 252–282). Hershey, PA: IGI Global. doi:10.4018/978-1-4666-1839-8.ch011

Li, X., Ortiz, P. J., Browne, J., Franklin, D., Oliver, J. Y., Geyer, R., & Chong, F. T. et al. (2012). A study of reusing smartphones to augment elementary school education. *International Journal of Handheld Computing Research*, *3*(2), 73–92. doi:10.4018/jhcr.2012040105

Lin, C., Chu, L., & Hsu, H. (2013). Study on the performance and exhaust emissions of motorcycle engine fuelled with hydrogen-gasoline compound fuel. *International Journal of Technology and Human Interaction, 8*(3), 69–81. doi:10.4018/jthi.2012070107

Lingarchani, A. (2011). Environmental challenges in mobile services. In B. Unhelkar (Ed.), *Handbook of research on green ICT: Technology, business and social perspectives* (pp. 355–363). Hershey, PA: IGI Global. doi:10.4018/978-1-61692-834-6.ch025

Lingarchani, A. (2012). Environmental challenges in mobile services. In *Wireless technologies: Concepts, methodologies, tools and applications* (pp. 1891–1899). Hershey, PA: IGI Global. doi:10.4018/978-1-61350-101-6.ch710

Loeser, F., Erek, K., & Zarnekow, R. (2013). Green IT strategies: A conceptual framework for the alignment of information technology and corporate sustainability strategy. In P. Ordóñez de Pablos (Ed.), *Green technologies and business practices: An IT approach* (pp. 58–95). Hershey, PA: IGI Global. doi:10.4018/978-1-4666-1972-2.ch004

Loi, N. K. (2013). Sustainable land use and watershed management in response to climate change impacts: Overview and proposed research techniques. In *Geographic information systems: Concepts, methodologies, tools, and applications* (pp. 2080–2101). Hershey, PA: IGI Global. doi:10.4018/978-1-4666-2038-4.ch124

Lucignano, C., Squeo, E. A., Guglielmotti, A., & Quadrini, F. (2013). Recycling of waste epoxy-polyester powders for foam production. In J. Davim (Ed.), *Dynamic methods and process advancements in mechanical, manufacturing, and materials engineering* (pp. 91–101). Hershey, PA: IGI Global. doi:10.4018/978-1-4666-1867-1.ch005

Lynar, T. M. Simon, Herbert, R. D., & Chivers, W. J. (2012). Clustering obsolete computers to reduce e-waste. In J. Wang (Ed.), Societal impacts on information systems development and applications (pp. 24-34). Hershey, PA: IGI Global. doi:10.4018/978-1-4666-0927-3.ch002

Mahbub, P., Ayoko, G., Egodawatta, P., Yigitcanlar, T., & Goonetilleke, A. (2011). Traffic and climate change impacts on water quality. In I. Management Association (Ed.), Green technologies: Concepts, methodologies, tools and applications (pp. 1804-1823). Hershey, PA: IGI Global. doi:10.4018/978-1-60960-472-1.ch716

Maillé, E., & Espinasse, B. (2011). Pyroxene: A territorial decision support system based on spatial simulators integration for forest fire risk management. *International Journal of Agricultural and Environmental Information Systems, 2*(2), 52–72. doi:10.4018/jaeis.2011070104

Maillé, E., & Espinasse, B. (2012). Pyroxene: A territorial decision support system based on spatial simulators integration for forest fire risk management. In P. Papajorgji & F. Pinet (Eds.), *New technologies for constructing complex agricultural and environmental systems* (pp. 244–264). Hershey, PA: IGI Global. doi:10.4018/978-1-4666-0333-2.ch014

Mallios, Z. (2012). Irrigation water valuation using spatial hedonic models in GIS environment. In J. Wang (Ed.), *Societal impacts on information systems development and applications* (pp. 308–320). Hershey, PA: IGI Global. doi:10.4018/978-1-4666-0927-3.ch020

Manikas, I., Ieromonachou, P., & Bochtis, D. (2014). Environmental sustainability initiatives in the agrifood supply chain. In Z. Andreopoulou, V. Samathrakis, S. Louca, & M. Vlachopoulou (Eds.), *E-innovation for sustainable development of rural resources during global economic crisis* (pp. 221–232). Hershey, PA: IGI Global. doi:10.4018/978-1-4666-4550-9.ch016

Manou, D., & Papathanasiou, J. (2013). Exploring the development of new tourism activities in the municipality of Kerkini by using the area's natural resources sustainably, municipality of Kerkini, Greece. In J. Papathanasiou, B. Manos, S. Arampatzis, & R. Kenward (Eds.), *Transactional environmental support system design: Global solutions* (pp. 172–175). Hershey, PA: IGI Global. doi:10.4018/978-1-4666-2824-3.ch012

Maragkogianni, A., Papaefthimiou, S., & Zopounidis, C. (2013). Emissions trading schemes in the transportation sector. In A. Jean-Vasile, T. Adrian, J. Subic, & D. Dusmanescu (Eds.), *Sustainable technologies, policies, and constraints in the green economy* (pp. 269–289). Hershey, PA: IGI Global. doi:10.4018/978-1-4666-4098-6.ch015

Maragkogianni, A., Papaefthimiou, S., & Zopounidis, C. (2014). Emissions trading schemes in the transportation sector. In *Sustainable practices: Concepts, methodologies, tools and applications* (pp. 65–85). Hershey, PA: IGI Global. doi:10.4018/978-1-4666-4852-4.ch004

Marino, D. J., Castro, E. A., Massolo, L., Mueller, A., Herbarth, O., & Ronco, A. E. (2011). Characterization of polycyclic aromatic hydrocarbon profiles by multivariate statistical analysis. *International Journal of Chemoinformatics and Chemical Engineering, 1*(2), 1–14. doi:10.4018/ijcce.2011070101

Marino, D. J., Castro, E. A., Massolo, L., Mueller, A., Herbarth, O., & Ronco, A. E. (2013). Characterization of polycyclic aromatic hydrocarbon profiles by multivariate statistical analysis. In A. Haghi (Ed.), *Methodologies and applications for chemoinformatics and chemical engineering* (pp. 102–116). Hershey, PA: IGI Global. doi:10.4018/978-1-4666-4010-8.ch008

Marshall, A. (2011). The middle ground for nuclear waste management: Social and ethical aspects of shallow storage. *International Journal of Technoethics, 2*(2), 1–13. doi:10.4018/jte.2011040101

Mbzibain, A. (2013). The effect of farmer capacities, farm business resources and perceived support of family, friends and associational networks on intentions to invest in renewable energy ventures in the UK. *International Journal of Applied Behavioral Economics, 2*(3), 43–58. doi:10.4018/ijabe.2013070104

Mbzibain, A. (2014). The effect of farmer capacities, farm business resources and perceived support of family, friends and associational networks on intentions to invest in renewable energy ventures in the UK. In *Sustainable practices: Concepts, methodologies, tools and applications* (pp. 1072–1088). Hershey, PA: IGI Global. doi:10.4018/978-1-4666-4852-4.ch059

McKnight, K. P., Messina, J. P., Shortridge, A. M., Burns, M. D., & Pigozzi, B. W. (2011). Using volunteered geographic information to assess the spatial distribution of West Nile Virus in Detroit, Michigan. *International Journal of Applied Geospatial Research, 2*(3), 72–85. doi:10.4018/jagr.2011070105

Mengel, M. A. (2011). Constructing an experience in a virtual green home. In G. Vincenti & J. Braman (Eds.), *Multi-user virtual environments for the classroom: Practical approaches to teaching in virtual worlds* (pp. 285–301). Hershey, PA: IGI Global. doi:10.4018/978-1-60960-545-2.ch018

Miidla, P. (2011). Data envelopment analysis in environmental technologies. In V. Olej, I. Obršálová, & J. Krupka (Eds.), *Environmental modeling for sustainable regional development: System approaches and advanced methods* (pp. 242–259). Hershey, PA: IGI Global. doi:10.4018/978-1-60960-156-0.ch012

Miidla, P. (2013). Data envelopment analysis in environmental technologies. In *Industrial engineering: Concepts, methodologies, tools, and applications* (pp. 625–642). Hershey, PA: IGI Global. doi:10.4018/978-1-4666-1945-6.ch036

Militano, L., Molinaro, A., Iera, A., & Petkovics, Á. (2013). A game theoretic approach to guarantee fairness in cooperation among green mobile network operators. *International Journal of Business Data Communications and Networking, 9*(3), 1–15. doi:10.4018/jbdcn.2013070101

Miller, W., & Birkeland, J. (2011). Green energy. In *Green technologies: Concepts, methodologies, tools and applications* (pp. 1–16). Hershey, PA: IGI Global. doi:10.4018/978-1-60960-472-1.ch101

Miralles, A., Pinet, F., & Bédard, Y. (2012). Describing spatio-temporal phenomena for environmental system development: An overview of today's needs and solutions. In P. Papajorgji & F. Pinet (Eds.), *New technologies for constructing complex agricultural and environmental systems* (pp. 211–226). Hershey, PA: IGI Global. doi:10.4018/978-1-4666-0333-2.ch012

Misso, R. (2011). Sustainable governance in the integrated system "environment-agriculture–health" through ICTs. In Z. Andreopoulou, B. Manos, N. Polman, & D. Viaggi (Eds.), *Agricultural and environmental informatics, governance and management: Emerging research applications* (pp. 87–101). Hershey, PA: IGI Global. doi:10.4018/978-1-60960-621-3.ch005

Mitroi, M. R., Fara, L., & Moraru, A. G. (2013). Organic solar cells modeling and simulation. In L. Fara & M. Yamaguchi (Eds.), *Advanced solar cell materials, technology, modeling, and simulation* (pp. 120–137). Hershey, PA: IGI Global. doi:10.4018/978-1-4666-1927-2.ch008

Mochal, T., & Krasnoff, A. (2013). GreenPM®: The basic principles for applying an environmental dimension to project management. In G. Silvius & J. Tharp (Eds.), *Sustainability integration for effective project management* (pp. 39–57). Hershey, PA: IGI Global. doi:10.4018/978-1-4666-4177-8.ch003

Montgomery, M. C., & Chakraborty, J. (2013). Social vulnerability to coastal and inland flood hazards: A comparison of GIS-based spatial interpolation methods. *International Journal of Applied Geospatial Research, 4*(3), 58–79. doi:10.4018/jagr.2013070104

Moreno, I. S., & Xu, J. (2013). Energy-efficiency in cloud computing environments: Towards energy savings without performance degradation. In S. Aljawarneh (Ed.), *Cloud computing advancements in design, implementation, and technologies* (pp. 18–36). Hershey, PA: IGI Global. doi:10.4018/978-1-4666-1879-4.ch002

Morris, J. Z., & Thomas, K. D. (2013). Implementing biosand filters in rural Honduras: A case study of his hands mission international in Copán, Honduras. In H. Muga & K. Thomas (Eds.), *Cases on the diffusion and adoption of sustainable development practices* (pp. 468–496). Hershey, PA: IGI Global. doi:10.4018/978-1-4666-2842-7.ch017

Mu, Z., Jing, L., Xiaohong, Z., Lei, T., Xiao-na, F., & Shan, C. (2011). Study on low-carbon economy model and method of Chinese tourism industry. *International Journal of Applied Logistics*, *2*(2), 69–102. doi:10.4018/jal.2011040105

Mu, Z., Jing, L., Xiaohong, Z., Lei, T., Xiao-na, F., & Shan, C. (2013). Study on low-carbon economy model and method of chinese tourism industry. In Z. Luo (Ed.), *Technological solutions for modern logistics and supply chain management* (pp. 284–317). Hershey, PA: IGI Global. doi:10.4018/978-1-4666-2773-4.ch018

Mudhoo, A., & Lin, Z. (2012). Phytoremediation of nickel: Mechanisms, application and management. In N. Ekekwe & N. Islam (Eds.), *Disruptive technologies, innovation and global redesign: Emerging implications* (pp. 173–195). Hershey, PA: IGI Global. doi:10.4018/978-1-4666-0134-5.ch010

Murugesan, S. (2011). Strategies for greening enterprise IT: Creating business value and contributing to environmental sustainability. In B. Unhelkar (Ed.), *Handbook of research on green ICT: Technology, business and social perspectives* (pp. 51–64). Hershey, PA: IGI Global. doi:10.4018/978-1-61692-834-6.ch004

Nagni, M., & Ventouras, S. (2013). Implementation of UML schema in relational databases: A case of geographic information. *International Journal of Distributed Systems and Technologies*, *4*(4), 50–60. doi:10.4018/ijdst.2013100105

Nair, S. R., & Ndubisi, N. O. (2013). Entrepreneurial values, environmental marketing and customer satisfaction: Conceptualization and propositions. In N. Ndubisi & S. Nwankwo (Eds.), *Enterprise development in SMEs and entrepreneurial firms: Dynamic processes* (pp. 257–269). Hershey, PA: IGI Global. doi:10.4018/978-1-4666-2952-3.ch013

Nourani, V., Roumianfar, S., & Sharghi, E. (2013). Using hybrid ARIMAX-ANN model for simulating rainfall - runoff - sediment process case study: Aharchai Basin, Iran. *International Journal of Applied Metaheuristic Computing*, *4*(2), 44–60. doi:10.4018/jamc.2013040104

Obara, S. (2011). Fuel reduction effect of the solar cell and diesel engine hybrid system with a prediction algorithm of solar power generation. In *Green technologies: Concepts, methodologies, tools and applications* (pp. 815–839). Hershey, PA: IGI Global. doi:10.4018/978-1-60960-472-1.ch414

Oktay, D. (2014). Sustainable urbanism revisited: A holistic framework based on tradition and contemporary orientations. In *Sustainable practices: Concepts, methodologies, tools and applications* (pp. 1723–1741). Hershey, PA: IGI Global. doi:10.4018/978-1-4666-4852-4.ch095

Olanrewaju, O. S. (2014). Evolving sustainable green ship technology. In O. Olanrewaju, A. Saharuddin, A. Ab Kader, & W. Wan Nik (Eds.), *Marine technology and sustainable development: Green innovations* (pp. 127–145). Hershey, PA: IGI Global. doi:10.4018/978-1-4666-4317-8.ch008

Olanrewaju, O. S. (2014). Risk requirement for multi-hybrid renewable energy for marine system. In O. Olanrewaju, A. Saharuddin, A. Ab Kader, & W. Wan Nik (Eds.), *Marine technology and sustainable development: Green innovations* (pp. 83–95). Hershey, PA: IGI Global. doi:10.4018/978-1-4666-4317-8.ch005

Olanrewaju, O. S., & Kader, A. S. (2014). Applying the safety and environmental risk and reliability model (SERM) for Malaysian Langat River collision aversion. In O. Olanrewaju, A. Saharuddin, A. Ab Kader, & W. Wan Nik (Eds.), *Marine technology and sustainable development: Green innovations* (pp. 193–225). Hershey, PA: IGI Global. doi:10.4018/978-1-4666-4317-8.ch013

Olej, V., & Hájek, P. (2011). Air quality modeling by fuzzy sets and IF-sets. In V. Olej, I. Obršálová, & J. Krupka (Eds.), *Environmental modeling for sustainable regional development: System approaches and advanced methods* (pp. 118–143). Hershey, PA: IGI Global. doi:10.4018/978-1-60960-156-0.ch006

Omer, A. M. (2012). Renewable energy and sustainable development. In P. Vasant, N. Barsoum, & J. Webb (Eds.), *Innovation in power, control, and optimization: Emerging energy technologies* (pp. 95–136). Hershey, PA: IGI Global. doi:10.4018/978-1-61350-138-2.ch003

Omer, A. M. (2014). Cooling and heating with ground source energy. In *Sustainable practices: Concepts, methodologies, tools and applications* (pp. 261–278). Hershey, PA: IGI Global. doi:10.4018/978-1-4666-4852-4.ch014

Ondieki, C. M. (2013). Hydrology and integrated water resource management for sustainable watershed management in Kenya. In H. Muga & K. Thomas (Eds.), *Cases on the diffusion and adoption of sustainable development practices* (pp. 352–375). Hershey, PA: IGI Global. doi:10.4018/978-1-4666-2842-7.ch013

Ozbakir, B. A. (2012). Urban environmental applications of GIScience: Challenges and new trends. In O. Ercoskun (Ed.), *Green and ecological technologies for urban planning: Creating smart cities* (pp. 192–211). Hershey, PA: IGI Global. doi:10.4018/978-1-61350-453-6.ch011

Ozbakir, B. A. (2014). Urban environmental applications of GIScience: Challenges and new trends. In *Sustainable practices: Concepts, methodologies, tools and applications* (pp. 602–620). Hershey, PA: IGI Global. doi:10.4018/978-1-4666-4852-4.ch034

Oztaysi, B., Isik, M., & Ercan, S. (2013). Multi-criteria decision aid for sustainable energy prioritization using fuzzy axiomatic design. *International Journal of Energy Optimization and Engineering*, *2*(1), 1–20. doi:10.4018/ijeoe.2013010101

Palantzas, G., Naniopoulos, A., & Koutitas, C. (2014). Management of environmental issues in port activities: The Hellenic caste study. *International Journal of Information Systems and Supply Chain Management, 7*(1), 40–55. doi:10.4018/ijisscm.2014010103

Pang, L., & Zhao, J. (2013). An empirical study on China's regional carbon emissions of agriculture. *International Journal of Asian Business and Information Management, 4*(4), 67–77. doi:10.4018/ijabim.2013100105

Papajorgji, P., Pinet, F., Miralles, A., Jallas, E., & Pardalos, P. (2012). Modeling: A central activity for flexible information systems development in agriculture and environment. In P. Papajorgji & F. Pinet (Eds.), *New technologies for constructing complex agricultural and environmental systems* (pp. 286–310). Hershey, PA: IGI Global. doi:10.4018/978-1-4666-0333-2.ch016

Papaspyropoulos, K. G., Christodoulou, A. S., Blioumis, V., Skordas, K. E., & Birtsas, P. K. (2011). The improvement of environmental performance in the nonprofit sector through informatics. In Z. Andreopoulou, B. Manos, N. Polman, & D. Viaggi (Eds.), *Agricultural and environmental informatics, governance and management: Emerging research applications* (pp. 359–376). Hershey, PA: IGI Global. doi:10.4018/978-1-60960-621-3.ch019

Pappis, C. P. (2011). Frameworks of policy making under climate change. In C. Pappis (Ed.), *Climate change, supply chain management and enterprise adaptation: Implications of global warming on the economy* (pp. 271–308). Hershey, PA: IGI Global. doi:10.4018/978-1-61692-800-1.ch009

Paquette, S. (2011). Applying knowledge management in the environmental and climate change sciences. In D. Schwartz & D. Te'eni (Eds.), *Encyclopedia of knowledge management* (2nd ed., pp. 20–26). Hershey, PA: IGI Global. doi:10.4018/978-1-59904-931-1.ch003

Pechanec, V., & Vávra, A. (2013). Education portal on climate change with web GIS client. *Journal of Cases on Information Technology, 15*(1), 51–68. doi:10.4018/jcit.2013010104

Perl-Vorbach, E. (2012). Communicating environmental information on a company and inter-organizational level. In *Regional development: Concepts, methodologies, tools, and applications* (pp. 914–932). Hershey, PA: IGI Global. doi:10.4018/978-1-4666-0882-5.ch505

Perry, J., Paas, L., Arreola, M. E., Santer, E., Sharma, N., & Bellali, J. (2011). Promoting e-governance through capacity development for the global environment. In *Green technologies: Concepts, methodologies, tools and applications* (pp. 980–1010). Hershey, PA: IGI Global. doi:10.4018/978-1-60960-472-1.ch423

Pessoa, M., Fernandes, E., Nascimento de Queiroz, S., Ferracini, V., Gomes, M., & Dornelas de Souza, M. (2011). Mathematical-modelling simulation applied to help in the decision-making process on environmental impact assessment of agriculture. In H. do Prado, A. Barreto Luiz, & H. Filho (Eds.), *Computational methods for agricultural research: Advances and applications* (pp. 199–233). Hershey, PA: IGI Global. doi:10.4018/978-1-61692-871-1.ch011

Peters, E. J. (2013). Promoting rainwater harvesting (RWH) in small island developing states (SIDS): A case in the Grenadines. In H. Muga & K. Thomas (Eds.), *Cases on the diffusion and adoption of sustainable development practices* (pp. 403–438). Hershey, PA: IGI Global. doi:10.4018/978-1-4666-2842-7.ch015

Peters, E. J. (2014). Promoting rainwater harvesting (RWH) in small island developing states (SIDS): A case in the Grenadines. In Sustainable practices: Concepts, methodologies, tools and applications (pp. 1657-1682). Hershey, PA: IGI Global. doi:10.4018/978-1-4666-4852-4.ch092

Ploberger, C. (2011). A critical assessment of environmental degeneration and climate change: A multidimensional (political, economic, social) challenge for China's future economic development. *International Journal of Applied Logistics*, *2*(2), 1–16. doi:10.4018/jal.2011040101

Ploberger, C. (2013). A critical assessment of environmental degeneration and climate change: A multidimensional (political, economic, social) challenge for China's future economic development. In Z. Luo (Ed.), *Technological solutions for modern logistics and supply chain management* (pp. 212–229). Hershey, PA: IGI Global. doi:10.4018/978-1-4666-2773-4.ch014

Ploberger, C. (2013). China's environmental issues, a domestic challenge with regional and international implications. *International Journal of Applied Logistics*, *4*(3), 47–61. doi:10.4018/jal.2013070104

Polat, E. (2012). An approach for land-use suitability assessment using decision support systems, AHP and GIS. In O. Ercoskun (Ed.), *Green and ecological technologies for urban planning: Creating smart cities* (pp. 212–233). Hershey, PA: IGI Global. doi:10.4018/978-1-61350-453-6.ch012

Polat, E. (2013). An approach for land-use suitability assessment using decision support systems, AHP and GIS. In *Data mining: Concepts, methodologies, tools, and applications* (pp. 2153–2173). Hershey, PA: IGI Global. doi:10.4018/978-1-4666-2455-9.ch110

Pülzl, H., & Wydra, D. (2013). The evaluation of the implementation of sustainability norms: An exercise for experts or citizens? In E. Carayannis (Ed.), *Creating a sustainable ecology using technology-driven solutions* (pp. 32–45). Hershey, PA: IGI Global. doi:10.4018/978-1-4666-3613-2.ch003

Pusceddu, C. (2012). Grenelle environment project: An institutional tool for building collaborative environmental policies at a national level. In M. Tortora (Ed.), *Sustainable systems and energy management at the regional level: Comparative approaches* (pp. 348–364). Hershey, PA: IGI Global. doi:10.4018/978-1-61350-344-7.ch018

Puškaric, A., Subic, J., & Bekic, B. (2013). Regionalization as a factor of agriculture development of the Republic of Serbia. *International Journal of Sustainable Economies Management*, *2*(1), 46–54. doi:10.4018/ijsem.2013010105

Rafferty, J. M. (2012). Design of outdoor and environmentally integrated learning spaces. In M. Keppell, K. Souter, & M. Riddle (Eds.), *Physical and virtual learning spaces in higher education: Concepts for the modern learning environment* (pp. 51–70). Hershey, PA: IGI Global. doi:10.4018/978-1-60960-114-0.ch004

Rahim, R. E., & Rahman, A. A. (2014). Green IT capability and firm's competitive advantage. *International Journal of Innovation in the Digital Economy*, *5*(1), 41–49. doi:10.4018/ijide.2014010104

Raj, P. P., & Azeez, P. A. (2012). Public on conserving an urban wetland: A case from Kerala, India. In E. Carayannis (Ed.), *Sustainable policy applications for social ecology and development* (pp. 1–7). Hershey, PA: IGI Global. doi:10.4018/978-1-4666-1586-1.ch001

Rasulev, B., Leszczynska, D., & Leszczynski, J. (2014). Nanoparticles: Towards predicting their toxicity and physico-chemical properties. In Nanotechnology: Concepts, methodologies, tools, and applications (pp. 1071-1089). Hershey, PA: IGI Global. doi:10.4018/978-1-4666-5125-8.ch049

Rene, E. R., Behera, S. K., & Park, H. S. (2012). Predicting adsorption behavior in engineered floodplain filtration system using backpropagation neural networks. In S. Kulkarni (Ed.), *Machine learning algorithms for problem solving in computational applications: Intelligent techniques* (pp. 179–194). Hershey, PA: IGI Global. doi:10.4018/978-1-4666-1833-6.ch011

Rene, E. R., López, M. E., Park, H. S., Murthy, D. V., & Swaminathan, T. (2012). ANNs for identifying shock loads in continuously operated biofilters: Application to biological waste gas treatment. In M. Khan & A. Ansari (Eds.), *Handbook of research on industrial informatics and manufacturing intelligence: Innovations and solutions* (pp. 72–103). Hershey, PA: IGI Global. doi:10.4018/978-1-4666-0294-6.ch004

Rene, E. R., López, M. E., Veiga, M. C., & Kennes, C. (2011). Artificial neural network modelling for waste: Gas and wastewater treatment applications. In B. Igelnik (Ed.), *Computational modeling and simulation of intellect: Current state and future perspectives* (pp. 224–263). Hershey, PA: IGI Global. doi:10.4018/978-1-60960-551-3.ch010

Rivas, A. A., Kahn, J. R., Freitas, C. E., Hurd, L. E., & Cooper, G. (2013). The role of payments for ecological services in the sustainable development and environmental preservation of the rainforest: A case study of Barcelos, Amazonas, BR. *International Journal of Social Ecology and Sustainable Development, 4*(3), 13–27. doi:10.4018/jsesd.2013070102

Rodrigues dos Anjos, M., & Schulz, M. (2013). Investigation of deforestation of environmental protection areas of Madeira River permanent preservation areas in Rondônia Amazon, Brazil. In E. Carayannis (Ed.), *Creating a sustainable ecology using technology-driven solutions* (pp. 335–343). Hershey, PA: IGI Global. doi:10.4018/978-1-4666-3613-2.ch023

Rojas-Mora, J., Josselin, D., Aryal, J., Mangiavillano, A., & Ellerkamp, P. (2013). The weighted fuzzy barycenter: Definition and application to forest fire control in the PACA region. *International Journal of Agricultural and Environmental Information Systems, 4*(4), 48–67. doi:10.4018/ijaeis.2013100103

Rolim da Paz, A., Uvo, C., Bravo, J., Collischonn, W., & Ribeiro da Rocha, H. (2011). Seasonal precipitation forecast based on artificial neural networks. In H. do Prado, A. Barreto Luiz, & H. Filho (Eds.), *Computational methods for agricultural research: Advances and applications* (pp. 326–354). Hershey, PA: IGI Global. doi:10.4018/978-1-61692-871-1.ch016

Romano, B., & Zullo, F. (2013). Models of urban land use in Europe: Assessment tools and criticalities. *International Journal of Agricultural and Environmental Information Systems, 4*(3), 80–97. doi:10.4018/ijaeis.2013070105

Rosen, M., Krichevsky, T., & Sharma, H. (2011). Strategies for a sustainable enterprise. In B. Unhelkar (Ed.), *Handbook of research on green ICT: Technology, business and social perspectives* (pp. 1–28). Hershey, PA: IGI Global. doi:10.4018/978-1-61692-834-6.ch001

Roussey, C., Pinet, F., & Schneider, M. (2013). Representations of topological relations between simple regions in description logics: From formalization to consistency checking. *International Journal of Agricultural and Environmental Information Systems, 4*(2), 50–69. doi:10.4018/jaeis.2013040105

Rushforth, R., & Phillips, C. F. (2012). Gathering under a green umbrella: collaborative rainwater harvesting at the University of Arizona. In E. Carayannis (Ed.), *Sustainable policy applications for social ecology and development* (pp. 139–149). Hershey, PA: IGI Global. doi:10.4018/978-1-4666-1586-1.ch011

Ruutu, J., Nurminen, J. K., & Rissanen, K. (2013). Energy efficiency of mobile device recharging. *International Journal of Handheld Computing Research, 4*(1), 59–69. doi:10.4018/jhcr.2013010104

Saïdi, S., Camara, A., Gazull, L., Passouant, M., & Soumaré, M. (2013). Lowlands mapping in forest Guinea. *International Journal of Agricultural and Environmental Information Systems, 4*(1), 20–34. doi:10.4018/jaeis.2013010102

Salewicz, K. A., Nakayama, M., & Bruch, C. (2011). Building capacity for better water decision making through internet-based decision support systems. In *Green technologies: Concepts, methodologies, tools and applications* (pp. 466–492). Hershey, PA: IGI Global. doi:10.4018/978-1-60960-472-1.ch306

Salter, S. J. (2011). When low-carbon means low-cost: Putting lessons from nature to work in our cities. *International Journal of Social Ecology and Sustainable Development, 2*(4), 12–25. doi:10.4018/jsesd.2011100102

Salter, S. J. (2013). When low-carbon means low-cost: putting lessons from nature to work in our cities. In E. Carayannis (Ed.), *Creating a sustainable ecology using technology-driven solutions* (pp. 282–295). Hershey, PA: IGI Global. doi:10.4018/978-1-4666-3613-2.ch019

Salvadó, J. A., López, J. E., & Martín de Castro, G. (2012). Social innovation, environmental innovation, and their effect on competitive advantage and firm performance. In F. Nobre, D. Walker, & R. Harris (Eds.), *Technological, managerial and organizational core competencies: Dynamic innovation and sustainable development* (pp. 89–104). Hershey, PA: IGI Global. doi:10.4018/978-1-61350-165-8.ch006

Saroar, M. M., & Routray, J. K. (2013). Desert in Bengal Delta- Changes in landscape, changes in livelihood: Can diffusion and adoption of sustainable adaptation make a difference? In H. Muga & K. Thomas (Eds.), *Cases on the diffusion and adoption of sustainable development practices* (pp. 83–117). Hershey, PA: IGI Global. doi:10.4018/978-1-4666-2842-7.ch004

Saroar, M. M., & Routray, J. K. (2014). Desert in Bengal Delta-Changes in landscape, changes in livelihood: Can diffusion and adoption of sustainable adaptation make a difference? In *Sustainable practices: Concepts, methodologies, tools and applications* (pp. 1414–1441). Hershey, PA: IGI Global. doi:10.4018/978-1-4666-4852-4.ch079

Schmehl, M., Eigner-Thiel, S., Ibendorf, J., Hesse, M., & Geldermann, J. (2012). Development of an information system for the assessment of different bioenergy concepts regarding sustainable development. In *Regional development: Concepts, methodologies, tools, and applications* (pp. 274–292). Hershey, PA: IGI Global. doi:10.4018/978-1-4666-0882-5.ch206

Schröter, M., Jakoby, O., Olbrich, R., Eichhorn, M., & Baumgärtner, S. (2011). Remote sensing of bush encroachment on commercial cattle farms in semi-arid rangelands in Namibia. In V. Olej, I. Obršálová, & J. Krupka (Eds.), *Environmental modeling for sustainable regional development: System approaches and advanced methods* (pp. 327–343). Hershey, PA: IGI Global. doi:10.4018/978-1-60960-156-0.ch016

Schultz, R. A. (2014). Environmentalism and sustainability. In *Technology versus ecology: Human superiority and the ongoing conflict with nature* (pp. 180–212). Hershey, PA: IGI Global. doi:10.4018/978-1-4666-4586-8.ch010

Schultz, R. A. (2014). More about modern technology. In *Technology versus ecology: Human superiority and the ongoing conflict with nature* (pp. 145–158). Hershey, PA: IGI Global. doi:10.4018/978-1-4666-4586-8.ch008

Schultz, R. A. (2014). The role of science and technology. In *Technology versus ecology: Human superiority and the ongoing conflict with nature* (pp. 213–230). Hershey, PA: IGI Global. doi:10.4018/978-1-4666-4586-8.ch011

Sdrali, D., Galanis, N., Goussia-Rizou, M., & Abeliotis, K. (2014). Are Greek high school students environmental citizens? A cluster analysis approach. *International Journal of Information Systems and Social Change*, *5*(1), 16–29. doi:10.4018/ijissc.2014010102

Selmaoui-Folcher, N., Flouvat, F., Gay, D., & Rouet, I. (2012). Spatial pattern mining for soil erosion characterization. In P. Papajorgji & F. Pinet (Eds.), *New technologies for constructing complex agricultural and environmental systems* (pp. 190–210). Hershey, PA: IGI Global. doi:10.4018/978-1-4666-0333-2.ch011

Shahid, M., Mishra, H., Mishra, H. K., Tripathi, T., Khan, H. M., Sobia, F., & Singh, A. (2012). Pharmaco-ecomicrobiology and its potential role in medical and environmental sciences. In T. Gasmelseid (Ed.), *Pharmacoinformatics and drug discovery technologies: Theories and applications* (pp. 291–302). Hershey, PA: IGI Global. doi:10.4018/978-1-4666-0309-7.ch018

Shakir, I., Ali, Z., Rana, U. A., Nafady, A., Sarfraz, M., Al-Nashef, I., & Kang, D. et al. (2014). Nanostructured materials for the realization of electrochemical energy storage and conversion devices: Status and prospects. In M. Bououdina & J. Davim (Eds.), *Handbook of research on nanoscience, nanotechnology, and advanced materials* (pp. 376–413). Hershey, PA: IGI Global. doi:10.4018/978-1-4666-5824-0.ch015

Sharma, P., Hussain, N., Das, M. R., Deshmukh, A. B., Shelke, M. V., Szunerits, S., & Boukherroub, R. (2014). Metal oxide-graphene nanocomposites: Synthesis to applications. In M. Bououdina & J. Davim (Eds.), *Handbook of research on nanoscience, nanotechnology, and advanced materials* (pp. 196–225). Hershey, PA: IGI Global. doi:10.4018/978-1-4666-5824-0.ch010

Sidorov, E., & Ritschelová, I. (2011). Economic performance and environmental quality at the regional level an approach to modeling depletion adjusted macro aggregates for the Czech coal mining regions. In V. Olej, I. Obršálová, & J. Krupka (Eds.), *Environmental modeling for sustainable regional development: System approaches and advanced methods* (pp. 281–302). Hershey, PA: IGI Global. doi:10.4018/978-1-60960-156-0.ch014

Silva, M. R., & McLellan, S. L. (2012). Environmental and social impact of stormwater outfalls at Lake Michigan beaches. In E. Carayannis (Ed.), *Sustainable policy applications for social ecology and development* (pp. 150–165). Hershey, PA: IGI Global. doi:10.4018/978-1-4666-1586-1.ch012

Snyder, A. (2014). Environmental protection agency. In J. Krueger (Ed.), *Cases on electronic records and resource management implementation in diverse environments* (pp. 363–377). Hershey, PA: IGI Global. doi:10.4018/978-1-4666-4466-3.ch022

Somavat, P., & Namboodiri, V. (2012). Information and communication technology revolution and global warming. In W. Hu & N. Kaabouch (Eds.), *Sustainable ICTs and management systems for green computing* (pp. 23–44). Hershey, PA: IGI Global. doi:10.4018/978-1-4666-1839-8.ch002

Spanu, V., & McCall, M. K. (2013). Eliciting local spatial knowledge for community-based disaster risk management: Working with cybertracker in Georgian caucasus. *International Journal of E-Planning Research, 2*(2), 45–59. doi:10.4018/ijepr.2013040104

Srinivasa, K. R, H., Sunil, K. N., B, A., Abhishek, S., Harish, R. C., & Anil, K. M. (2012). Weather nowcasting using environmental sensors integrated to the mobile. In A. Kumar, & H. Rahman (Eds.), Mobile computing techniques in emerging markets: Systems, applications and services (pp. 183-203). Hershey, PA: IGI Global. doi:10.4018/978-1-4666-0080-5.ch007

Stewart, A. W. (2014). Integrating sustainability within higher education. In K. Thomas & H. Muga (Eds.), *Handbook of research on pedagogical innovations for sustainable development* (pp. 369–382). Hershey, PA: IGI Global. doi:10.4018/978-1-4666-5856-1.ch017

Stewart, C. O., & Rhodes, C. (2014). Global warming as a socioscientific controversy. In R. Hart (Ed.), *Communication and language analysis in the public sphere* (pp. 276–289). Hershey, PA: IGI Global. doi:10.4018/978-1-4666-5003-9.ch016

Suaire, R., Durickovic, I., Simonnot, M., & Marchetti, M. (2013). Monitoring of road deicers in a retention pond. *International Journal of Measurement Technologies and Instrumentation Engineering, 3*(1), 39–47. doi:10.4018/ijmtie.2013010104

Subic, J., & Jelocnik, M. (2013). Economic and environmental aspects of controlled vegetable production within the region of Danube basin. In A. Jean-Vasile, T. Adrian, J. Subic, & D. Dusmanescu (Eds.), *Sustainable technologies, policies, and constraints in the green economy* (pp. 39–62). Hershey, PA: IGI Global. doi:10.4018/978-1-4666-4098-6.ch003

Tabrizi, A., & Sanguinetti, P. (2013). Case study: Evaluation of renewable energy strategies using building information modeling and energy simulation. *International Journal of 3-D Information Modeling (IJ3DIM), 2*(4), 25-37. doi:10.4018/ij3dim.2013100103

Taylor, R., Barron, E., & Eames, K. A. (2014). Embedding sustainability learning: Robustness in changing circumstances - Perspectives from a United Kingdom (UK) higher education institution (HEI). In K. Thomas & H. Muga (Eds.), *Handbook of research on pedagogical innovations for sustainable development* (pp. 641–671). Hershey, PA: IGI Global. doi:10.4018/978-1-4666-5856-1.ch033

Thiell, M., & Zuluaga, J. P. (2013). Is it feasible to implement green logistics in emerging markets? *International Journal of Applied Logistics*, *4*(1), 1–13. doi:10.4018/jal.2013010101

Thongmak, M. (2013). A systematic framework for sustainable ICTs in developing countries. *International Journal of Information Technologies and Systems Approach*, *6*(1), 1–19. doi:10.4018/jitsa.2013010101

Torrens, F., & Castellano, G. (2012). Cluster origin of solvent features of fullerenes, single-wall carbon nanotubes, nanocones, and nanohorns. In E. Castro & A. Haghi (Eds.), *Nanoscience and advancing computational methods in chemistry: Research progress* (pp. 1–57). Hershey, PA: IGI Global. doi:10.4018/978-1-4666-1607-3.ch001

Torrens, F., & Castellano, G. (2014). Cluster origin of solvent features of fullerenes, single-wall carbon nanotubes, nanocones, and nanohorns. In I. Management Association (Ed.), Nanotechnology: Concepts, methodologies, tools, and applications (pp. 262-318). Hershey, PA: IGI Global. doi:10.4018/978-1-4666-5125-8.ch011

Touza, L. L., & López-Gunn, E. (2012). Climate change policies—Mitigation and adaptation at the local level: The case of the city of Madrid (Spain). In M. Tortora (Ed.), *Sustainable systems and energy management at the regional level: Comparative approaches* (pp. 261–287). Hershey, PA: IGI Global. doi:10.4018/978-1-61350-344-7.ch014

Trautmann, N. M., & McLinn, C. M. (2012). Using online data for student investigations in biology and ecology. In A. Olofsson & J. Lindberg (Eds.), *Informed design of educational technologies in higher education: Enhanced learning and teaching* (pp. 80–100). Hershey, PA: IGI Global. doi:10.4018/978-1-61350-080-4.ch005

Trisurat, Y., Shrestha, R. P., & Alkemade, R. (2011). Linkage between biodiversity, land use informatics and climate change. In Y. Trisurat, R. Shrestha, & R. Alkemade (Eds.), *Land use, climate change and biodiversity modeling: Perspectives and applications* (pp. 1–22). Hershey, PA: IGI Global. doi:10.4018/978-1-60960-619-0.ch001

Trivedi, B. (2011). Developing environmentally responsible business strategies: A research perspective. *International Journal of Green Computing*, *2*(1), 47–57. doi:10.4018/jgc.2011010105

Trivedi, B. (2013). Developing environmentally responsible business strategies: A research perspective. In K. Ganesh & S. Anbuudayasankar (Eds.), *International and interdisciplinary studies in green computing* (pp. 47–57). Hershey, PA: IGI Global. doi:10.4018/978-1-4666-2646-1.ch005

Tsalapata, H., Alimsi, R., & Heidmann, O. (2012). Environmental education through envkids didactical framework and ICT tools. In Z. Lu (Ed.), *Learning with mobile technologies, handheld devices, and smart phones: Innovative methods* (pp. 147–161). Hershey, PA: IGI Global. doi:10.4018/978-1-4666-0936-5.ch009

Tsalapata, H., Alimsi, R., & Heidmann, O. (2014). Environmental education through envkids didactical framework and ICT tools. In *Sustainable practices: Concepts, methodologies, tools and applications* (pp. 1492–1504). Hershey, PA: IGI Global. doi:10.4018/978-1-4666-4852-4.ch083

Turek, A. (2013). Sustainable agriculture: between sustainable development and economic competitiveness. In A. Jean-Vasile, T. Adrian, J. Subic, & D. Dusmanescu (Eds.), *Sustainable technologies, policies, and constraints in the green economy* (pp. 219–235). Hershey, PA: IGI Global. doi:10.4018/978-1-4666-4098-6.ch012

Turgut, E. T., & Rosen, M. A. (2012). Emission assessment of aviation. In E. Abu-Taieh, A. El Sheikh, & M. Jafari (Eds.), *Technology engineering and management in aviation: Advancements and discoveries* (pp. 20–72). Hershey, PA: IGI Global. doi:10.4018/978-1-60960-887-3.ch002

Twesigye, C. K. (2011). Application of remote sensing technologies and geographical information systems in monitoring environmental degradation in the Lake Victoria watershed, East Africa. In *Green technologies: Concepts, methodologies, tools and applications* (pp. 653–677). Hershey, PA: IGI Global. doi:10.4018/978-1-60960-472-1.ch405

Uchida, S., Hayashi, K., Sato, M., & Hokazono, S. (2011). Construction of agri-environmental data using computational methods: The case of life cycle inventories for agricultural production systems. In H. do Prado, A. Barreto Luiz, & H. Filho (Eds.), *Computational methods for agricultural research: advances and applications* (pp. 412–433). Hershey, PA: IGI Global. doi:10.4018/978-1-61692-871-1.ch019

Unhelkar, B. (2011). Green enterprise architecture using environmental intelligence. *International Journal of Green Computing*, *2*(1), 58–65. doi:10.4018/jgc.2011010106

Unhelkar, B., & Trivedi, B. (2011). A framework for environmentally responsible business strategies. In B. Unhelkar (Ed.), *Handbook of research on green ICT: Technology, business and social perspectives* (pp. 214–232). Hershey, PA: IGI Global. doi:10.4018/978-1-61692-834-6.ch014

Urban, M. J., Marker, E., & Falvo, D. A. (2012). An interdisciplinary exploration of the climate change issue and implications for teaching STEM through inquiry. In L. Lennex & K. Nettleton (Eds.), *Cases on inquiry through instructional technology in math and science* (pp. 523–550). Hershey, PA: IGI Global. doi:10.4018/978-1-4666-0068-3.ch019

Urooj, S., Hussain, A., & Srivastava, N. (2013). Biodiesel production from algal blooms: A step towards renewable energy generation & measurement. *International Journal of Measurement Technologies and Instrumentation Engineering*, *2*(3), 60–71. doi:10.4018/ijmtie.2012070106

Uyttersprot, I., & Vermeir, I. (2014). Should I recycle or not? Effects of attitude strength and social pressure. In A. Kapoor & C. Kulshrestha (Eds.), *Dynamics of competitive advantage and consumer perception in social marketing* (pp. 292–308). Hershey, PA: IGI Global. doi:10.4018/978-1-4666-4430-4.ch012

Varga, L., Camci, F., Boxall, J., Toossi, A., Machell, J., Blythe, P. T., & Taylor, C. (2013). Transforming critical infrastructure: Matching the complexity of the environment to policy. *International Journal of E-Planning Research*, *2*(3), 38–49. doi:10.4018/ijepr.2013070104

Viaggi, D., & Raggi, M. (2011). Estimation of irrigation water demand on a regional scale combining positive mathematical programming and cluster analysis in model calibration. In V. Olej, I. Obršálová, & J. Krupka (Eds.), *Environmental modeling for sustainable regional development: System approaches and advanced methods* (pp. 204–220). Hershey, PA: IGI Global. doi:10.4018/978-1-60960-156-0.ch010

Wang, H. (2011). GHG emissions from the international goods movement by ships and the adaptation funding distribution. In Z. Luo (Ed.), *Green finance and sustainability: Environmentally-aware business models and technologies* (pp. 274–290). Hershey, PA: IGI Global. doi:10.4018/978-1-60960-531-5.ch015

Wang, H., & Ghose, A. K. (2011). Green strategic alignment: Aligning business strategies with sustainability objectives. In B. Unhelkar (Ed.), *Handbook of research on green ICT: Technology, business and social perspectives* (pp. 29–41). Hershey, PA: IGI Global. doi:10.4018/978-1-61692-834-6.ch002

Wang, S., Ku, C., & Chu, C. (2013). Sustainable campus project: Potential for energy conservation and carbon reduction education in Taiwan. *International Journal of Technology and Human Interaction, 8*(3), 19–30. doi:10.4018/jthi.2012070103

Wang, Y. (2014). Texted environmental campaign in China: A case study of new media communication. In J. Pelet & P. Papadopoulou (Eds.), *User behavior in ubiquitous online environments* (pp. 19–43). Hershey, PA: IGI Global. doi:10.4018/978-1-4666-4566-0.ch002

Wani, A. H., Amin, M., Shahnaz, M., & Shah, M. A. (2012). Antimycotic activity of nanoparticles of MgO, FeO and ZnO on some pathogenic fungi. *International Journal of Manufacturing, Materials, and Mechanical Engineering, 2*(4), 59–70. doi:10.4018/ijmmme.2012100105

Wani, A. H., Amin, M., Shahnaz, M., & Shah, M. A. (2014). Antimycotic activity of nanoparticles of MgO, FeO and ZnO on some pathogenic fungi. In Nanotechnology: Concepts, methodologies, tools, and applications (pp. 1289-1299). Hershey, PA: IGI Global. doi:10.4018/978-1-4666-5125-8.ch059

Wati, Y., & Koo, C. (2011). A new recommendation for green IT strategies: A resource-based perspective. In Z. Luo (Ed.), *Green finance and sustainability: Environmentally-aware business models and technologies* (pp. 153–175). Hershey, PA: IGI Global. doi:10.4018/978-1-60960-531-5.ch008

Williamson, T. B., Hauer, G. K., & Luckert, M. K. (2011). Economic concepts, methods, and tools for risk analysis in forestry under climate change. In V. Olej, I. Obršálová, & J. Krupka (Eds.), *Environmental modeling for sustainable regional development: System approaches and advanced methods* (pp. 303–326). Hershey, PA: IGI Global. doi:10.4018/978-1-60960-156-0.ch015

Wong, Y. M. (2014). Fair share of supply chain responsibility for low carbon manufacturing. In Z. Luo (Ed.), *Smart manufacturing innovation and transformation: interconnection and intelligence* (pp. 303–332). Hershey, PA: IGI Global. doi:10.4018/978-1-4666-5836-3.ch012

Wu, J., & Haasis, H. (2013). Integration of knowledge management approach to the planning stage of freight villages: Towards sustainable development. *International Journal of Applied Logistics, 4*(2), 46–65. doi:10.4018/jal.2013040104

Yamaguchi, M., & Fara, L. (2013). New trends in solar cells. In L. Fara & M. Yamaguchi (Eds.), *Advanced solar cell materials, technology, modeling, and simulation* (pp. 1–21). Hershey, PA: IGI Global. doi:10.4018/978-1-4666-1927-2.ch001

Yi, L. (2011). E-business/ICT and carbon emissions. In O. Bak & N. Stair (Eds.), *Impact of e-business technologies on public and private organizations: Industry comparisons and perspectives* (pp. 214–232). Hershey, PA: IGI Global. doi:10.4018/978-1-60960-501-8.ch013

Yi, L. (2013). E-business/ICT and carbon emissions. In *Industrial engineering: Concepts, methodologies, tools, and applications* (pp. 1833–1852). Hershey, PA: IGI Global. doi:10.4018/978-1-4666-1945-6.ch098

Younessi, D. (2011). Sustainable business value. In B. Unhelkar (Ed.), *Handbook of research on green ICT: Technology, business and social perspectives* (pp. 98–115). Hershey, PA: IGI Global. doi:10.4018/978-1-61692-834-6.ch007

Yu, T., Lenzen, M., & Dey, C. (2011). Large-scale computational modeling for environmental impact assessment. In V. Olej, I. Obršálová, & J. Krupka (Eds.), *Environmental modeling for sustainable regional development: system approaches and advanced methods* (pp. 1–17). Hershey, PA: IGI Global. doi:10.4018/978-1-60960-156-0.ch001

Zhu, Y., & Heath, T. (2012). Technologies in urban design practice: Integrating environmental design considerations. In O. Ercoskun (Ed.), *Green and ecological technologies for urban planning: Creating smart cities* (pp. 133–152). Hershey, PA: IGI Global. doi:10.4018/978-1-61350-453-6.ch008

Zoltáni, T. A. (2013). Carbon as an emerging tool for risk management. *International Journal of Applied Logistics*, *4*(4), 51–69. doi:10.4018/ijal.2013100104

Compilation of References

Adedemeji, M. (2013). What is academic humility. *The Alma Matter*. Retrieved July 29, 2016, from http://mahfouzadedimeji.com/what-is-academic-humility/

Adeleke, D. S., & Emeahara, E. N. (2016). Relationship between information literacy and use of electronic information resources by postgraduate students of the university of Ibadan. *Library Philosophy and Practice*. Retrieved from http://digitalcommons.unl.edu/libphilprac/1381

Advanced Learner's Dictionary of Current English. (1952). Oxford: Oxford University Press.

Agu, N. N., Omenyi, A. S., & Odimegwu, C. (2015). Evaluation of doctorate dissertation in Nigerian universities: do faculties provide and use explicit criteria/rubrics? *IJTIE: International Journal of Technology and Inclusive Education*, *4*(1), 565-569.

Almind, T., & Ingwersen, P. (1997). Informetric analyses on the world wide web: Methodological approaches to 'webometrics'. *The Journal of Documentation*, *53*(4), 60–64. doi:10.1108/EUM0000000007205

American Library Association (ALA). (1989). ALA Presidential Committee on Information Literacy Final Report. Chicago: American Library Association.

American Medical Association Foundation Health Literacy. (2003). *Help Your Patients Understand*. CME Programme.

Amsaveni, N., & Ramesh, K. (2016). Mapping of research productivity in forensic science: A scientometric analysis. *Journal of Advances in Library and Information Science*, *5*(4), 372–378.

Anilkumar, N. & Rajaram, S. (2013). Theses submitted by doctoral students of physical research laboratory, India: a citation analysis. *Serials Review*, *39*(2), 114–120.

Anyi, K. W. U., Zainab, A. N., & Anuar, N. B. (2009). Bibliometric studies on single journals: A review. *Malaysian Journal of Library and Information Science*, *14*(1), 17–55.

Arunachalam, S. (2003). Use of SCI-based publication counts. *Current Science*, *85*(10), 1391–1392.

Ashbaugh, D. R. (2000). *Quantitative-Qualitative Friction Ridge Analysis: An introduction to basic and advanced ridgeology*. Boca Raton: CRC Press.

Ashok, J., & Gang, K. S. (1992). Laser research in India: Scientific study and model projection. *Scientometrics*, *23*(3), 395–415. doi:10.1007/BF02029806

Ashokkumar, N., Radhakrishnan, N., & Mahesh, K. C. (2008). Recent trends in ophthalmology– A citation analysis. *PEARL–. Journal of Library and Information Science*, *2*(2), 13–16.

Austin, M. (2014) Humility is a trait worth having. *Psychology Today*. Retrieved 14 August, 2016. from https://www.psychologytoday.com/blog/ethics-everyone/201206/humility

Awuni, J. A. (2016). Students' perspective of effective supervision of graduate programmes in Ghana. *American Journal of Educational Research*, *4*(2), 163–169.

Balarabe, Y. (2009). PhD as minimum qualification for academic staff in Nigerian universities: a policy of self deception. *Modern Ghana*. Retrieved March 12, 2016, from http//:www.modernghana/phd-as-minimum-qualification-for-academic-staff-in-nigeria-universities.htm.

Beaver, D., & Rosen, R. (1978). Studies in scientific collaboration: Part I--The professional origins of scientific co-authorship. *Scientometrics*, *1*(1), 65–84. doi:10.1007/BF02016840

Bentham Open Publisher. (2017). Retrieved March 27, 2017 from http://www.benthamopen.com

Beri, G. C. (1993). *Marketing Research*. Tata: McGraw-Hill.

Berkman, N. D., Davis, T. C., & McCormack, L. (2010). Health Literacy: What Is It? *Journal of Health Communication*, *15*(2), 9–19. doi:10.1080/10810730.2010.499985 PMID:20845189

Betageri, S. S. (2015). Bibliometric Analysis of contributions in the journal of dairying, foods and home sciences. *Journal of Advances in Library and Information Science*, *4*(1), 86–89.

Bettencourta, L. M. A., & Kaur, J. (2011). Evolution and structure of sustainability science. *Proceedings of the National Academy of Sciences of the United States of America*, *108*(49), 19540–19545. doi:10.1073/pnas.1102712108 PMID:22114186

Bhaskaran, C. (2013). Research productivity of Alagappa University during 1999-2011: A bibliometric study. *DESIDOC Journal of Library and Information Technology*, *33*(3), 236–242. doi:10.14429/djlit.33.3.4609

Bhattacharya, G. (1978). *Information Science: A unified view through systems approach*. Calcutta: IASLIC.

Bibina, C. B., & Kabir, S. H. (2016). Understanding the factors related to digital literacy: A survey among science-research scholars of university of Kerala. *Journal of Knowledge & Communication Management*, *6*(2), 93–104. doi:10.5958/2277-7946.2016.00008.5

Bioline International. (2017). Retrieved March 27, 2017 from http://www.bioline.org.br

Bird, D. K. (2009). The use of questionnaires for acquiring information on public perception of natural hazards and risk mitigation –a review of current knowledge and practice. *Natural Hazards and Earth System Sciences*, *9*(4), 1307–1325. doi:10.5194/nhess-9-1307-2009

Black, S. (2012). Frequently cited journals in forensic psychology. *Psychological Reports*, *110*(1), 276–282. doi:10.2466/09.13.17.PR0.110.1.276-282 PMID:22489393

Bormann, L. D. (2007). What do we know about the h index? *Journal of the American Society for Information Science and Technology*, *58*(9), 1381–1385. doi:10.1002/asi.20609

Börner, K., Chen, C., & Boyack, K. (2003). Visualizing Knowledge Domains. In B. Cronin (Ed.), *Annual Review of Information Science & Technology*. Medford, NJ: American Society for Information Science and Technology. doi:10.1002/aris.1440370106

Bothakur, P. (2015). Citation analysis of theses and dissertations in chemistry submitted to the LNB, library, Dibrugarh university, 2009-13. *International Journal of Research in Library Science*, *1*(2), 33–41.

Brookes, B. C. (1990). Biblio-, sciento-, infor-metrics??? What are we talking about. In L. Egghe & R. Rousseau (Eds.), *Informetrics 89/90* (pp. 31–43). Amsterdam: Elsevier Science Publishers B.V.

Bryman, A., & Bell, E. (2010). *Business research methods*. USA: Oxford University Press.

Burchell, B., & Marsh, C. (1992). The effect of questionnaire length on survey response. *Quality & Quantity*, *26*(3), 233–244. doi:10.1007/BF00172427

Card, S., Mackinlay, J., & Shneiderman, B. (1999). *Readings in Information Visualization: Using Vision to Think*. San Francisco: Morgan Kaufmann.

Carizzo-Sainero, G. (2000). Towards a concept of bibliometrics. *Journal of Spanish Research on Information Science*, *1*(2), 5986.

Castillo-Pérez, J. J., Muñoz-Valera, L., García-Gómez, F., & Mejía-Aranguré, J. M. (2015). Bibliometric analysis of scientific output on influenza in Mexico 2000-2012. *Revista Medica del Instituto Mexicano del Seguro Social*, *53*, 294–301. PMID:25984614

Catalano, A., & Philips, S. R. (2016). Information literacy and retention: A case study of the value of the library. *Evidence Based Library and Information Practice*, *11*(4), 2–13. doi:10.18438/B82K7W

Centers for Disease Control and Prevention. (2016, April 13). *CDC concludes* zika *causes microcephaly and other birth defects*. News Room Release.

Charrel, R. N., Leparc-Goffart, I., Pas, S., de Lamballerie, X., Koopmans, M., & Reusken, C. (2016). State of knowledge on zika virus for an adequate laboratory response. *Bulletin of the World Health Organization*. doi:10.2471/BLT.16.171207

Chastain, M. (2016). National Institutes of Health: zika virus is a pandemic. Breitbart; Retrieved 13 February 2016.

Chen, C., Paul, R., & O'Keefe, B. (2001). Fitting the jigsaw of citation: Information visualization in domain analysis. *Journal of the American Society for Information Science and Technology*, *52*(4), 315–330. doi:10.1002/1532-2890(2000)9999:9999<::AID-ASI1074>3.0.CO;2-2

Childers, T. L., & Ferrell, O. C. (1979). Response rates and perceived questionnaire length in mail surveys. *JMR, Journal of Marketing Research*, *16*(3), 429–431. doi:10.2307/3150720

Chiu, W.-T., & Ho, Y.-S. (2005). Bibliometric analysis of homoeopathy research during the period of 1991 to 2003. *Scientometrics*, *63*(1), 3–23. doi:10.1007/s11192-005-0201-7

Chiu, W.-T., Huang, J.-S., & Ho, Y.-S. (2004). Bibliometric analysis of Severe Acute Respiratory Syndrome-related research in the beginning stage. *Scientometrics*, *61*(1), 69–77. doi:10.1023/B:SCIE.0000037363.49623.28

Christian, G. E. (2008). *Open access initiative and developing world*. Retrieved March 25, 2017 from https://ssrn.com/abstract=1304665

Colquhoun, H. L., Levac, D., O'Brien, K. K., Straus, S., Tricco, A. C., Perrier, L., & Moher, D. et al. (2014). Scoping reviews: Time for clarity in definition, methods, and reporting. *Journal of Clinical Epidemiology*, *67*(12), 1291–1294. doi:10.1016/j.jclinepi.2014.03.013 PMID:25034198

Cronin, B., & Atkins, H. B. E. (2000). *The Web of Knowledge: A Festschrift in Honor of Eugene Garfield*. ASIST.

Das, A. K. (2008). *Open Access to Knowledge and Information: Scholarly Literature and Digital Library Initiatives -The South Asian Scenario*. Retrieved March 25, 2017 from http://www.portal.unesco.org/ci/en/files/.../12075628443open_access.../open_access_book_en.pd

De Forest, P. R., Gaensslen, R. E., & Lee, H. C. (1983). *Forensic Science: an introduction to criminalistics*. New York: McGraw-Hill Book Company.

Delirrad, M., Rashidi, A., & Karimi, S. (2012). A bibiliometric analysis of Toxicology publications of Iran and Turkey in ISI web of science. *Iranian Journal of Toxicology*, *6*(19), 735–745.

Derr, D. L. (1983). A conceptual analysis of information needs. *Information Processing & Management*, *19*(5), 273–278. doi:10.1016/0306-4573(83)90001-8

Dervin, B. (1983). An overview of sense-making research: Concepts, methods and results to date. In *Proceedings of the International Communication Association Annual Meeting,* Dallas, TX.

Deutsch, A. (1977). *The Librarians glossary*. New Delhi: Indian Book Company.

Devi, B. M., & Sankar, J. V. S. (2014). Information use pattern of researchers in commerce: A citation analysis of doctoral dissertations. *Library Philosophy and Practice*. Retrieved from http://digitalcommons.unl.edu/libphilprac/1105

Devi, B. M. (2013). Bibliometric study of core journals in toxicology. *E-Science World*, *5*(3-4), 13–23.

Devi, S. (2010). Medlars: Health information system in library. *University News*, *48*(40), 14–22.

Dharmapalan, B. (2014). To be successful in science one needs to follow a disciplined way of life – Dr. Ignacimuthu. *Scientific Reports*, *51*(9), 42–44.

Dhiman, C. (2016). Professors must have PhDs, at least. *The Daily Star*, November 13 edition, 23-25.

DOAJ. (2017). Retrieved March 27, 2017 from http://www.doaj.org

DOAR. (2017). Retrieved March 25, 2017 from http://www.opendoar.org

Dolfman, M. (1973). The concept of health: An historic and analytic examination. *The Journal of School Health*, *43*(8), 491–497. doi:10.1111/j.1746-1561.1973.tb03740.x PMID:4583082

Doyle, A. C. (1905). *The Great Keinplatz Experiment*. New York: George H. Doran Co.

DRDO Publications. (2017). Retrieved April 20, 2017 from http://www.publications.drdo.gov.in

DRTC. (2017). Retrieved March 25, 2017 from http://www.drtc.isibang.ac.in

Dubos, R. (2007). *Saunders Comprehensive Veterinary Dictionary* (3rd ed.). Elsevier.

Dutt, B., & Nikam, K. (2014). Scientometrics of collaboration pattern in solar cell research in India. *Annals of Library and Information Studies*, *61*(1), 65–73.

Educational Broadcasting Corporation. (2004). Constructivism as a paradigm for teaching and learning. *Concept to Classroom* Retrieved March 14, 2016, from http://www.thirteen.org/edonline/concept2class/constructivism/

Egghe, L., & Rousseau, R. (1990). *Introduction to Informetrics: Quantitative Methods in Library, Documentation and Information Science*. New York: Elsevier.

Ekundayo, S. (2014). Maggots in our university kola nut. *The Guardian Newspaper*, March 9 edition, p.11.

Encyclopedia Americana (International ed.). (1981). Grolier Inc.

e-PG Pathshala@Inflibnet. (2017). Retrieved March 25, 2017 from http://www.epgp.inflibnet.ac.in

Etzkowitza, H., & Leydesdorffb, L. (2000). The dynamics of innovation: From national systems and 'mode2' to a triple helix of university-industry-government relations. *Research Policy*, *29*(2), 109–123. doi:10.1016/S0048-7333(99)00055-4

European University – Industry Network. (2007). Teaching and research synergy in the context of university – industry cooperation. Eindhoven: ZKP. Chevalier de Seyn Publishers.

Exline, J., & Hill, P. (2012, May). Humility: A consistent and robust predictor of generosity. *The Journal of Positive Psychology*, *7*(3), 208–218. doi:10.1080/17439760.2012.671348

Fairthorne, R. A. (1969). Empirical hyperbolic distributions (Bradford-Zipf-Madelbrot) for bibliometrics description and prediction. *The Journal of Documentation*, *25*(4), 319–343. doi:10.1108/eb026481

Farooq, R. (2016a). Role of structural equation modeling in scale development. *Journal of Advances in Management Research*, *13*(1), 75–91. doi:10.1108/JAMR-05-2015-0037

Felder, R. M. (1994). The myth of the superhuman professor. *Journal of Engineering Education*, *82*(2), 105–110. doi:10.1002/j.2168-9830.1994.tb01087.x

Fulder, S. (1996). *The Handbook of Alternative and Complementary Medicine*. London: Hodder and Stoughton.

Galton, F. (1892). *Fingerprints*. London: Macmillan.

Gann, R. (1986). *The Health Information Hand Book: Resources for self care*. England: Gower.

Garfield (1994). Scientography: Mapping the tracks of science. *Current Contents: Social and Behavioural Sciences,* *7*(45), 5-10.

Garfield, E. (1979). *Citation Indexing: Its theory and application in Science, Technology and Humanities*. USA: Wiley Inter Science Publications.

Garfield, E. (1994). Scientography: Mapping the tracks of science. *Current Contents: Social and Behavioural Sciences*, *7*(45), 5–10.

Gašparac, P. (2006). The role and relevance of bibliographic citation databases. *Biochemia Medica*, *16*(2), 93–102. doi:10.11613/BM.2006.009

Gevitz, N. (1993). Unorthodox medical theories. In W. F. Bynum & R. Porter (Eds.), *Companion Encyclopaedia of the History of Medicine*. London: Routledge.

Gill, P., Jeffreys, A. J., & Werrett, D. J. (1985). Forensic application of DNA 'fingerprints'. *Nature*, *318*(6046), 577–579. doi:10.1038/318577a0 PMID:3840867

Gopikuttan, A., & Aswathy, S. (2014). Publication productivity of university of Kerala: A scientometric view. *DESIDOC Journal of Library and Information Technology*, *34*(2), 131–139. doi:10.14429/djlit.34.4280

Graham, E. C. (1993). Encyclopaedia of Scientific Terms (Vol. 2). New Delhi: Crest Publishing House.

Gualdi-Russo, E., & Fonti, G. (2013). Recent trend and perspectives in forensic anthropology: A bibliometric analysis. *Collegium Antropologicum*, *37*(2), 595–599. PMID:23941009

Gupta, B.M. (2012). Scientometric analysis of pakistan's S &T resource output. *Annals of library and information studies, 59*(1), 25-38.

Gupta, A. B., & Har, K. (2011). Mapping of AIDS/HIV research in India: A scientometrics analysis of publications output during 1999-2008. *Collnet Journal of Scientometrics and Information Management*, *5*(2), 185–203. doi:10.1080/09737766.2011.10700912

Gupta, B. M., & Bhattcharya, S. (2004). A bibliometric approach to word & mapping the dynamics of science and technology. *DESIDOC Bulletin of Information Technology*, *24*(1), 3–8. doi:10.14429/dbit.24.1.3616

Gupta, B. M., & Dhawan, S. M. (2008). A scientometric analysis of S & T publications output by India during 1985-2002. *DESIDOC Journal of Library and Information &. Información Tecnológica*, *28*(2), 73–85.

Haiai, Z. (1994). A Bibliometric study on medicine Chinese tradition in medicine database. *Scientometrics*, *31*(3), 241–250. doi:10.1007/BF02016874

Health Metrics Network Framework and Standards for Country Health Information Systems. (2008). WHO.

Healthfield, S. M. (2016). What seniority means at work. *The Balance*. Retrieved March 7, 2016, from https://www.thebalance.com/what-seniority-means-at-work-1919372

Heilig, L., &VoB, S. (2014). A scientometric analysis of cloud computing literature. *IEEE Transactions on Cloud Computing, 2*(3), 266-278.

He, Q. (1999). Knowledge discovery through co-word analysis. *Library Trends*, *48*(1), 133–159.

Hertzel, D. H. (1987). Bibliometrics, history of development of ideas in". In A. Kent (Ed.), Encyclopedia of Library and Information Science (Vol. 42, pp. 144-219). New York: Marcell Dekker.

Herzog, A. R., & Bachman, J. G. (1981). Effects of questionnaire length on response quality. *Public Opinion Quarterly*, *45*(4), 549–559. doi:10.1086/268687

High Wire Press. (2017). Retrieved March 27, 2017 from High Wire Press: http://www.highwire.org

Holmes, O. (2015). Dissertation Diaries: Lessons in academic humility. *Society for Personality and Social Psychology*, November.

Hume, D. (1983). *An enquiry into the principles of morals*. Indianapolis: Hackett.

IETE. (2017). Retrieved March 25, 2017 from http://www. iete-elan.ac.in

Indian Academy of Sciences. (2017). Retrieved March 27, 2017 from Indian Academy of Science: http://www.ias.ac.in

Indian Agricultural Research Institute. (2017). Retrieved March 27, 2017 from http://www. isas.org.in

Indian Journals. *Com*. (2017). Retrieved March 27, 2017 from http://www.indianjournals.com

Indian National Science Academy. (2017). Retrieved March 27, 2017 from http://www.insa.nic.in

IndMED. (2017). Retrieved March 27, 2017 from http://www.indmed.nic.in

INFLIBNET Institution Repository. (2017). Retrieved March 25, 2017 from http://www.ir.inflibnet.ac.in

Infoport@ INFLIBNET. (2017). Retrieved March 25, 2017 from http://www.infoport.inflibnet.ac.in/

Ingwersen, P., & Christensen, F. H. (1997). Data set isolation for bibliometric online analyses of research publications: Fundamental methodological issues. *Journal of the American Society for Information Science*, *48*(3), 205–217. doi:10.1002/(SICI)1097-4571(199703)48:3<205::AID-ASI3>3.0.CO;2-0

Inman, K., & Rudin, N. (2001). *Principles and Practice of Criminalistics: The profession of Forensic Science*. Boca Raton: CRC Press.

Iyer, L. M., Balaji, S., Koonin, E. V., & Aravind, L. (2006). Evolutionary genomics of nucleo-cytoplasmic large DNA viruses. *Virus Research*, *117*(1), 156–184. doi:10.1016/j.virusres.2006.01.009 PMID:16494962

Jain, P. (2012). Promoting open access to research in academic libraries. Retrieved March 30, 2017, from http://unllib.unl.edu/LPP/

James, S. H., & Nordby, J. J. (2003). *Forensic Science: An introduction to Scientific and investigative techniques*. Boca Raton: CRC Press.

Jansen, D., Wald, A., Franke, K., Schmoch, U., & Schubert, T. (2007). Third party research funding and performance in research. On the effects of institutional conditions on research performance of teams. *Kolner Zeitschrift Fur Soziloogie Und Sozialpsychoogie*, *59*(1), 125–149. doi:10.1007/s11577-007-0006-1

Jeyasekar, J. J., & Saravanan, P. (2013). Journal of forensic sciences: a bibliometric study for the period 2006 to 2010. In *Proceedings of the Second National Conference of Scientometrics and Knowledge Management, Dharwad* (CD-ROM version).

Jeyasekar, J. J., & Saravanan, P. (2014a). Scientometric analysis of the journal medicine, science and the law from 2001 to 2012. In *Proceedings of UGC Sponsored National Seminar on Application of Quantitative and Qualitative Indicators for the Excellence of Scholarly Communication, Annamalai Nagar,* Annamalai Nagar (pp. 308-313).

Jeyasekar, J.J., & Saravanan, P. (2014d). A scientometric analysis of global forensic science research publications. *Library Philosophy & Practice*.

Jeyasekar, J.J., & Saravanan, P. (2016). Science maps of global and Indian wildlife forensics: a comparative map. *Library Philosophy and Practice.*

Jeyasekar, J. J., & Saravanan, P. (2012a). Scientometric study of forensic science: a study based on *SCOPUS* database. In *Proceedings of the UGC Sponsored National Seminar on Scientometrics and Informetrics, Annamalai Nagar* (pp. 75-78).

Jeyasekar, J. J., & Saravanan, P. (2012b). Scientometric analysis of Indian forensic science literature based on ICI database. *Journal of Library Advancements*, *2*(1), 1–5.

Jeyasekar, J. J., & Saravanan, P. (2014b). A scientometric analysis of the Journal of Forensic Identification. *KELPRO Bulletin*, *18*(2), 34–47.

Jeyasekar, J. J., & Saravanan, P. (2014c). A scientometric portrait of the journal Digital Investigation. *Journal of Advances in Library and Information Science*, *3*(2), 155–162.

Jeyasekar, J. J., & Saravanan, P. (2015a). Indian forensic science research literature: A bibliometric study of its growth, authorship and publication patterns. *SRELS Journal of Information Management*, *52*(1), 67–75.

Jeyasekar, J. J., & Saravanan, P. (2015b). Mapping forensic odontology literature using open source bibliographies and software: a case study. In S. Thanuskodi (Ed.), *Handbook of Research in Inventive Digital Tools for Collection Management and Development in Modern Libraries*. Philadelphia: IGI Global. doi:10.4018/978-1-4666-8178-1.ch011

Jeyshankar, R., Babu, B. R., & Rajendran, P. (2011). Research output of CSIR-central electro chemical research institute (CECRI): A study. *Annals Library Information Studies*, *58*(4), 301–306.

Jeyshankar, R., & Vellaichamy, A. (2015). A bibliometric analysis of global literature on cervical cancer. *International Journal of Information Dissemination and Technology*, *4*, 298–302.

Jeyshankar, R., & Vellaichamy, A. (2015). Research productivity of lung cancer by Indian scientist during 1984 – 2013. *Productivity*, *56*, 24–35.

Jeyshankar, R., & Vellaichamy, A. (2016). Scientometric analysis of autism research output during 2007-2011. *SRELS Journal of Information Management*, *53*(1), 55–64. doi:10.17821/srels/2016/v53i1/86768

Johnstone, S. (2005). *International institute for educational planning,Open educational resources and open content for higher education background note*. UNESCO International institute for educational planning. Retrieved March 25, 2017 from http://www.unesco.org/iiep/virtualuniversity/media/forum/oer_forum_session_1_note.pdf

Jones, A. W. (1993). The impact of forensic science journals. *Forensic Science International*, *62*(3), 172–178. doi:10.1016/0379-0738(93)90206-P PMID:12742682

Jones, A. W. (1999). The Impact of alcohol and alcoholism among substance abuse journals. *Alcohol and Alcoholism*, *34*(1), 25–34. doi:10.1093/alcalc/34.1.25 PMID:10075398

Jones, A. W. (2002). JAT's impact factor--room for improvement? journal of analytical toxicology. *Journal of Analytical Toxicology*, *26*(1), 2–5. doi:10.1093/jat/26.1.2 PMID:11888013

Jones, A. W. (2003). Impact factors of forensic science and toxicology journals: What do the numbers really mean? *Forensic Science International, 133*(1-2), 1–8. doi:10.1016/S0379-0738(03)00042-2 PMID:12742682

Jones, A. W. (2004). Impact of JAT publications 1981-2003: The most prolific authors and the most highly cited articles. *Journal of Analytical Toxicology, 28*(7), 541–545. doi:10.1093/jat/28.7.541 PMID:16078379

Jones, A. W. (2005a). Creme de la creme in forensic science and legal medicine the most highly cited articles, authors and journals 1981-2003. *International Journal of Legal Medicine, 119*(2), 59–65. doi:10.1007/s00414-004-0512-x PMID:15645298

Jones, A. W. (2005b). Which articles and which topics in the forensic sciences are most highly cited? *Science & Justice, 45*(4), 175–182. doi:10.1016/S1355-0306(05)71661-0 PMID:16686268

Jones, A. W. (2007). The distribution of forensic journals, reflections on authorship practices, peer-review and role of the impact factor. *Forensic Science International, 165*(2-3), 115–128. doi:10.1016/j.forsciint.2006.05.013 PMID:16784827

Jose, C. S., & Casado, E. (1983). Bibliometric analysis pf cheese literature covered by the food science and technology. *Journal of Dairy Science, 66*(7), 1528–1535. doi:10.3168/jds.S0022-0302(83)81969-9

Junquera, B., & Mitre, M. (2007). Value of bibliometric analysis for research policy: A case study of Spanish research into innovation and technology management. *Scientometrics, 71*(3), 443–454. doi:10.1007/s11192-007-1689-9

Kademani, B. S.; Kalyane, V. L. & Kumar, V. (2000). Scientometric portrait of Vikram Ambalal Sarabhai: a citation analysis. *SRELS Journal of Information Management*, 37(2).

Kanuk, L., & Berenson, C. (1975). Mail surveys and response rates: A literature review. *JMR, Journal of Marketing Research, 12*(4), 440–453. doi:10.2307/3151093

Kaplan, S. (2013). *Scholarliness developed.* London: Oxford University Press.

Katz, J. S., & Martin, B. R. (1997). What is research collaboration? *Research Policy, 26*(1), 1–18. doi:10.1016/S0048-7333(96)00917-1

Kavitha, E. S., & Sivaraj, K. (2014). Bibliometric study of Ph.D. thesis in commerce. *e-library science research journal, 2*(11), 1-9.

Keshava & Thimmaiah, B. N. (2015). Growth of poliomyelitis research literature during 1999-2014: A scientometric study. *Indian Journal of Research, 4*, 199–201.

Kessler, M. M. (1963). Bibliographic coupling between scientific papers. *American Documentation, 14*(1), 10–25. doi:10.1002/asi.5090140103

Khan, J. (2015). Use of information sources and need of information literacy among students in Aligarh Muslim university, Aligarh. *International Journal of Library and Information Science, 7*(1), 10–13. doi:10.5897/IJLIS2014.0447

Khan, N. A., & Ahangar, H. (2015). *Research productivity of government medical college Jammu: a Bibliometric Analysis.* Library Philosophy and Practice.

Kirk, P. L. (1953). *Crime Investigation*. New York: John Wiley.

Koonin, E. V., Senkevich, T. G., & Dolja, V. V. (2006). The ancient virus world and evolution of cells. *Biology Direct*, *1*(1), 29. doi:10.1186/1745-6150-1-29 PMID:16984643

Kothari, C. R., & Garg, G. (2014). *Research methodology: Methods and techniques* (3rd ed.). New Delhi: New age international publishers.

Kumar, A. H., Dora, M., & Desai, A. (2015). A Bibliometrics profile of Gujarat university, Ahmedabad during 2004-2013. *DESIDOC Journal of Library and Information Technology*, *35*(1), 9–16. doi:10.14429/djlit.35.1.7699

Kumar, S., & Garg, K. C. (2005). Scientometrics of computer science research in India and China. *Scientometrics*, *64*(2), 121–132. doi:10.1007/s11192-005-0244-9

Kumbar, M. (2009). Growth and impact of research output of University of Mysore, 1996-2006: A case study. *Annals Library Information Studies*, *55*(3), 185–195.

Kumbar, P., & Biradar, N. (2015). Research trends in forensic science: A study of scientometric analysis. *International Journal of Research in Library Science*, *1*(2), 42–48.

Kumbar, P., & Biradar, N. (2015). Research trends in forensic science; A study of scientometric analysis. *International Journal of Research in Library Science*, *1*(2), 42–48.

Lahiri, R. (1996). Citation Analysis: As a tool for collection Development in Libraries. Management of Libraries: Concepts and practices. New Delhi: Ess Ess publication.

Lathman, B. (2007) Sampling: What is it? *Quantitative Research Methods,* Retrieved 15 October, 2014 from http://webpages.acs.ttu.edu/rlatham/Coursework/5377(Quant))/Sampling_Methodology_Paper.pdf

Leslie, S. (2000). The conflict between research and teaching. In H. Kline & M. Alder (Eds.), *Research and development*. New York: Prentice Hall.

Leung, W. C. (2001). How to design a questionnaire. *Student BMJ*, *9*(11), 187–189. PMID:11407060

Levac, D., Colquhoun, H., & O'Brien, K. K. (2010). Scoping studies: Advancing the methodology. *Implementation Science; IS*, *5*(1), 69. doi:10.1186/1748-5908-5-69 PMID:20854677

Leydesdorff, L. (2001). *The Challenge of Scientometrics: The Development, Measurement, and Self-Organization of Scientific Communications*. Parkland: Universal-Publishers.

Line, M. B. (1974). Draft definitions: Information and library needs wants, demands and users. *Aslib Proceedings*, *26*(2), 87. doi:10.1108/eb050451

Lippeveld, T., Sauerborn, R., Bodart, C., & World Health Organization. (2000). Design and Implementation of Health Information Systems. Geneva: WHO.

Loudon, I. (2006). A brief history of homoeopathy. *Journal of the Royal Society of Medicine*, *99*(12), 607–610. doi:10.1258/jrsm.99.12.607 PMID:17139061

Lund, E., & Gram, I. T. (1998). Response rate according to title and length of questionnaire. *Scandinavian Journal of Public Health*, *26*(2), 154–160. doi:10.1080/14034949850153680 PMID:9658516

MacLeod, K. (1987). The seniority principle. Is it discriminatory? Kingston: IRC: Queens' University.

Mafa, O., & Mapolisa, T. (2014). Supervisors' experiences in supervising postgraduate education students' dissertations and theses at the Zimbabwe Open University (ZOU). *International Journal of Asian Social Sciences*, *2*(10), 1685–1697.

Mahapatra, G. (2009). *Bibliometric Studies in the Internet Era*. New Delhi: Indiana Publishing House.

Mahapatra, M. (1985). On the validity of the theory of exponential growth of scientific literature. In *Proceedings of the 15th IASLIC conference, Bangalore* (pp. 61-70).

Maher, R.C. (2010). Authoritative sources. *EE417*. Retrieved online March 9, 2016 from http//:authoritative-sources/ee417.fall2010/htm

Malone, R. W., Homan, J., Callahan, M. V., Glasspool-Malone, J., Damodaran, L., Schneider, A. D. B., & Wilson, J. et al. (2016). Zika virus: Medical countermeasure development challenges. *PLoS Neglected Tropical Diseases*, *10*(3), e0004530. doi:10.1371/journal.pntd.0004530 PMID:26934531

Martinez-Pulgarin, D. F., Acevedo-Mendoza, W. F., Cardona-Ospina, J., Rodriguez-Morales, A. J., & Paniz-Mondolfi, A. E. (2016). A bibliometric analysis of global zika research. *Travel Medicine and Infectious Disease*, *14*(1), 55–57. doi:10.1016/j.tmaid.2015.07.005 PMID:26257029

Maslow. (1970). *Motivation and Personality* (2nd ed.). New York: Harper and Row.

Matthews, A. (2013). Physics publication productivity in South African universities. *Scientometrics*, *95*(1), 69–86. doi:10.1007/s11192-012-0842-2

Mbagaya, G. M., & Odhiambo, M. O. (2005). Mother's health seeking behaviour during child illness in a rural western Kenya community. *African Health Sciences*, *5*(4), 322–327. PMID:16615844

McCain, K. W. (1990). Mapping authors in intellectual space: A technical overview. *Journal of the American Society for Information Science*, *41*(6), 433–443. doi:10.1002/(SICI)1097-4571(199009)41:6<433::AID-ASI11>3.0.CO;2-Q

MDPI. (2017). Retrieved March 30, 2017 from MDPI: http://www.mdpi.com

Medical Library Association (MLA). (2003). The Medical Library Association Task Force on Health Information Literacy. July 23, 2003. Retrieved January 7, 2011 from http://www.mlanet.org/resources/healthlit/ define.html

Medknow Publishing. (2017). Retrieved March 30, 2017 from http://www.medknow.com

Meera & Sahu, S.K. (2014). Research output of university college of medical science, university of Delhi: A bibliometric study. *COLLNET Journal of Scientometrics and Information Management, 8*(2), 401-418.

Mishra, P. N., & Gowsami, N.U. (2013). Impact of grant aid project at CSIR national matellurgical laboratory, Indian: A Bibliometric study. *SRELS journal of information management, 50*(4), 387-412.

Mookerjee, B. P. (1998). *Citation Indexing in Subject Indexing Systems: Concepts, Methods and Techniques*. Calcutta: IASLIC.

Morland, N. (1950). *An Outline of Scientific Criminology*. New York: Philosophical Library.

Mukhereje, M. (2012). Improving Women's Health. *HealthAction*, *25*(4), 4–7.

Mulla, K. R. (2012). Identifying and mapping the information science and scientometrics analysis studies in India (2005-2009): A bibliometric study. *Library Philosophy and Practice*, (June).

Murray, C., Ke, W., & Börner, K. (2006). Mapping Scientific Disciplines and Author Expertise Based on Personal Bibliography Files. In *Proceedings of the KDVis Symposium at Information Visualisation*, London, UK.

Nalimov, V. V., & Mulchenko, M. (1969). *Scientometrics: A study of science as information process. Macoow*: Nauka.

Nanda, B. B., & Tewari, R. K. (2001). *Forensic Science in India: A vision for the twenty-first century*. New Delhi: Select Publishers.

Narayana, G. J. (1991). *Library and Information Management*. New delhi: Prentice Hall of India Pvt. Ltd.

National Digital Library of India (NDL). (2017). Retrieved April 20, 2017 from https://www.ndl.iitkpg.ac.in/

Nederhof, A. J. (2008). Policy impact of bibliometric rankings of research performance of departments and individuals in economics. *Scientometrics*, *74*(1), 163–174. doi:10.1007/s11192-008-0109-0

Niazi, M., & Hussain, A. (2011). Agent-based computing from multi-agent systems to agent-based models: A visual survey. *Scientometrics*, *89*(2), 479–486. doi:10.1007/s11192-011-0468-9

NISCAIR. (2017). Retrieved March 25, 2017 from http:// www.niscair.res.in

Noyons, E. C. M., & van Raan, A. F. J. (1998). Monitoring Scientific Developments from a Dynamic Perspective: Self-Organized Structuring to Map Neural Network Research. *Journal of the American Society for Information Science*, *49*(1), 68–81.

NPTEL. (2017). Retrieved March 25, 2017 from http://www.nptel.ac.in

O.A. Library. (2017). Retrieved March 30, 2017 from http://www.oalib.com

O'Leary, P. D. (2016). Surviving oral exam. Retrieved March 14, 2016, from https://www.cs.umd.edu/~oleary/gradstudy/node7.html

Ofiaja, A. (2014). Why elders deserve only earned respect. *This is Africa*. Retrieve March 14, 2016, from http://www.why-elders-deserve-only-earned-respect/thisisafrica.com

Olayinka, A. I. (2014). My stewardship. Ibadan: Office of the Deputy Vice-Chancellor (Academic, University of Ibadan.

Osterburg, J. W. (1968). *The Crime Laboratory: Case studies of scientific criminal investigation*. Bloomington: Indiana University Press.

Papreja, S., & Shukla, A. (2016). Dental clinics of North America 2004-2014: A bibliometric study. *Library Herald*, *54*(1), 49–51. doi:10.5958/0976-2469.2016.00004.X

PDFDRIVE. (2017). Retrieved October 14, 2017 from https://www.pdfdrive.net/

Pigott, D. M., Golding, N., Mylne, A., Huang, Z., & Henry, A. J. (2014). Mapping the zoonotic niche of Ebola virus disease in Africa. *eLife, 04395*. doi:10.7554/eLife.04395 PMID:25201877

PLOS. (2017). Retrieved March 30, 2017 from PLOS: https://www.plos.org

Pope, G. U. (1886). *The Sacred Kurral of Tiruvalluva Nayanar*. New Delhi: Asian Education Services.

Price, D. (1963). *Little Science, Big Science*. New York: Columbia University Press.

Prince, G. & Saravanan, P. (2015). A study on awareness and perception towards open access resources among the users in the higher educational institutions in Kanyakumari district. *International Journal of Next Generation Library and Technologies, 1*(3), 01-09.

Prince, G. & Saravanan, P. (2015). A study on awareness and perception towards open access resources among the users in the higher educational institutions in Kanyakumari District. *International Journal of Next Generation Library and Technologies, 1*(3), 01-09.

Pritchard, A. (1969). Statistical bibliography or bibliometrics. *The Journal of Documentation, 25*(4), 348–349.

Pyrek, K. M. (2007). *Forensic Science Under Siege: The challenges of forensic laboratories and the medico-legal investigation system*. Amsterdam: Elsevier.

Radhakrishnan, N. & Velmurugan, C. (2015). Scholarly publications by the faculty members of Periyar University, Salem, India: A scientometric approach. *Library Philosophy and Practice*.

Radha, P. (2012). Empowering women for healthy families. *HealthAction, 25*(4), 8–10.

Raja, S., & Balasubramani, R. (2011). Plasmodium falciparum research publication in India: A scientometric analysis. *European Journal of Scientific Research, 56*(3), 294–300.

Rajinikanth, A. (2009). A bibliometrics analysis of surveying engineering literature. *Indian Journal of Information Science and Services, 3*(1), 31–35.

Ranganathan, S. R. (1969). Librametry and its scope. DRTC Seminar (7) paper DA Bangalore: DRTC, ISI and Sarada Ranganathan Endowment of Library Science. Reprinted in: *JISSI: The International Journal of Scientometrics and Informetrics, 1*(1), 15-21.

Ranganathan, S. R. (1963). *Documentation and its facts*. Bombay: Asia Publishing House.

Rao, R. (1983). *Qualitative methods in library and Information science*: Newdelhi: Wiley eastern.

Rao, G. M. (2014). *Food Science and Technology*. New Delhi: New Age Publishers.

Rao, I. K. R., Raghavan, K. S., & Bhargav, K. N. (2017). Quantitative analysis of scientific literature in forensic science. *COLLNET Journal of Scientometrics and Information Management, 11*(1), 59–72. doi:10.1080/09737766.2016.1251561

Rasmussen, S. A., Jamieson, D. J., Honein, M. A., & Petersen, L. R. (2016). Zika virus and birth defects - reviewing the evidence for causality. *The New England Journal of Medicine, 374*(20), 1981–1987. Retrieved15April2016. doi:10.1056/NEJMsr1604338 PMID:27074377

Ravichandra Rao, I. K. (1985). Bibliometric Model: An entropy approach and their application. *IASLIC Special Publication, 25*, 1–8.

Reifz, J. M. (2004). *Dictionary for Library and Information Science*. USA: Libraries Unlimited.

ROAR. (2017). Retrieved March 25, 2017 from http://www.roar.eprints.org

Rockman, I. F. (2003). Integrating information literacy into the learning outcomes of academic disciplines, A critical 21st century issue. *College & Research Libraries News*, *64*(9), 612–615.

Rolison, J. (1998). Health information for the teenage years, what do they want to know? *Information Research, 3*(2). Retrieved from http://information.net/lr/3 3/papaer42html

Rowley, J. (2014). Designing and using research questionnaires. *Management Research Review*, *37*(3), 308–330. doi:10.1108/MRR-02-2013-0027

Roy, P. M. (1983). Towards a theory of citing in citation analysis studies. Developing horizons in Library and Information Science, 1.

Saferstein, R. (1995). *Criminalistics: an introduction to forensic science* (7th ed.). New Jersey: Prentice-Hall.

Sahu, D. K., & Parmar, R. C. (2006). Open access in the developing world: Regaining the lost impact In *Workshop on electronic publishing and open access: Developing country perspectives*, Bangalore, India, *November 2-3*.

Salauddin, N. (2015). Usage of information communication technology application in national physical laboratory library, Delhi. *International Journal of Next Generation Library and Technologies*, *1*(1), 1–8.

Saravanan, P., & Mary, A. L. (2007). Teachers approach to internet and online information resources. *IASLIC Bulletin*, *52*(3), 150–158.

Saunders, M. N. (2005). *Research Methods for Business Students, 3/e*. Pearson Education India.

Saunders, M. N. (2011). *Research Methods for Business Students, 5/e*. Pearson Education India.

Sauvageau, A., Desnoyers, S., & Godin, A. (2009). Mapping the literature in forensic sciences: A bibliometric study of North-American journals from 1980 to 2005. *The Open Forensic Science Journal*, *2*(1), 41–46. doi:10.2174/1874402800902010041

Scott, D. (2009, June 3). After 50 years of teaching, professor sees Ph.D. oral exam as the golden hour. Stanford Report. Retrieves March 14, 2016, from http://news.stanford.edu/news/2009/june3/scott-060309.html

Sengupta, I. N. (1985). Ranganathan's Philosophy and Librametrics. In T. S. Rajagopalan (Ed.), *Ranganathan's Philosophy: Assessment, impact, and relevance* (pp. 447–458). New Delhi: Vikas Publishing House.

Sensabaugh, G. F. (1986). Forensic Science Research: Who Does it and Where is it Going? In G. Davies (Ed.), *Forensic Science* (2nd ed., pp. 129–140). Washington, DC: American Chemical Society.

Sethi, M. & Rao, E.S. (2005). *Food Science*. New Delhi: New Age International.

Shafi, S. M. (2002). Citation Analysis of Ph.D. Thesis: A Study of Doctoral Theses submitted during, 1980-2000 in Natural Science. *Library Philosophy and Practice*.

Sheikh, K., & Mattingly, S. (1981). Investigating non-response bias in mail surveys. *Journal of Epidemiology and Community Health, 35*(4), 293–296. doi:10.1136/jech.35.4.293 PMID:6461711

Shen, J., Wu, D., Achhpiliya, V., Bieber, M., & Hiltz, R. (2004). Participatory learning approach: A research agenda. *Semantic Scholar, 1*(6), 1–13.

Shodhganga@INFLIBNET. (2017). Retrieved March 25, 2017 from http://www.shodhganga.inflibnet.ac.in

Shores, L. (1954). *Basic reference sources*. Chicago: American Library Association.

Sikka, V., Chattu, V., Popli, R. K., Galwankar, S. C., Kelkar, D., Sawicki, S. G., & Papadimos, T. J. et al. (2016). The emergence of zika virus as a global health security threat: A review and a consensus statement of the INDUSEM Joint working Group (JWG). *Journal of Global Infectious Diseases, 8*(1), 3–15. doi:10.4103/0974-777X.176140 PMID:27013839

Small, H. (1999). Visualizing science by citation mapping. *Journal of the American Society for Information Science, 50*(9), 799–813. doi:10.1002/(SICI)1097-4571(1999)50:9<799::AID-ASI9>3.0.CO;2-G

Smith, M. M. (2013). *Humble leadership: The research shows it's a competitive advantage. Talent management and human resources*. New York: TLNT.

Soper, M. E. (1990). *The librarian's thesaurus: A concise guide to library and information terms*. Chicago: American Library Association.

Standler, R. B. (2004). Evaluating credibility of information on the internet. Retrieved January 30, 2017 from http://www.rbs0.com

Stehlik, T. (2008). *Final report: The teaching-research nexus in the division of education, arts and social sciences*. Sydney: University of Australia.

Stoney, D. A. (1984). Evaluation of associative evidence: Choosing the relevant question. *Journal - Forensic Science Society, 24*(5), 473–482. doi:10.1016/S0015-7368(84)72326-7

Stoney, D. A. (1991). What made us ever think we could individualise using statistics? *Journal - Forensic Science Society, 3*(2), 197–199. doi:10.1016/S0015-7368(91)73138-1 PMID:1940832

Straits, K. J. E., Bird, D. M., Tsinajinnie, E., Espinoza, J., Goodkind, J., Spencer, O., & Willging, C. et al.The Guiding Principles Workgroup. (2012). *Guiding principles for engaging in research with native American communities (Version 1)*. Washington: UNM Center for Rural and Community Behavioral Health & Albuquerque Area Southwest Tribal Epidemiology Center.

Suber, P. (2012). Open Access. In Open Access (pp. 4-10). Cambridge: MIT Press.

Subramanyam, K. (1983). Bibliometric study of research collaboration: A review. *Journal of Information Science, 6*, 33–38.

Sudhamani, N. (2008). Revitalizing Primary Health Care through Women's Empowerment. *HealthAction, 21*(12), 48.

Sudhierpillai, K. G. & Dileepkumar, V. (2010) Scientometric study of doctoral dissertations in biochemistry in the university of Kerala, India. *Library philosophy and practice*, June.

Sudhierpillai, K. G. (2007). Journal citation in physics doctoral dissertation of Indian institute of science. *Annals of library and Information Studies, 54*(4), 177-184.

SudhierPillai, K.G. & Priyalekshmi, V. (2013). Research publication trend among the scientists of Central Tuber Crops Research Institute (CTCRI), Thiruvananathapuram: A scientometric studies. *Annals of Library and Information Studies*, *60*(1), 7–14.

Surwase, G., Mohan, L., Kademani, B. S., & Bhanumurthy, K. (2014). Research trends on food preservation: A scientometric analysis. *DESIDOC Journal of Library and Information Technology*, *34*(3), 257–264. doi:10.14429/djlit.34.5979

Tabatabaei-Malazy, O., Ramezani, A., Atlsai, R., Larijani, B., & Abdollahi, M. (2016). Scientometric study of academic publications on antioxidative herbal medicines in type 2 diabetes mellitus. *Journal of Diabetes and Metabolic Disorders*, *15*(1), 48. Retrieved 13 October 2017 from doi:10.1186/s40200-016-0273-3 PMID:27785446

Tague-Sutcliffe, J. (1992). An Introduction to Informetrics. *Information Processing & Management*, *28*(1), 13. doi:10.1016/0306-4573(92)90087-G

The United Nations. (1948Universal Declaration of Human Rights (UDHR). *General Assembly resolution 217A*.

Thirumagal, A. (2012). Scientific publications of Manonmaniam Sundaranar university, Tirunelveli: Scientometric analysis. *Library Philosophy and Practice*.

Thirumagal, A., & SethuKumari., N.S. (2013). Mapping of scholarly research in cloud computing: A bibliometric study. *SERLS journal of information management, 50*(5).

Tripathi, H. K., & Garg, K. C. (2014). Scientometrics of Indian crop science research as reflected by the coverage in Scopus, CABI and ISA databases during 2008-2010. *Annals of Library and Information Studies*, *61*(1), 41–48.

Tryon, R. C., & Bailey, D. E. (1970). *Cluster analysis*. New York: McGraw-Hill Book Co.

Ugboma, M.U. (2010). Provision and use of legal information among civil rights groups in the oil producing communities of Nigeria. *Library Philosophy and Practice.*

Umbecco, E. (2015). *How to write a thesis*. New York: Academic Emergence Worldpress.

UNESCO. (2015). *Introduction to open access*. Retrieved March 25, 2017 from http://www.unesdoc.unesco.org/images/0023/002319/231920e.pdf

University of Edinburgh. (2016). *Code of practice for supervisors and research students*. Edinburgh: University of Edinburgh.

University of Otago. (2011). *Pacific research protocols*. Otago: University of Otago.

van Raan, A. F. J. (2004). Sleeping beauties in science. *Scientometrics*, *59*(3), 461–466. doi:10.1023/B:SCIE.0000018543.82441.f1

Van Raan, A. F. S., & Van Leeuwen, T. H. N. (2002). Assessment of the scientific basic of interdisciplinary applied research, and Application of bibliometric methods in nutrition and food research. *Research Policy*, *31*, 611–632. doi:10.1016/S0048-7333(01)00129-9 vanRaan, A. F. J. (1991). Fractal geometry of information space as represented by co-citation clustering. *Scientometrics*, *20*(3), 439–449. doi:10.1007/BF02019764

Vellaichamy, A. (2016). A scientometric analysis of 20 years literature on Ebola virus. *International Journal of Innovative Knowledge Concepts.*, *4*, 255–262.

Vellaichamy, A., & Jeyshankar, R. (2014). *Anemia research in India: a bibliometric analysis of publications output during 1993–2013.* Library Philosophy and Practice.

Vellaichamy, A., & Jeyshankar, R. (2015). Dengue research in India and China: A comparative study using bibliometrics. *International Journal of Library Science and Information Management*, *1*, 1–9.

Vellaichamy, A., & Jeyshankar, R. (2015). Myeloma research: A bibliometric assessment of world publications output during 2008-2012. *International Journal of Digital Library Services*, *5*, 13–25.

Velmurugan, C., & Radhakrishnan, N. (2015). A scientometric analysis of research papers published on pharmacognosy as reflected in the Web of Science. *Advances in Pharmacognosy and Phytomedicine*, *1*(1), 27–40.

Velmurugan, C., & Radhakrishnan, N. (2015). Scientometric observations of authorship trends and collaborative research on DESIDOC journal of library and information technology. *Collnet Journal of Scientometrics and Information Management*, *9*(2), 193–204. doi:10.1080/09737766.2015.1069957

Velmurugan, C., & Radhakrishnan, N. (2015). Webology journal: A scientometric profile. *International Journal of Information Dissemination and Technology*, *5*(2), 137.

Vijay, R. K. (2005). Bibliometric study of research publication trends among, Indian food scientists and technologists. *Annals of Library and Information Studies*, *52*(3), 77–81.

Waddell, L. A., & Greig, J. D. (2016). Scoping review of the Zika virus literature. *PLoS One*, *11*(5), e0156376. doi:10.1371/journal.pone.0156376 PMID:27244249

Webster's new world dictionary of the American language. (1960). Cleveland: World Publishing Company.

Weinstock, N. (n.d.). [Citation indexes]. In Encyclopedia of Library and Information Science (Vol. 15). New York: Marcel Deccker.

Wellington, J. J. (2010). *Making supervision work for you: A student's guide*. London: Sage.

White, H. D., & McCain, K. W. (1989). Bibliometrics. In M. E. Williams (Ed.), *Annual Review of Information Science and Technology* (Vol. 24, pp. 119–186). Amsterdam: Elsevier Science Publishers B.V.

White, H. D., & McCain, K. W. (1989). Bibliometrics. In M. E. Williams (Ed.), *Annual Review on Information Science and Technology* (Vol. 24). Amsterdam, Netherlands: Elsevier Science Publishers.

Wilson. (1981). On user studies and information needs, *Journal of Documentation, 37*(1), 3-15.

Wilson, C. S. (1999). Informetrics. *Annual Review of Information Science & Technology*, *34*, 107–247.

Wilson, T. D. (1981). On user studies and Information Needs. *The Journal of Documentation*, *37*(1), 5–6. doi:10.1108/eb026702

World Health Organization. (1946). *Preamble to the constitution of World Health Organization* as *adapted by the International Health Conferences.*

Worthingtong, E. L. (2007). The Paradox of Humility. *Greater good: The science of a meaningful Life*. Retrieved July 29, 2016, from http://greatergood.berkeley.edu/article/item/paradox_humility

Wulder, M. (2014). A practical guide to the use of selected multivariate statistics. Retrieved 15 July 2014 from http://www.psych.yorku.ca/lab/psy6140/DataScreeningChecklist.pdf

Young, I., Waddell, L., Sanchez, J., Wilhelm, B., McEwen, S. A., & Rajic, A. (2014). The application of knowledge synthesis methods in agri-food public health: Recent advancements, challenges and opportunities. *Preventive Veterinary Medicine*, *113*(4), 339–355. doi:10.1016/j.prevetmed.2013.11.009 PMID:24485274

Zarafshan, H., Mohammadi, M. R., Motevalian, S. A., Abolhassani, F., Khaleghi, A., & Sharifi, V. (2016). Autism research in Iran: A scientometric study. *Iranian Journal of Psychiatry and Behavioral Sciences*, *11*(2), e7350. doi:10.5812/ijpbs.7350

Zyoud, S. H., Al-Jabi, S. W., Swelleh, W. M., & Awany, R. (2014). A bibilometric analysi of toxicology research productivity in Middle Eastern Arab countries during a 10 year period (2003-2012). *Health Research Policy and Systems*, *12*(4). PMID:24443999

About the Contributors

J. John Jeyasekar graduated from Bishop Heber College, Tiruchirapalli, India specializing in Library, Information and Documentation Science. He obtained MLIS Degree and also PG Diploma in Library Automation and Networking from Annamalai University, India. In addition to these degrees, he was awarded Diploma in French by the University of Madras, India and PhD in Science Mapping by the Manonmaniam Sundaranar University, India. He has more about thirty years of professional experience. He completed his internship under Prof.T.K.S.Iyengar at the Indian Institute of Science, Bangalore and started his career as a school librarian. At present he is the librarian of the Tamil Nadu State Forensic Sciences Department, India. He is a life member of the Society for Advancement of Library and Information Science (SALIS). He has 10 research papers in national and international journals and 2 book chapters to his credit. He has also presented 6 papers in national conferences.

P. Saravanan is presently working as a College Librarian (S.G) at Lekshmipuram College of Arts & Science, Neyyoor-629802, Tamilnadu, India. In addition to Masters degree in Library & Information Science, he also has Masters Degrees in Mathematics and Education. Before joining as a Librarian at Lekshmipuram College, he has worked as Lecturer in Mathematics and Hon. Lecturer in Mathematics at Kerala University Teacher Education Centers. He has qualified UGC-NET examination both in Library Science and Education, he has published a book titled 'Secret of Success in UGC-NET Examination, more than 25 articles in Journals and more than 40 papers in conferences as his scholarly output. Moreover, he has successfully completed two UGC-Minor Research Project in Library Science, and 4 scholars have completed Ph.D in Library & Information Science under his guidance.

* * *

R. Gnana Bai is presently serving as a library and information professional in Scott Christian College (Autonomous), Nagercoil, Kanyakumari District, India. Before joining as a library professional, she worked as an Accountant in Scott Christian College, affiliated to Manonmaniam Sundaranar University, Tirunelveli, India. She has a masters degree and M.Phil. degree in commerce and also PhD in Library and Information Science. She has attended more than fifteen seminars and has published two research articles. She has 20 years of experience in library field.

Floribert Patrick C. Endong (Ph.D) is a research consultant in the humanities and social sciences. His areas of interest include sexuality, culture, religion, gender, media communication, religious communication and international relations. He is author of numerous peer-reviewed articles and book chapters in the above-mentioned areas.

Rayees Farooq is a Senior Research Fellow (SRF) under the Maulana Azad National Fellowship of the University Grants Commission (UGC), India. He is pursuing a PhD on "Knowledge Management Orientation and its Relationship with Business Performance" at the Lovely Professional University, Phagwara (India). He has an MBA and MPhil in management. He participated in the 2nd IIM-A Summer School, 2013 at the Indian Institute of Management, Ahmedabad, and attended the Doctoral Students Colloquium at IIM, Kozhikode during the XVI Annual Convention of the Strategic Management Forum (SMF) in May 2013. He has Nine publications in peer-reviewed journals, and has a good command of multivariate data analysis tools. Rayees Farooq can be contacted at: rayeesfarooq@rediffmail.com

S. Hariharan is presently working as a Librarian at Marthandam College of Enginering and Technology, Marthandam, Kanyakumari District, Tamilnadu India. In addition to Master of Library and Information Science, he also has a Master's Degree in History. He is having experience more than 10 years in Librarianship and his area of interest is e-resources. He has also published more five articles in Journals.

E.S. Kavitha obtained her MLIS and Ph,D from Annamalai University,Tamilnadu.She has been working as Assistant Professor at Periyar University, Periyar palkalai nagar, Salem -636011, Tamilnadu since 2013. She has eight years experience as Librarian at Sri Vasavi College. She serves more than eleven years in the LIS profession. She has published many articles in National and International Journals and conferences. She has attended number of Seminars, workshop and training programmes. She has also organized two workshops in LIS. Her areas of interest are User study, Scientometrics, ICT. She is a life member of SALIS.

A. Lawrence Mary is presently working as a Research Co-ordinator at Loyola Institute of Technology and Science, Thovalai, Nagercoil, Tamilnadu. She has 25 years experience as Librarian. She is formerly a College Librarian of TDMNS College, T. Kallikulam affiliated to Manonmaniam Sundaranar University, Tirunelveli. At her credit, she has published more than 10 books, more than 50 articles and about 50 articles in seminar proceedings. She has organised number of seminars and conferences. Under her guidance, more than 35 students completed their M.Phil dissertations and 18 scholars have completed Ph.D. Moreover, she is the founder president of KDLF (Kanyakumari District Librarians Federation) and chief-editor of 'Journal of Library Advancements'.

Jahir Hussain Haneefa Mohamed obtained Ph.D in Library and Information Science in 2014. He also holds M.Phil in Library and Information Science in addition to Master Degree. Presently, he is working as an Assistant Librarian, Waljat College of Applied Sciences, KOM-Rusayl, Sultanate of Oman. Previously he worked as an Assistant Librarian at Saudi ARAMCO – Kingdom of Saudi Arabia. His area of interest is Library Automation. He has published number of articles in journals and presented papers in Seminars and Conferences.

Usharani Munusamy works as Junior Scientific officer in Tamil Nadu State Forensic Sciences Department. Has research experience in HSPs, cell signaling, and environmental stress adaptation.

V. S. Prasanth is presently working as Head of the Department of History, Lekshmipuram College of Arts & Science, Neyyoor-629802, Tamilnadu, India. He is interested in Library Science, in particular on quantification of information. He is also a recognised research supervisor for Ph.D Programme in History. He has presented more than 15 articles in refereed journals and presented more than 20 papers in seminars, and conferences. He has also participated number of seminars on Library Science.

Natarajan Radhakrishnan is currently working for Periyar University, Salem, India. He has published number of research papers which are indexed in Scopus and Web of science. He had worked in India and UAE. He has guided for M.Phil and Ph.D students.

Rajshree Raghunath is a postgraduate in Forensic Sciences and holding a PhD degree in Biotechnology. She has put in 25 years of service in the Tamil Nadu State Forensic Sciences Department India, in the field of biological sciences.

G. Ramadhas is presently working as a Deputy Librarian at Noorul Islam Center for Higher Education (Deemed University), Kumaracoil, Thuckalay -629175, Tamilnadu. He is actively involved in research, published articles in refereed journals and presented papers in conference at various level. His area of interest is user study, and digital library.

A.S. Suman Sankar, M.D. (Hom.), is presently working as a Professor in Repertory at Sarda Krishna Homoeopathic Medical College, Kulasekaram -629161, Tamilnadu. His area of specialization is Repertory in Homoeopathy and is interested in plant research and public health. He is also an eminent physician in Homoeopathy. He is supervising PG projects and STSH programmes of CCH (Central Council of Homoeopathy, New Delhi).

P. Sheela is presently working as college librarian at N.V.K.S.D.College of Education, Attoor, Kanyakumari District. She has qualified U.G.C.NET examination in Library Science. She id handling the elective paper "Library and Information Science". She has publication in various reputed education journals and has presented paper in many national, International conferences and seminars.

S. Lawyed Stephen is currently working as Chief Librarian of Noorul Islam Center for Higher Education (Deemed University), Thuckaly, Tamilnadu, India. He completed his MLIS course from University of Madras in regular stream. He also holds M.Phil in Library Science from Alagappa University, Karikudi and PhD from Manonmaniam Sundaranar University, Tirunelveli. He is also a Recipient of S.R.Ranganthan Award from University of Madras and Best Librarian Award for the year 2006.

N. V. Sugathan, M.D (Hom.) is presently working as Professor of Practice of Medicine at Sarada Krishna Homoeopathic Medical College, Kulasekaram, Tamilnadu. He had more than 20 years experience in teaching and practicing Homoeopathic practice. He is also serving as Deputy Medical Superintendent of Sarada Krishna Homoeopathic Medical College. He has also served as various capacities such as member and examiner in the The Tamilnadu Dr. M.G.R. Medical University Chennai, and as member of Central Council of Homoeopathy, New Delhi.

M. Vijayakumar, M.A. MLISC., M.Phil, Ph.D is presently working as a Librarian at St. Xavier's Catholic College of Engineering, Nagercoil-3 for the last 20 years. He has also served as Librarian at The Indian Engineering College, Vadakankulam, Tirunelveli District and at Kamarajar Institute of Technology, Virudhunagar. His area of specialisation is Digital Library. He is also the Secretary of Kanyakumari District Librarians Federation (KDLF). He has presented number of papers in conferences and more than 5 articles in refereed journals.

Index

A

B

C

D

E

F

G

H

I

J

K

L

M

N

O

Q

R

S

T

U

V

W

Y

Z

www.ingramcontent.com/pod-product-compliance
Lightning Source LLC
LaVergne TN
LVHW060305011026
849161LV00011B/2773